Physics I For Dummies®

2nd Edition

by Steven Holzner, PhD

Wiley Publishing, Inc.

Physics I For Dummies®, 2nd Edition

Published by
Wiley Publishing, Inc.
111 River St.
Hoboken, NJ 07030-5774
www.wiley.com

Published by Wiley Publishing, Inc., Indianapolis, Indiana

Published simultaneously in Canada

For general information on our other products and services, please contact our Customer Care Department within the U.S. at 877-762-2974, outside the U.S. at 317-572-3993, or fax 317-572-4002.

For technical support, please visit www.wiley.com/techsupport.

Wiley also publishes its books in a variety of electronic formats. Some content that appears in print may not be available in electronic books.

Library of Congress Control Number: 2011926317

ISBN: 978-0-470-90324-7

Manufactured in the United States of America

10 9 8 7 6 5 4 3 2 1

About the Author

Steven Holzner is an award-winning author of 94 books, which have sold more than 2 million copies and have been translated into 18 languages. He served on the Physics faculty at Cornell University for more than a decade, teaching both Physics 101 and Physics 102. Dr. Holzner received his PhD in physics from Cornell and performed his undergrad work at MIT, where he has also served as a faculty member.

Dedication

To Nancy.

Author's Acknowledgments

Any book such as this one is the work of many people besides the author. I'd like to thank my acquisitions editor, Stacy Kennedy, and everyone else who had a hand in the book's contents, including Tracy Barr, Danielle Voirol, Joel Bryan, Eric Hedin, and Neil Clark. Thank you, everyone.

Publisher's Acknowledgments

We're proud of this book; please send us your comments at http://dummies.custhelp.com. For other comments, please contact our Customer Care Department within the U.S. at 877-762-2974, outside the U.S. at 317-572-3993, or fax 317-572-4002.

Some of the people who helped bring this book to market include the following:

Acquisitions, Editorial, and Media Development

Editors: Tracy Barr, Danielle Voirol

Acquisitions Editor: Stacy Kennedy

Assistant Editor: David Lutton

Editorial Program Coordinator: Joe Niesen

Technical Editors: Joel Bryan, PhD; Eric Hedin, PhD

Editorial Managers: Jennifer Erhlich, Senior Editorial Manager; Carmen Krikorian, Editorial Supervisor and Reprint Editor

Editorial Assistants: Jennette ElNaggar, Rachelle S. Amick

Cover Photos: © iStockphoto.com / Marco Martins

Cartoons: Rich Tennant (www.the5thwave.com)

Composition Services

Project Coordinator: Katherine Crocker

Layout and Graphics: Corrie Socolovitch, Christin Swinford, Laura Westhuis

Proofreaders: Laura Albert, Laura Bowman

Indexer: Infodex Indexing Services, Inc.

Special Help

Neil Clark, Krista Fanning

Publishing and Editorial for Consumer Dummies

Diane Graves Steele, Vice President and Publisher, Consumer Dummies

Kristin Ferguson-Wagstaffe, Product Development Director, Consumer Dummies

Ensley Eikenburg, Associate Publisher, Travel

Kelly Regan, Editorial Director, Travel

Publishing for Technology Dummies

Andy Cummings, Vice President and Publisher, Dummies Technology/General User

Composition Services

Debbie Stailey, Director of Composition Services

Contents at a Glance

Table of Contents

Part III: Manifesting the Energy to Work 161

Chapter 9: Getting Some Work Out of Physics 163

Chapter 10: Putting Objects in Motion: Momentum and Impulse 187

Introduction

Physics is what it's all about. What *what's* all about? Everything. Physics is present in every action around you. And because physics is everywhere, it gets into some tricky places, which means it can be hard to follow. Studying physics can be even worse when you're reading some dense textbook that's hard to follow.

For most people who come into contact with physics, textbooks that land with 1,200-page *whumps* on desks are their only exposure to this amazingly rich and rewarding field. And what follows are weary struggles as the readers try to scale the awesome bulwarks of the massive tomes. Has no brave soul ever wanted to write a book on physics from the *reader's* point of view? One soul is up to the task, and here I come with such a book.

About This Book

Physics I For Dummies, 2nd Edition, is all about physics from your point of view. I've taught physics to many thousands of students at the university level, and from that experience, I know that most students share one common trait: confusion. As in, "I'm confused about what I did to deserve such torture."

This book is different. Instead of writing it from the physicist's or professor's point of view, I wrote it from the reader's point of view. After thousands of one-on-one tutoring sessions, I know where the usual book presentation of this stuff starts to confuse people, and I've taken great care to jettison the top-down kinds of explanations. You don't survive one-on-one tutoring sessions for long unless you get to know what really makes sense to people — what they want to see from *their* points of view. In other words, I designed this book to be crammed full of the good stuff — and *only* the good stuff. You also discover unique ways of looking at problems that professors and teachers use to make figuring out the problems simple.

Conventions Used in This Book

Some books have a dozen conventions that you need to know before you can start. Not this one. All you need to know is that variables and new terms appear in italics, like *this,* and that vectors — items that have both a magnitude and a direction — appear in **bold.** Web addresses appear in `monofont`.

What You're Not to Read

I provide two elements in this book that you don't have to read at all if you're not interested in the inner workings of physics — sidebars and paragraphs marked with a Technical Stuff icon.

Sidebars provide a little more insight into what's going on with a particular topic. They give you a little more of the story, such as how some famous physicist did what he did or an unexpected real-life application of the point under discussion. You can skip these sidebars, if you like, without missing any essential physics.

The Technical Stuff material gives you technical insights into a topic, but you don't miss any information that you need to do a problem. Your guided tour of the world of physics won't suffer at all.

Foolish Assumptions

In writing this book, I made some assumptions about you:

- You have no or very little prior knowledge of physics.
- You have some math prowess. In particular, you know algebra and a little trig. You don't need to be an algebra pro, but you should know how to move items from one side of an equation to another and how to solve for values.
- You want physics concepts explained clearly and concisely, and you want examples that let you see those concepts in action.

How This Book Is Organized

The natural world is, well, *big*. And to handle it, physics breaks the world down into different parts. The following sections present the various parts you see in this book.

Part I: Putting Physics into Motion

You usually start your physics journey with motion, because describing motion — including acceleration, velocity, and displacement — isn't very difficult. You have only a few equations to deal with, and you can get them under your belt in no time at all. Examining motion is a great way to understand how physics works, both in measuring and in predicting what's going on.

Part II: May the Forces of Physics Be with You

"For every action, there is an equal and opposite reaction." Ever heard that one? The law (and its accompanying implications) comes up in this part. Without forces, the motion of objects wouldn't change at all, which would make for a very boring world. Thanks to Sir Isaac Newton, physics is particularly good at explaining what happens when you apply forces. You also take a look at the motion of fluids.

Part III: Manifesting the Energy to Work

If you apply a force to an object, moving it around and making it go faster, what are you really doing? You're doing work, and that work becomes the kinetic energy of that object. Together, work and energy explain a whole lot about the whirling world around you, which is why I dedicate Part III to these topics.

Part IV: Laying Down the Laws of Thermodynamics

What happens when you stick your finger in a candle flame and hold it there? You get a burned finger, that's what. And you complete an experiment in heat transfer, one of the topics you see in Part IV, which is a roundup of thermodynamics — the physics of heat and heat flow. You also see how heat-based engines work, how ice melts, how the ideal gas behaves, and more.

Part V: The Part of Tens

The Parts of Tens is made up of fast-paced lists of ten items each. You discover all kinds of amazing topics here, like some far-out physics — everything from black holes and the Big Bang to wormholes in space and the smallest distance you can divide space into — as well as some famous scientists whose contributions made a big difference in the field.

Icons Used in This Book

You come across some icons that call attention to certain tidbits of information in this book. Here's what the icons mean:

This icon marks information to remember, such as an application of a law of physics or a particularly juicy equation.

When you run across this icon, be prepared to find a shortcut in the math or info designed to help you understand a topic better.

This icon highlights common mistakes people make when studying physics and solving problems.

This icon means that the info is technical, insider stuff. You don't have to read it if you don't want to, but if you want to become a physics pro (and who doesn't?), take a look.

Where to Go from Here

You can leaf through this book; you don't have to read it from beginning to end. Like other *For Dummies* books, this one was designed to let you skip around as you like. This is your book, and physics is your oyster. You can jump into Chapter 1, which is where all the action starts; you can head to Chapter 2 for a discussion of the necessary algebra and trig you should know; or you can jump in anywhere you like if you know exactly what topic you want to study. And when you're ready for more-advanced topics, from electromagnetism to relativity to nuclear phsics, you can check out *Physics II For Dummies*.

Part I
Putting Physics into Motion

"Physics explains motion. Like the acceleration of this glass moving backward creating a velocity that results in a displacement. Any weenie knows that.

In this part . . .

Part I is designed to give you an introduction to the ways of physics. Motion is one of the easiest physics topics to work with, and you can become a motion meister with just a few equations. This part also arms you with foundational info on math and measurement to show how physics equations describe the world around you. Just plug in the numbers, and you can make calculations that astound your peers.

Chapter 1

Using Physics to Understand Your World

In This Chapter

- Recognizing the physics in your world
- Understanding motion
- Handling the force and energy around you
- Getting hot under the collar with thermodynamics

Physics is the study of the world and universe around you. Luckily, the behavior of the matter and energy — the stuff of this universe — is not completely unruly. Instead, it strictly obeys laws, which physicists are gradually revealing through the careful application of the *scientific method,* which relies on experimental evidence and sound rigorous reasoning. In this way, physicists have been uncovering more and more of the beauty that lies at the heart of the workings of the universe, from the infinitely small to the mind-bogglingly large.

Physics is an all-encompassing science. You can study various aspects of the natural world (in fact, the word *physics* is derived from the Greek word *physika,* which means "natural things"), and accordingly, you can study different fields in physics: the physics of objects in motion, of energy, of forces, of gases, of heat and temperature, and so on. You enjoy the study of all these topics and many more in this book. In this chapter, I give an overview of physics — what it is, what it deals with, and why mathematical calculations are important to it — to get you started.

What Physics Is All About

Many people are a little on edge when they think about physics. For them, the subject seems like some highbrow topic that pulls numbers and rules out of thin air. But the truth is that physics exists to help you make sense of the world. Physics is a human adventure, undertaken on behalf of everyone, into the way the world works.

At its root, physics is all about becoming aware of your world and using mental and mathematical models to explain it. The gist of physics is this: You start by making an observation, you create a model to simulate that situation, and then you add some math to fill it out — and voilà! You have the power to predict what will happen in the real world. All this math exists to help you see what happens and why.

In this section, I explain how real-world observations fit in with the math. The later sections take you on a brief tour of the key topics that comprise basic physics.

Observing the world

You can observe plenty going on around you in your complex world. Leaves are waving, the sun is shining, light bulbs are glowing, cars are moving, computer printers are printing, people are walking and riding bikes, streams are flowing, and so on. When you stop to examine these actions, your natural curiosity gives rise to endless questions such as these:

- Why do I slip when I try to climb that snow bank?
- How distant are other stars, and how long would it take to get there?
- How does an airplane wing work?
- How can a thermos flask keep hot things warm *and* keep cold things cool?
- Why does an enormous cruise ship float when a paper clip sinks?
- Why does water roll around when it boils?

Any law of physics comes from very close observation of the world, and any theory that a physicist comes up with has to stand up to experimental measurements. Physics goes beyond qualitative statements about physical things — "If I push the child on the swing harder, then she swings higher," for example. With the laws of physics, you can predict precisely how high the child will swing.

Making predictions

Physics is simply about modeling the world (although an alternative viewpoint claims that physics actually uncovers the truth about the workings of the world; it doesn't just model it). You can use these mental models to describe how the world works: how blocks slide down ramps, how stars form and shine, how black holes trap light so it can't escape, what happens when cars collide, and so on.

When these models are first created, they sometimes have little to do with numbers; they just cover the gist of the situation. For example, a star is made up of this layer and then that layer, and as a result, this reaction takes place, followed by that one. And pow! — you have a star. As time goes on, those models become more numeric, which is where physics students sometimes start having problems. Physics class would be a cinch if you could simply say, "That cart is going to roll down that hill, and as it gets toward the bottom, it's going to roll faster and faster." But the story is more involved than that — not only can you say that the cart is going to go faster, but in exerting your mastery over the physical world, you can also say how much faster it'll go.

There's a delicate interplay between theory, formulated with math, and experimental measurements. Often experimental measurements not only verify theories but also suggest ideas for new theories, which in turn suggest new experiments. Both feed off each other and lead to further discovery.

Many people approaching this subject may think of math as something tedious and overly abstract. However, in the context of physics, math comes to life. A quadratic equation may seem a little dry, but when you're using it to work out the correct angle to fire a rocket at for the perfect trajectory, you may find it more palatable! Chapter 2 explains all the math you need to know to perform basic physics calculations.

Reaping the rewards

So what are you going to get out of physics? If you want to pursue a career in physics or in an allied field such as engineering, the answer is clear: You'll need this knowledge on an everyday basis. But even if you're not planning to embark on a physics-related career, you can get a lot out of studying the subject. You can apply much of what you discover in an introductory physics course to real life:

- In a sense, all other sciences are based upon physics. For example, the structure and electrical properties of atoms determine chemical reactions; therefore, all of chemistry is governed by the laws of physics. In fact, you could argue that everything ultimately boils down to the laws of physics!
- Physics does deal with some pretty cool phenomena. Many videos of physical phenomena have gone viral on YouTube; take a look for yourself. Do a search for "non-Newtonian fluid," and you can watch the creeping, oozing dance of a cornstarch/water mixture on a speaker cone.

- More important than the applications of physics are the problem-solving skills it arms you with for approaching any kind of problem. Physics problems train you to stand back, consider your options for attacking the issue, select your method, and then solve the problem in the easiest way possible.

Observing Objects in Motion

Some of the most fundamental questions you may have about the world deal with objects in motion. Will that boulder rolling toward you slow down? How fast do you have to move to get out of its way? (Hang on just a moment while I get out my calculator. . . .) Motion was one of the earliest explorations of physics.

When you take a look around, you see that the motion of objects changes all the time. You see a motorcycle coming to a halt at a stop sign. You see a leaf falling and then stopping when it hits the ground, only to be picked up again by the wind. You see a pool ball hitting other balls in just the wrong way so that they all move without going where they should. Part I of this book handles objects in motion — from balls to railroad cars and most objects in between. In this section, I introduce motion in a straight line, rotational motion, and the cyclical motion of springs and pendulums.

Measuring speed, direction, velocity, and acceleration

Speeds are big with physicists — how fast is an object going? Thirty-five miles per hour not enough? How about 3,500? No problem when you're dealing with physics. Besides speed, the direction an object is going is important if you want to describe its motion. If the home team is carrying a football down the field, you want to make sure they're going in the right direction.

When you put speed and direction together, you get a vector — the velocity vector. Vectors are a very useful kind of quantity. Anything that has both size and direction is best described with a *vector.* Vectors are often represented as arrows, where the length of the arrow tells you the magnitude (size), and the direction of the arrow tells you the direction. For a velocity vector, the length corresponds to the speed of the object, and the arrow points in the direction the object is moving. (To find out how to use vectors, head to Chapter 4.)

Everything has a velocity, so velocity is great for describing the world around you. Even if an object is at rest with respect to the ground, it's still on the Earth, which itself has a velocity. (And if everything has a velocity, it's no wonder physicists keep getting grant money — somebody has to measure all that motion.)

If you've ever ridden in a car, you know that velocity isn't the end of the story. Cars don't start off at 60 miles per hour; they have to accelerate until they get to that speed. Like velocity, acceleration has not only a magnitude but also a direction, so acceleration is a vector in physics as well. I cover speed, velocity, and acceleration in Chapter 3.

Round and round: Rotational motion

Plenty of things go round and round in the everyday world — CDs, DVDs, tires, pitchers' arms, clothes in a dryer, roller coasters doing the loop, or just little kids spinning from joy in their first snowstorm. That being the case, physicists want to get in on the action with measurements. Just as you can have a car moving and accelerating in a straight line, its tires can rotate and accelerate in a circle.

Going from the linear world to the rotational world turns out to be easy, because there's a handy physics *analog* (which is a fancy word for "equivalent") for everything linear in the rotational world. For example, distance traveled becomes angle turned. Speed in meters per second becomes angular speed in angle turned per second. Even linear acceleration becomes rotational acceleration.

So when you know linear motion, rotational motion just falls in your lap. You use the same equations for both linear and angular motion — just different symbols with slightly different meanings (angle replaces distance, for example). You'll be looping the loop in no time. Chapter 7 has the details.

Springs and pendulums: Simple harmonic motion

Have you ever watched something bouncing up and down on a spring? That kind of motion puzzled physicists for a long time, but then they got down to work. They discovered that when you stretch a spring, the force isn't constant. The spring pulls back, and the more you pull the spring, the stronger it pulls back.

So how does the force compare to the distance you pull a spring? The force is directly proportional to the amount you stretch the spring: Double the amount you stretch the spring, and you double the amount of force with which the spring pulls back.

Physicists were overjoyed — this was the kind of math they understood. Force proportional to distance? Great — you can put that relationship into an equation, and you can use that equation to describe the motion of the object tied to the spring. Physicists got results telling them just how objects tied to springs would move — another triumph of physics.

This particular triumph is called *simple harmonic motion*. It's *simple* because force is directly proportional to distance, and so the result is simple. It's *harmonic* because it repeats over and over again as the object on the spring bounces up and down. Physicists were able to derive simple equations that could tell you exactly where the object would be at any given time.

But that's not all. Simple harmonic motion applies to many objects in the real world, not just things on springs. For example, pendulums also move in simple harmonic motion. Say you have a stone that's swinging back and forth on a string. As long as the arc it swings through isn't too high, the stone on a string is a pendulum; therefore, it follows simple harmonic motion. If you know how long the string is and how big of an angle the swing covers, you can predict where the stone will be at any time. I discuss simple harmonic motion in Chapter 13.

When Push Comes to Shove: Forces

Forces are a particular favorite in physics. You need forces to get motionless things moving — literally. Consider a stone on the ground. Many physicists (except, perhaps, geophysicists) would regard it suspiciously. It's just sitting there. What fun is that? What can you measure about that? After physicists had measured its size and mass, they'd lose interest.

But kick the stone — that is, apply a force — and watch the physicists come running over. Now something is happening — the stone started at rest, but now it's moving. You can find all kinds of numbers associated with this motion. For instance, you can connect the force you apply to something to its mass and get its acceleration. And physicists love numbers, because numbers help describe what's happening in the physical world.

Physicists are experts in applying forces to objects and predicting the results. Got a refrigerator to push up a ramp and want to know if it'll go? Ask a physicist. Have a rocket to launch? Same thing.

Absorbing the energy around you

You don't have to look far to find your next piece of physics. (You never do.) As you exit your house in the morning, for example, you may hear a crash up the street. Two cars have collided at a high speed, and locked together, they're sliding your way. Thanks to physics (and more specifically, Part III of this book), you can make the necessary measurements and predictions to know exactly how far you have to move to get out of the way.

Having mastered the ideas of energy and momentum helps at such a time. You use these ideas to describe the motion of objects with mass. The energy of motion is called *kinetic energy,* and when you accelerate a car from 0 to 60 miles per hour in 10 seconds, the car ends up with plenty of kinetic energy.

Where does the kinetic energy come from? It comes from *work,* which is what happens when a force moves an object through a distance. The energy can also come from *potential energy,* the energy stored in the object, which comes from the work done by a particular kind of force, such as gravity or electrical forces. Using gasoline, for example, an engine does work on the car to get it up to speed. But you need a force to accelerate something, and the way the engine does work on the car, surprisingly, is to use the force of friction with the road. Without friction, the wheels would simply spin, but because of a frictional force, the tires impart a force on the road. For every force between two objects, there is a reactive force of equal size but in the opposite direction. So the road also exerts a force on the car, which causes it to accelerate.

Or say that you're moving a piano up the stairs of your new place. After you move up the stairs, your piano has potential energy, simply because you put in a lot of work against gravity to get the piano up those six floors. Unfortunately, your roommate hates pianos and drops yours out the window. What happens next? The potential energy of the piano due to its height in a gravitational field is converted into kinetic energy, the energy of motion. You decide to calculate the final speed of the piano as it hits the street. (Next, you calculate the bill for the piano, hand it to your roommate, and go back downstairs to get your drum set.)

That's heavy: Pressures in fluids

Ever notice that when you're 5,000 feet down in the ocean, the pressure is different from at the surface? Never been 5,000 feet beneath the ocean waves? Then you may have noticed the difference in pressure when you dive into a swimming pool. The deeper you go, the higher the pressure is because of the weight of the water above you exerting a force downward. *Pressure* is just force per area.

Got a swimming pool? Any physicists worth their salt can tell you the approximate pressure at the bottom if you tell them how deep the pool is. When working with fluids, you have all kinds of other quantities to measure, such as the velocity of fluids through small holes, a fluid's density, and so on. Once again, physics responds with grace under pressure. You can read about forces in fluids in Chapter 8.

Feeling Hot but Not Bothered: Thermodynamics

Heat and cold are parts of your everyday life. Ever take a look at the beads of condensation on a cold glass of water in a warm room? Water vapor in the air is being cooled when it touches the glass, and it condenses into liquid water. The condensing water vapor passes thermal energy to the glass, which passes thermal energy to the cold drink, which ends up getting warmer as a result.

Thermodynamics can tell you how much heat you're radiating away on a cold day, how many bags of ice you need to cool a lava pit, and anything else that deals with heat energy. You can also take the study of thermodynamics beyond planet Earth. Why is space cold? In a normal environment, you radiate heat to everything around you, and everything around you radiates heat back to you. But in space, your heat just radiates away, so you can freeze.

Radiating heat is just one of the three ways heat can be transferred. You can discover plenty more about heat, whether created by a heat source like the sun or by friction, through the topics in Part IV.

Chapter 2

Reviewing Physics Measurement and Math Fundamentals

In This Chapter

- Mastering measurements (and keeping them straight as you solve equations)
- Accounting for significant digits and possible error
- Brushing up on basic algebra and trig concepts

Physics uses observations and measurements to make mental and mathematical models that explain how the world (and everything in it) works. This process is unfamiliar to most people, which is where this chapter comes in.

This chapter covers some basic skills you need for the coming chapters. I cover measurements and scientific notation, give you a refresher on basic algebra and trigonometry, and show you which digits in a number to pay attention to — and which ones to ignore. Continue on to build a physics foundation, solid and unshakable, that you can rely on throughout this book.

Measuring the World around You and Making Predictions

Physics excels at measuring and predicting the physical world — after all, that's why physics exists. Measuring is the starting point — part of observing the world so you can then model and predict it. You have several different measuring sticks at your disposal: some for length, some for mass or weight, some for time, and so on. Mastering those measurements is part of mastering physics.

Using systems of measurement

To keep like measurements together, physicists and mathematicians have grouped them into *measurement systems.* The most common measurement system you see in introductory physics is the meter-kilogram-second (MKS) system, referred to as SI (short for *Système International d'Unités,* the International System of Units), but you may also come across the foot-pound-second (FPS) system. Table 2-1 lists the primary units of measurement in the MKS system, along with their abbreviations.

Table 2-1 Units of Measurement in the MKS System

Measurement	*Unit*	*Abbreviation*
Length	meter	m
Mass	kilogram	kg
Time	second	s
Force	newton	N
Energy	joule	J
Pressure	pascal	Pa
Electric current	ampere	A
Magnetic flux density	tesla	T
Electric charge	coulomb	C

Because different measurement systems use different standard lengths, you can get several different numbers for one part of a problem, depending on the measurement you use. For example, if you're measuring the depth of the water in a swimming pool, you can use the MKS measurement system, which gives you an answer in meters, or the less common FPS system, in which case you determine the depth of the water in feet. The point? When working with equations, stick with the same measurement system all the way through the problem. If you don't, your answer will be a meaningless hodgepodge, because you're switching measuring sticks for multiple items as you try to arrive at a single answer. Mixing up the measurements causes problems — imagine baking a cake where the recipe calls for 2 cups of flour, but you use 2 liters instead.

From meters to inches and back again: Converting between units

Physicists use various measurement systems to record numbers from their observations. But what happens when you have to convert between those systems? Physics problems sometimes try to trip you up here, giving you the data you need in mixed units: centimeters for this measurement but meters for that measurement — and maybe even mixing in inches as well. Don't be fooled. You have to convert *everything* to the same measurement system before you can proceed. How do you convert in the easiest possible way? You use conversion factors, which I explain in this section.

Using conversion factors

To convert between measurements in different measuring systems, you can multiply by a conversion factor. A *conversion factor* is a ratio that, when you multiply it by the item you're converting, cancels out the units you don't want and leaves those that you do. The conversion factor must equal 1.

Here's how it works: For every relation between units — for example, 24 hours = 1 day — you can make a fraction that has the value of 1. If, for example, you divide both sides of the equation 24 hours = 1 day by 1 day, you get

$$\frac{24\text{ hours}}{1\text{ day}}=1$$

Suppose you want to convert 3 days to hours. You can just multiply your time by the preceding fraction. Doing so doesn't change the value of the time because you're multiplying by 1. You can see that the unit of *days* cancels out, leaving you with a number of hours:

$$\frac{3\text{ days}}{1}\times\frac{24\text{ hours}}{1\text{ day}}=\frac{3\text{ }\cancel{\text{days}}}{1}\times\frac{24\text{ hours}}{1\text{ }\cancel{\text{day}}}=72\text{ hours}$$

Words such as *days, seconds,* and *meters* act like the variables x and y in that if they're present in both the numerator and the denominator, they cancel each other out.

To convert the other way — hours into days, in this example — you simply use the same original relation, 24 hours = 1 day, but this time divide both sides by 24 hours to get

$$1 = \frac{1 \text{ day}}{24 \text{ hours}}$$

Then multiply by this fraction to cancel the units from the bottom, which leaves you with the units on the top.

Consider the following problem. Passing the state line, you note that you've gone 4,680 miles in exactly three days. Very impressive. If you went at a constant speed, how fast were you going? Speed is just as you may expect — distance divided by time. So you calculate your speed as follows:

$$\frac{4{,}680 \text{ miles}}{3 \text{ days}} = 1{,}560 \text{ miles/day}$$

Your answer, however, isn't exactly in a standard unit of measure. You have a result in miles per day, which you write as miles/day. To calculate miles per hour, you need a conversion factor that knocks *days* out of the denominator and leaves *hours* in its place, so you multiply by *days/hour* and cancel out *days:*

$$\frac{\text{miles}}{\cancel{\text{day}}} \times \frac{\cancel{\text{days}}}{\text{hour}} = \frac{\text{miles}}{\text{hour}}$$

Your conversion factor is *days/hour*. When you multiply by the conversion factor, your work looks like this:

$$\frac{1{,}560 \text{ miles}}{1 \text{ day}} \times \frac{1 \text{ day}}{24 \text{ hours}}$$

Note that because there are 24 hours in a day, the conversion factor equals exactly 1, as all conversion factors must. So when you multiply 1,560 miles/day by this conversion factor, you're not changing anything — all you're doing is multiplying by 1.

When you cancel out *days* and multiply across the fractions, you get the answer you've been searching for:

$$\frac{1{,}560 \text{ miles}}{1 \cancel{\text{ day}}} \times \frac{1 \cancel{\text{ day}}}{24 \text{ hours}} = 65 \text{ miles/hour}$$

So your average speed is 65 miles per hour, which is pretty fast considering that this problem assumes you've been driving continuously for three days.

Looking at the units when numbers make your head spin

Want an inside trick that teachers and instructors often use to solve physics problems? Pay attention to the units you're working with. I've had thousands of one-on-one problem-solving sessions with students in which we worked on homework problems, and I can tell you that this is a trick that instructors use all the time.

As a simple example, say you're given a distance and a time, and you have to find a speed. You can cut through the wording of the problem immediately because you know that distance (for example, meters) divided by time (for example, seconds) gives you speed (meters/second). Multiplication and division are reflected in the units. So, for example, because a rate like speed is given as a distance divided by a time, the units (in MKS) are meters/second. As another example, a quantity called *momentum* is given by velocity (meters/second) multiplied by mass (kilograms); it has units of kg·m/s.

As the problems get more complex, however, more items are involved — say, for example, a mass, a distance, a time, and so on. You find yourself glancing over the words of a problem to pick out the numeric values and their units. Have to find an amount of energy? Energy is mass times distance squared over time squared, so if you can identify these items in the question, you know how they're going to fit into the solution and you won't get lost in the numbers.

The upshot is that units are your friends. They give you an easy way to make sure you're headed toward the answer you want. So when you feel too wrapped up in the numbers, check the units to make sure you're on the right path. But remember: You still need to make sure you're using the right equations!

You don't *have* to use a conversion factor; if you instinctively know that you need to divide by 24 to convert from miles per day to miles per hour, so much the better. But if you're ever in doubt, use a conversion factor and write out the calculations, because taking the long road is far better than making a mistake. I've seen far too many people get everything in a problem right except for this kind of simple conversion.

Eliminating Some Zeros: Using Scientific Notation

Physicists have a way of getting their minds into the darndest places, and those places often involve really big or really small numbers. Physics has a way of dealing with very large and very small numbers; to help reduce clutter and make them easier to digest, it uses *scientific notation*.

In scientific notation, you write a number as a decimal (with only one digit before the decimal point) multiplied by a power of ten. The power of ten (10 with an exponent) expresses the number of zeroes. To get the right power of ten for a vary large number, count all the places in front of the decimal point, from right to left, up to the place just to the right of the first digit (you don't include the first digit because you leave it in front of the decimal point in the result).

For example, say you're dealing with the average distance between the sun and Pluto, which is about 5,890,000,000,000 meters. You have a lot of meters on your hands, accompanied by a lot of zeroes. You can write the distance between the sun and Pluto as follows:

$$5{,}890{,}000{,}000{,}000 \text{ meters} = 5.89 \times 10^{12} \text{ meters}$$

The exponent is 12 because you count 12 places between the end of 5,890,000,000,000 (where a decimal would appear in the whole number) and the decimal's new place after the 5.

Scientific notation also works for very small numbers, such as the one that follows, where the power of ten is negative. You count the number of places, moving left to right, from the decimal point to just after the first nonzero digit (again leaving the result with just one digit in front of the decimal):

$$0.0000000000000000005339 \text{ meters} = 5.339 \times 10^{-19} \text{ meters}$$

Using unit prefixes

Scientists have come up with a handy notation that helps take care of variables that have very large or very small values in their standard units. Say you're measuring the thickness of a human hair and find it to be 0.00002 meters thick. You could use scientific notation to write this as 2×10^{-5} meters (20×10^{-6} meters), or you could use the unit prefix μ, which stands for *micro:* 20 μm. When you put μ in front of any unit, it represents 10^{-6} times that unit.

A more familiar unit prefix is *k*, as in *kilo,* which represents 10^3 times the unit. For example the kilometer, km, is 10^3 meters, which equals 1,000 meters. The following table shows other common unit prefixes that you may see.

Unit Prefix	*Exponent*
mega (M)	10^6
kilo (k)	10^3
centi (c)	10^{-2}
milli (m)	10^{-3}
micro (μ)	10^{-6}
nano (n)	10^{-9}
pico (p)	10^{-12}

If the number you're working with is larger than ten, you have a positive exponent in scientific notation; if it's smaller than one, you have a negative exponent. As you can see, handling super large or super small numbers with scientific notation is easier than writing them all out, which is why calculators come with this kind of functionality already built in.

Here's a simple example: How does the number 1,000 look in scientific notation? You'd like to write 1,000 as 1.0 times ten to a power, but what is the power? You'd have to move the decimal point of 1.0 three places to the right to get 1,000, so the power is three:

$$1{,}000 = 1.0 \times 10^3$$

Checking the Accuracy and Precision of Measurements

Accuracy and precision are important when making (and analyzing) measurements in physics. You can't imply that your measurement is more precise than you know it to be by adding too many significant digits, and you have to account for the possibility of error in your measurement system by adding a ± when necessary. This section delves deeper into the topics of significant digits, precision, and accuracy.

Knowing which digits are significant

This section is all about how to properly account for the known precision of the measurements and carry that through the calculations, how to represent numbers in a way that is consistent with their known precision, and what to do with calculations that involve measurements with different levels of precision.

Finding the number of significant digits

In a measurement, *significant digits* (or *significant figures*) are those that were actually measured. Say you measure a distance with your ruler, which has millimeter markings. You can get a measurement of 10.42 centimeters, which has four significant digits (you estimate the distance between markings to get the last digit). But if you have a very precise micrometer gauge, then you can measure the distance to within one-hundredth of that, so you may measure the same thing to be 10.4213 centimeters, which has six significant digits.

By convention, zeroes that simply fill out values down to (or up to) the decimal point aren't considered significant. When you see a number given as 3,600, you know the the 3 and 6 are included because they're significant. However, knowing which, if any, of the zeros are significant can be tricky.

The best way to write a number so that you leave no doubt about how many significant digits there are is to use scientific notation. For example, if you read of a measurement of 1,000 meters, you don't know if there are one, two, three, or four significant figures. But if it were written as 1.0×10^3 meters, you would know that there are two significant figures. If the measurement were written as 1.000×10^3 meters, then you would know that there are four significant figures.

Rounding answers to the correct number of digits

When you do calculations, you often need to round your answer to the correct number of significant digits. If you include any more digits, you claim a precision that you don't really have and haven't measured.

For example, if someone tells you that a rocket traveled 10.0 meters in 7.0 seconds, the person is telling you that the distance is known to three significant digits and the seconds are known to two significant digits (the number of digits in each of the measurements). If you want to find the rocket's speed, you can whip out a calculator and divide 10.0 meters by 7.0 seconds to come up with 1.428571429 meters per second, which looks like a very precise measurement indeed. But the result is too precise — if you know your measurements to only two or three significant digits, you can't say you know the answer to ten significant digits. Claiming as such would be like taking a meter stick, reading down to the nearest millimeter, and then writing down an answer to the nearest ten-millionth of a millimeter. You need to round your answer.

The rules for determining the correct number of significant digits after doing calculations are as follows:

- **When you multiply or divide numbers:** The result has the same number of significant digits as the original number that has the fewest significant digits. In the case of the rocket, where you need to divide, the result should have only two significant digits (the number of significant digits in 7.0). The best you can say is that the rocket is traveling at 1.4 meters per second, which is 1.428571429 rounded to one decimal place.
- **When you add or subtract numbers:** Line up the decimal points; the last significant digit in the result corresponds to the right-most column where all numbers still have significant digits. If you have to add 3.6, 14, and 6.33, you'd write the answer to the nearest whole number — the 14 has no significant digits after the decimal place, so the answer shouldn't, either. You can see what I mean by taking a look for yourself:

$$\begin{array}{r} 3.6 \\ 14 \\ +6.33 \\ \hline 23.93 \end{array}$$

When you round the answer to the correct number of significant digits, your answer is 24.

When you round a number, look at the digit to the right of the place you're rounding to. If that right-hand digit is 5 or greater, round up. If it's 4 or less, round down. For example, you round 1.428 up to 1.43 and 1.42 down to 1.4.

Estimating accuracy

Physicists don't always rely on significant digits when recording measurements. Sometimes, you see measurements that use plus-or-minus signs to indicate possible error in measurement, as in the following:

5.36 ± 0.05 meters

The ± part (0.05 meters in the preceding example) is the physicist's estimate of the possible error in the measurement, so the physicist is saying that the actual value is between 5.36 + 0.05 (that is, 5.41) meters and 5.36 – 0.05 (that is, 5.31 meters), inclusive. Note that the possible error isn't the amount your measurement *differs* from the "right" answer; it's an indication of how precisely your apparatus can measure — in other words, how reliable your results are as a measurement.

Arming Yourself with Basic Algebra

Physics deals with plenty of equations, and to be able to handle them, you should know how to move the variables in them around. Note that algebra doesn't just allow you to plug in numbers and find values of different variables; it also lets you rearrange equations so you can make substitutions in other equations, and these new equations show different physics concepts. If you can follow along with the derivation of a formula in a physics book, you can get a better understanding of why the world works the way it does. That's pretty important stuff! Time to travel back to basic algebra for a quick refresher.

You need to be able to isolate different variables. For instance, the following equation tells you the distance, *s,* that an object travels if it starts from rest and accelerates at rate of *a* for a time, *t:*

$$s = \frac{1}{2}at^2$$

Now suppose the problem actually tells you the time the object is in motion and the distance it travels and asks you to calculate the object's acceleration. By rearranging the equation algebraically, you can solve for the acceleration:

$$a = \frac{2s}{t^2}$$

In this case, you've multiplied both sides by 2 and divided both sides by t^2 to isolate the acceleration, *a,* on one side of the equation.

What if you have to solve for the time, *t?* By moving the number and variables around, you get the following equation:

$$t = \sqrt{\frac{2s}{a}}$$

Do you need to memorize all three of these variations on the same equation? Certainly not. You just memorize one equation that relates these three items — distance, acceleration, and time — and then rearrange the equation as needed. (If you need a review of algebra, get a copy of *Algebra I For Dummies* by Mary Jane Sterling [Wiley].)

Tackling a Little Trig

You need to know a little trigonometry, including the sine, cosine, and tangent functions, for physics problems. To find these values, start with a simple right triangle. Take a look at Figure 2-1, which displays a right triangle in all its glory, complete with labels I've provided for the sake of explanation. Note in particular the angle θ, which appears between one of the triangle's legs and the hypotenuse (the longest side, which is opposite the right angle). The side *y* is opposite θ, and the side *x* is adjacent to θ.

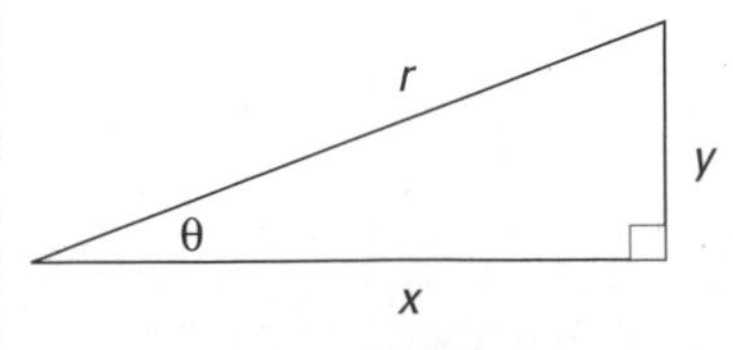

Figure 2-1: A labeled triangle that you can use to find trig values.

To find the trigonometric values of the triangle in Figure 2-1, you divide one side by another. Here are the definitions of sine, cosine, and tangent:

- $\sin\theta = \frac{y}{r}$
- $\cos\theta = \frac{x}{r}$
- $\tan\theta = \frac{y}{x}$

If you're given the measure of one angle and one side of the triangle, you can find all the other sides. Here are some other forms of the trig relationships — they'll probably become distressingly familiar before you finish any physics course, but you *don't* need to memorize them. If you know the preceding sine, cosine, and tangent equations, you can derive the following ones as needed:

- $x = r\cos\theta = \frac{y}{\tan\theta}$
- $y = r\sin\theta = x\tan\theta$
- $r = \frac{y}{\sin\theta} = \frac{x}{\cos\theta}$

To find the angle θ, you can go backward with the inverse sine, cosine, and tangent, which are written as $\sin^{-1}$, $\cos^{-1}$, and $\tan^{-1}$. Basically, if you input the sine of an angle into the $\sin^{-1}$ equation, you end up with the measure of the angle itself. Here are the inverses for the triangle in Figure 2-1:

- $\sin^{-1}\left(\frac{y}{r}\right) = \theta$
- $\cos^{-1}\left(\frac{x}{r}\right) = \theta$
- $\tan^{-1}\left(\frac{y}{x}\right) = \theta$

If you need a more in-depth refresher, check out *Trigonometry For Dummies,* by Mary Jane Sterling (Wiley).

Interpreting Equations as Real-World Ideas

After teaching physics to college students for many years, I'm very familiar with one of the biggest problems they face — getting lost in, and being intimidated by, the math.

Be a genius: Don't focus on the math

Richard Feynman was a famous Nobel Prize winner in physics who had a reputation during the 1950s and '60s of being an amazing genius. He later explained his method: He attached the problem at hand to a real-life scenario, creating a mental image, while others got caught in the math. When someone would show him a long derivation that had gone wrong, for example, he'd think of some physical phenomenon that the derivation was supposed to explain. As he followed along, he'd get to the point where he suddenly realized the derivation no longer matched what happened in the real world, and he'd say, "No, that's the problem." He was always right, which mystified people who, awe-struck, took him for a supergenius. Want to be a supergenius? Do the same thing: Don't let the math scare you.

Always keep in mind that the real world comes first and the math comes later. When you face a physics problem, make sure you don't get lost in the math; keep a global perspective about what's going on in the problem, because doing so helps you stay in control.

In physics, the ideas and observations of the physical world are the things that are important. Math operations are really only a simplified language for accurately describing what is going on. For example, here's a simple equation for speed:

$$v = \frac{s}{t}$$

In this equation, v is the speed, s is the distance, and t is the time. You can examine this equation's terms to see how this equation embodies simple common-sense notions of speed. Say that you travel a larger distance in the same amount of time. In that case, the right side of the equation must be larger, which means that your speed, on the left, is also greater. If you travel the same distance but it takes you more time, then the right side of this equation becomes smaller, which means that your speed is lower. The relationship between all the different components makes sense.

You can think of all the equations you come across in a similar way to make sure they make sense in the real world. If your equation behaves in a way that doesn't make physical sense, then you know that something must be wrong with the equation.

Bottom line: In physics, math is your friend. You don't need to get lost in it. Instead, you use it to formulate the problem and help guide you in its solution. Alone, each of these mathematical operations is very simple, but when you put them together, they're very powerful.

Chapter 3

Exploring the Need for Speed

In This Chapter

- Getting up to speed on displacement
- Dissecting different kinds of speed
- Going with acceleration
- Examining the link among acceleration, time, and displacement
- Connecting velocity, acceleration, and displacement

There you are in your Formula 1 racecar, speeding toward glory. You have the speed you need, and the pylons are whipping past on either side. You're confident that you can win, and coming into the final turn, you're far ahead. Or at least you think you are. Seems that another racer is also making a big effort, because you see a gleam of silver in your mirror. You get a better look and realize that you need to do something — last year's winner is gaining on you fast.

It's a good thing you know all about velocity and acceleration. With such knowledge, you know just what to do: You floor the gas pedal, accelerating out of trouble. Your knowledge of velocity lets you handle the final curve with ease. The checkered flag is a blur as you cross the finish line in record time. Not bad. You can thank your understanding of the issues in this chapter: displacement, velocity, and acceleration.

You already have an intuitive feeling for what I discuss in this chapter, or you wouldn't be able to drive or even ride a bike. Displacement is about where you are, speed is about how fast you're going, and anyone who's ever been in a car knows about acceleration. These characteristics of motion concern people every day, and physics has made an organized study of them. This knowledge has helped people to plan roads, build spacecraft, organize traffic patterns, fly, track the motion of planets, predict the weather, and even get mad in slow-moving traffic jams. Understanding movement is a vital part of understanding physics, and that's the topic of this chapter. Time to move on.

Going the Distance with Displacement

When something moves from Point A to Point B, displacement takes place in physics terms. In plain English, *displacement* is a distance in a particular direction.

Like any other measurement in physics (except for certain angles), displacement always has units — usually centimeters or meters. You may also use kilometers, inches, feet, miles, or even *light-years* (the distance light travels in one year, a whopper of a distance not fit for measuring with a meter stick: 5,865,696,000,000 miles, which is 9,460,800,000,000 kilometers or 9,460,800,000,000,000 meters).

In this section, I cover position and displacement in one to three dimensions.

Understanding displacement and position

You find displacement by finding the distance between an object's initial position and its final position. Say, for example, that you have a fine new golf ball that's prone to rolling around, shown in Figure 3-1. This particular golf ball likes to roll around on top of a large measuring stick. You place the golf ball at the 0 position on the measuring stick, as you see in Figure 3-1, diagram A.

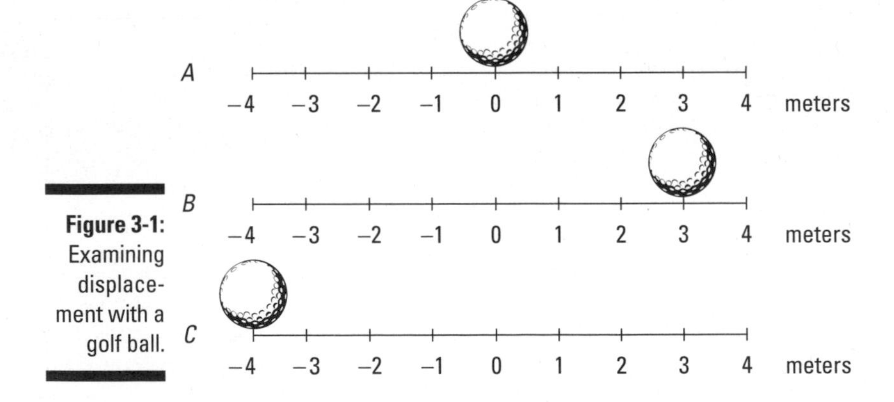

Figure 3-1: Examining displacement with a golf ball.

The golf ball rolls over to a new point, 3 meters to the right, as you see in Figure 3-1, diagram B. The golf ball has moved, so displacement has taken place. In this case, the displacement is just 3 meters to the right. Its initial position was 0 meters, and its final position is at +3 meters. The displacement is 3 meters.

In physics terms, you often see displacement referred to as the variable *s* (don't ask me why).

Scientists, being who they are, like to go into even more detail. You often see the term s_i, which describes *initial position,* (the *i* stands for *initial*). And you may see the term s_f used to describe *final position*.

In these terms, moving from diagram A to diagram B in Figure 3-1, s_i is at the 0-meter mark and s_f is at +3 meters. The displacement, *s,* equals the final position minus the initial position:

$$s = s_f - s_i$$
$$= 3\text{ m} - 0\text{ m} = 3\text{ m}$$

Displacements don't have to be positive; they can be zero or negative as well. If the positive direction is to the right, then a negative displacement means that the object has moved to the left.

In diagram C, the restless golf ball has moved to a new location, which is measured as –4 meters on the measuring stick. The displacement is given by the difference between the initial and final position. If you want to know the displacement of the ball from its position in diagram B, take the initial position of the ball to be $s_i = 3$ meters; then the displacement is given by

$$s = s_f - s_i$$
$$= -4\text{ m} - 3\text{ m} = -7\text{ m}$$

When working on physics problems, you can choose to place the origin of your position-measuring system wherever is convenient. The measurement of the position of an object depends on where you choose to place your origin; however, displacement from an initial position s_i to a final position s_f does not depend on the position of the origin because the displacement depends only on the *difference* between the positions, not the positions themselves.

Examining axes

Motion that takes place in the world isn't always in one dimension. Motion can take place in two or three dimensions. And if you want to examine motion in two dimensions, you need two intersecting meter sticks (or number lines), called *axes.* You have a horizontal axis — the *x*-axis — and a vertical axis — the *y*-axis. (For three-dimensional problems, watch for a third axis — the *z*-axis — sticking straight up out of the paper.)

Finding the distance

Take a look at Figure 3-2, where a golf ball moves around in two dimensions. The ball starts at the center of the graph and moves up to the right. In terms of the axes, the golf ball moves to +4 meters on the *x*-axis and +3 meters on the *y*-axis, which is represented as the point (4, 3); the *x* measurement comes first, followed by the *y* measurement: (*x*, *y*).

So what does this mean in terms of displacement? The change in the *x* position, Δx (Δ, the Greek letter delta, means "change in"), is equal to the final *x* position minus the initial *x* position. If the golf ball starts at the center of the graph — the origin of the graph, location (0, 0) — you have a change in the *x* location of

$$\begin{aligned}\Delta x &= x_f - x_i \\ &= 4\text{ m} - 0\text{ m} = 4\text{ m}\end{aligned}$$

The change in the *y* location is

$$\begin{aligned}\Delta y &= y_f - y_i \\ &= 3\text{ m} - 0\text{ m} = 3\text{ m}\end{aligned}$$

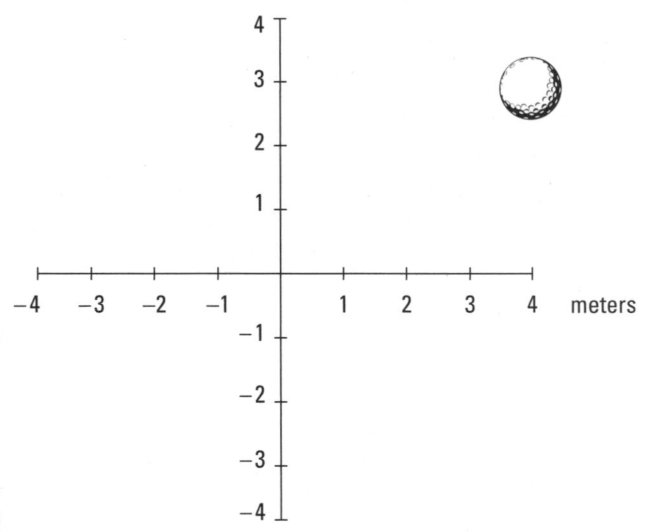

Figure 3-2: A ball moving in two dimensions.

If you're more interested in figuring out the magnitude (size) of the displacement than in the changes in the *x* and *y* locations of the golf ball, that's a different story. The question now becomes: How far is the golf ball from its starting point at the center of the graph?

Using the *distance formula* — which is just the Pythagorean theorem solved for the hypotenuse — you can find the *magnitude of the displacement* of the golf ball, which is the distance it travels from start to finish. The Pythagorean theorem states that the sum of the squares of the legs of a right triangle ($a^2 + b^2$) is equal to the square on the hypotenuse (c^2). Here, the legs of the triangle are Δx and Δy, and the hypotenuse is s. Here's how to work the equation:

$$\begin{aligned} s &= \sqrt{\Delta x^2 + \Delta y^2} \\ &= \sqrt{(4\text{ m})^2 + (3\text{ m})^2} \\ &= \sqrt{16\text{ m}^2 + 9\text{ m}^2} \\ &= \sqrt{25\text{ m}^2} \\ &= 5\text{ m} \end{aligned}$$

So in this case, the magnitude of the ball's displacement is exactly 5 meters.

Determining direction

You can find the direction of an object's movement from the values of Δx and Δy. Because these are just the legs of a right triangle, you can use basic trigonometry to find the angle of the ball's displacement from the *x*-axis. The tangent of this angle is simply given by

$$\tan\theta = \frac{\Delta y}{\Delta x}$$

Therefore, the angle itself is just the inverse tangent of that:

$$\begin{aligned} \theta &= \tan^{-1}\left(\frac{\Delta y}{\Delta x}\right) \\ &= \tan^{-1}\left(\frac{3\text{ m}}{4\text{ m}}\right) \\ &\approx 37^\circ \end{aligned}$$

The ball in Figure 3-2 has moved at an angle of 37° from the *x*-axis.

Speed Specifics: What Is Speed, Anyway?

There's more to the story of motion than just the actual movement. When displacement takes place, it happens in a certain amount of time. You may already know that speed is distance traveled per a certain amount of time:

$$\text{speed} = \frac{\text{distance}}{\text{time}}$$

For example, if you travel distance *s* in a time *t*, your speed, *v*, is

$$v = \frac{s}{t}$$

The variable *v* really stands for velocity, but true velocity also has a direction associated with it, whereas *speed* does not. For that reason, velocity is a vector (you usually see the velocity vector represented as **v** or $\vec{v}$. *Vectors* have both a magnitude (size) and a direction, so with velocity, you know not only how fast you're going but also in what direction. Speed is only a magnitude (if you have a certain velocity vector, in fact, the speed is the magnitude of that vector), so you see it represented by the term *v* (not in bold). You can read more about velocity and displacement as vectors in Chapter 4.

Just as you can measure displacement, you can measure the difference in time from the beginning to the end of the motion, and you usually see it written like this: $\Delta t = t_f - t_i$. Technically speaking (physicists love to speak technically), velocity is the change in position (displacement) divided by the change in time, so you can also represent it like this, if, say, you're moving along the *x*-axis:

$$v = \frac{\Delta x}{\Delta t} = \frac{x_f - x_i}{t_f - t_i}$$

Speed can take many forms, which you find out about in the following sections.

Reading the speedometer: Instantaneous speed

You already have an idea of what speed is; it's what you measure on your car's speedometer, right? When you're tooling along, all you have to do to see your speed is look down at the speedometer. There you have it: 75 miles per hour. Hmm, better slow it down a little — 65 miles per hour now. You're looking at your speed at this particular moment. In other words, you see your *instantaneous speed.*

Instantaneous speed is an important term in understanding the physics of speed, so keep it in mind. If you're going 65 mph right now, that's your instantaneous speed. If you accelerate to 75 mph, that becomes your instantaneous speed. Instantaneous speed is your speed at a particular instant of time. Two seconds from now, your instantaneous speed may be totally different.

Staying steady: Uniform speed

What if you keep driving 65 miles per hour forever? You achieve *uniform speed* in physics (also called *constant speed*). Uniform motion is the simplest speed variation to describe, because it never changes.

Uniform speed may be possible in the western portion of the United States, where the roads stay in straight lines for a long time and you don't have to change your speed. But uniform speed is also possible when you drive around a circle, too. Imagine driving around a racetrack; your velocity would change (because of the constantly changing direction), but your speed could remain constant as long as you keep your gas pedal pressed down the same amount. I discuss uniform circular motion in Chapter 7, but in this chapter, I stick to motion in straight lines.

Shifting speeds: Nonuniform motion

Nonuniform motion varies over time; it's the kind of speed you encounter more often in the real world. When you're driving, for example, you change speed often, and your changes in speed come to life in an equation like this, where v_f is your final speed and v_i is your initial speed:

$$\Delta v = v_f - v_i$$

The last part of this chapter is all about acceleration, which occurs in nonuniform motion. There, you see how changing speed is related to acceleration — and how you can accelerate even without changing speed!

Busting out the stopwatch: Average speed

Average speed is the total distance you travel divided by the total time it takes. Average speed is sometimes written as $\bar{v}$; a bar over a variable means *average* in physics terms.

Say, for example, that you want to pound the pavement from New York City to Los Angeles to visit your uncle's family, a distance of about 2,781 miles. If the trip takes you 4.000 days, what was your average speed? You divide the total distance by the change in time, so your average speed for the trip would be

$$\frac{2{,}781 \text{ miles}}{4.000 \text{ days}} \approx 695.3 \text{ miles/day}$$

This solution divides miles by days, so you come up with 695.3 miles per day. Not exactly a standard unit of measurement — what's that in miles per hour? To find it, you want to cancel *days* out of the equation and put in *hours* (see Chapter 2). Because a day is 24 hours, you can multiply this way (note that *days* cancels out, leaving miles over hours, or *miles per hour*):

$$\frac{2{,}781 \text{ miles}}{4.000 \text{ days}} \times \frac{1 \text{ day}}{24 \text{ hours}} \approx 28.97 \text{ miles/hour}$$

That's a better answer.

You can relate total distance traveled, s, with average speed, $\bar{v}$, and time, t, like this:

$$s = \bar{v}t$$

Contrasting average and instantaneous speed

Average speed differs from instantaneous speed, unless you're traveling in uniform motion (in which case your speed never varies). In fact, because average speed is the total distance divided by the total time, it may be very different from your instantaneous speed.

If you travel 2,781 miles in four days (a total of 96 hours), you go at an average speed of 28.97 miles per hour. That answer seems pretty slow, because when you're driving, you're used to going 65 miles per hour. You've calculated an average speed over the whole trip, obtained by dividing the total distance by the total trip time, which includes non-driving time. You may have stopped at a hotel several nights, and while you slept, your instantaneous speed was 0 miles per hour; yet even at that moment, your overall average speed was still 28.97 miles per hour!

Distinguishing average speed and average velocity

There is a difference between average speed and average velocity. Say, for example, that while you were driving in Ohio on your cross-country trip, you wanted to make a detour to visit your sister in Michigan after you dropped off a hitchhiker in Indiana. Your travel path may have looked like the straight lines in Figure 3-3 — first 80 miles to Indiana and then 30 miles to Michigan.

If you drove at an average speed or a uniform speed of 55 miles per hour and you had to cover $80 + 30 = 110$ miles, this trip took you 2.0 hours. But if you calculate the magnitude of the average velocity (by taking the distance between the starting point and the ending point, about 85 miles as the crow flies), you get

$$\frac{85 \text{ miles}}{2.0 \text{ hours}} \approx 43 \text{ miles/hour}$$

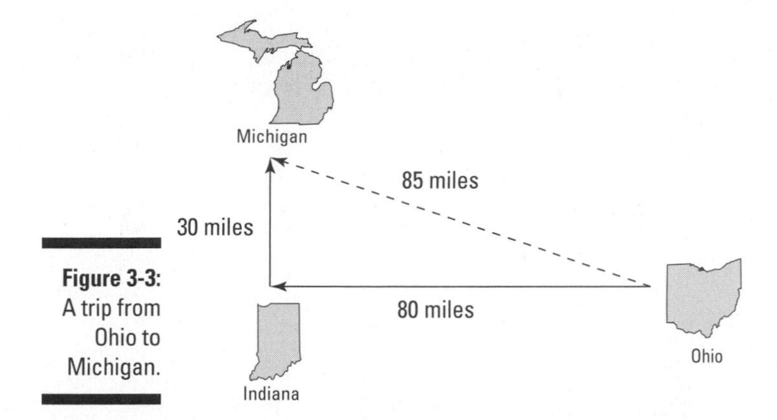

Figure 3-3: A trip from Ohio to Michigan.

The direction of the average velocity is just the direction between the start and end points. But if you're interested in your average speed along either of the two legs of the trip, you have to measure the time it takes for a leg and divide the length of that leg by that time to get the average speed.

To calculate the average speed over the whole trip, you look at the whole distance traveled, which is 80 + 30 = 110 miles, not just 85 miles. And 110 miles divided by 2.0 hours is 55 miles per hour; this is your average speed.

As another illustration of the difference between average speed and average velocity, consider the motion of the Earth around the sun. The Earth travels in its nearly circular orbit around the sun at an enormous average speed of something like 18 miles per second! However, if you consider one full revolution of the Earth, the Earth returns to its original position, relative to the sun, after one year. After one year, there's no displacement relative to the sun, so the Earth's average velocity over a year is zero, even though its average speed is enormous!

When considering motion, it's not only speed that counts but also direction. That's why velocity is important: It lets you record an object's speed and its direction. Pairing speed with direction enables you to handle cases like cross-country travel, where the direction can change.

Speeding Up (Or Down): Acceleration

Acceleration is a measure of how quickly your velocity changes. When you pass a parking lot's exit and hear squealing tires, you know what's coming next — someone is accelerating to cut you off. After he passes, he slows down right in front of you, forcing you to hit your brakes to slow down yourself. Good thing you know all about physics.

You may think that, with all this speeding up and slowing down, you'd use terms like *acceleration* and *deceleration.* Well, physics has no use for the term *deceleration,* because deceleration is just a particular kind of acceleration — one in which speed reduces.

Like speed, acceleration takes many forms that affect your calculations in various physics situations. In different physics problems, you have to take into account the direction of the acceleration (whether the acceleration is positive or negative in a particular direction), whether it's average or instantaneous, and whether it's uniform or nonuniform. This section tells you more about acceleration and explores its various forms.

Defining acceleration

In physics terms, *acceleration, a,* is the amount by which your velocity changes in a given amount of time, or

$$a = \frac{\Delta v}{\Delta t}$$

Given the initial and final velocities, v_i and v_f, and the initial and final times over which your speed changes, t_i and t_f, you can also write the equation like this:

$$a = \frac{\Delta v}{\Delta t} = \frac{v_f - v_i}{t_f - t_i}$$

Acceleration, like velocity, is actually a vector and is often written as **a,** in vector style (see Chapter 4). In other words, acceleration, like velocity but unlike speed, has a direction associated with it.

Determining the units of acceleration

You can calculate the units of acceleration easily enough by dividing velocity by time to get acceleration:

$$a = \frac{v_f - v_i}{t_f - t_i}$$

In terms of units, the equation looks like this:

$$a = \frac{v_f - v_i}{t_f - t_i} = \frac{\text{distance/time}}{\text{time}} = \text{distance/time}^2$$

Distance per time squared? Don't let that throw you. You end up with time squared in the denominator because you divide velocity by time. In other words, *acceleration* is the rate at which your velocity changes, because rates have time in the denominator. For acceleration, you see units of meters per second^2, centimeters per second^2, miles per second^2, feet per second^2, or even kilometers per hour^2.

It may be easier, for a given problem, to use units such as mph/s (miles per hour per second). This would be useful if the velocity in question had a magnitude of something like several miles per hour that changed typically over a number of seconds.

Looking at positive and negative acceleration

Just as for displacement and velocity, acceleration can be positive or negative. This section explains how positive and negative acceleration relate to changes in speed and direction.

Changing speed

The sign of the acceleration tells you whether you're speeding up or slowing down (depending on which direction you're traveling).

For example, say that you're driving at 75 miles per hour, and you see those flashing red lights in the rearview mirror. You pull over, taking 20 seconds to come to a stop. The officer appears by your window and says, "You were going 75 miles per hour in a 30-mile-per-hour zone." What can you say in reply?

You can calculate your rate of acceleration as you pulled over, which, no doubt, would impress the officer — look at you and your law-abiding tendencies! You whip out your calculator and begin entering your data. Remember that the acceleration is given in terms of the change in velocity divided by the change in time:

$$a = \frac{\Delta v}{\Delta t}$$

Plugging in the numbers, your calculations look like this:

$$\begin{aligned} a &= \frac{\Delta v}{\Delta t} \\ &= \frac{75 \text{ mph}}{20 \text{ s}} \\ &\approx 3.8 \text{ mph/s} \end{aligned}$$

Your acceleration was 3.8 mph/s. But that can't be right! You may already see the problem here; take a look at the original definition of acceleration:

$$a = \frac{\Delta v}{\Delta t} = \frac{v_f - v_i}{t_f - t_i}$$

Your final speed was 0 mph, and your original speed was 75 mph, so plugging in the numbers here gives you this acceleration:

$$\begin{aligned} a = \frac{\Delta v}{\Delta t} &= \frac{v_f - v_i}{t_f - t_i} \\ &= \frac{0 - 75 \text{ mph}}{20 \text{ s}} \\ &\approx -3.8 \text{ mph/s} \end{aligned}$$

In other words, –3.8 mph/s, *not* +3.8 mph/s — a big difference in terms of solving physics problems (and in terms of law enforcement). If you accelerated at +3.8 mph/s rather than –3.8 mph/s , you'd end up going 150 mph at the end of 20 seconds, not 0 mph. And that probably wouldn't make the cop very happy.

Now you have your acceleration. You can turn off your calculator and smile, saying, "Maybe I was going a little fast, officer, but I'm very law abiding. Why, when I heard your siren, I accelerated at –3.8 mph/s just in order to pull over promptly." The policeman pulls out his calculator and does some quick calculations. "Not bad," he says, impressed. And you know you're off the hook.

Accounting for direction

The sign of the acceleration depends on direction. If you slow down to a complete stop in a car, for example, and your original velocity was positive and your final velocity was 0, then your acceleration is negative because a positive velocity came down to 0. However, if you slow down to a complete stop in a car and your original velocity was negative and your final velocity was 0, then your acceleration would be positive because a negative velocity increased to 0.

Looking at positive and negative acceleration

When you hear that acceleration is going on in an everyday setting, you typically think that means the speed is increasing. However, in physics, that isn't always the case. An acceleration can cause speed to increase, decrease, and even stay the same!

Acceleration tells you the rate at which the velocity is changing. Because the velocity is a vector, you have to consider the changes to its magnitude and direction. The acceleration can change the magnitude and/or the direction of the velocity. Speed is only the magnitude of the velocity.

Here's a simple example that shows how a simple constant acceleration can cause the speed to increase and decrease in the course of an object's motion. Say you take a ball, throw it straight up in the air, and then catch it again. If you throw the ball upward with a speed of 9.8 m/s, the velocity has a magnitude of 9.8 m/s in the upward direction. Now the ball is under the influence of gravity, which, on the surface of the Earth, causes all free-falling objects to undergo a vertical acceleration of –9.8 m/s^2. This acceleration is negative because its direction is vertically downward.

With this acceleration, what's the velocity of the ball after 1.0 second ? Well, you know that

$$a = \frac{v_f - v_i}{t_f - t_i}$$

Rearrange this equation and plug in the numbers, and you find that the final velocity after 1.0 second is 0 meters/second:

$$\begin{aligned} v_f &= v_i + a(t_f - t_i) \\ &= 9.8 \text{ m/s} + (-9.8 \text{ m/s}^2)(1.0 \text{ s}) \\ &= 0 \text{ m/s} \end{aligned}$$

After 1.0 second, the ball has zero velocity because it's reached the top of its trajectory, just at the point where it's about to fall back down again. So the acceleration has actually slowed down the ball because it was going in the direction opposite the velocity.

Now see what happens as the ball falls back down to Earth. The ball has zero velocity, but the acceleration due to gravity accelerates the ball downward at a rate of –9.8 m/s^2. As the ball falls, it gathers speed before you catch it. What's its final velocity as you catch it, given that its initial velocity at the top of its trajectory is zero?

The time for the ball to fall back down to you is just the same as the time it took to reach the top of its trajectory, which is 1.0 second, so you can find the final velocity for this part of the ball's motion with this calcuation:

$$\begin{aligned} v_f &= v_i + a(t_f - t_i) \\ &= 0 \text{ m/s} + (-9.8 \text{ m/s}^2)(1.0 \text{ s}) \\ &= -9.8 \text{ m/s} \end{aligned}$$

So the final velocity is 9.8 meters/second directed straight downward. The magnitude of this velocity — that is, the speed of the ball — is 9.8 meters/second. The acceleration increases the speed of the ball as it falls because the acceleration is in the same direction as the velocity for this part of the ball's trajectory.

When you work with physics problems, bear in mind that acceleration can speed up or slow down an object, depending on the direction of the acceleration and the velocity of the object. Don't simply assume that just because something is accelerating its speed must be increasing. (By the way, if you want to see an example of how an acceleration can leave the speed of an object unchanged, take a look at the circular motion topic covered in Chapter 7.)

Examining average and instantaneous acceleration

Just as you can examine average and instantaneous speeds and velocities, you can also examine average and instantaneous acceleration. *Average acceleration* is the ratio of the change in velocity to the change in time. You calculate average acceleration, also written as $\bar{a}$, by taking the final velocity, subtracting the initial velocity, and dividing the result by the total time (final time minus the initial time):

$$\bar{a} = \frac{v_f - v_i}{t_f - t_i}$$

At any given point, the acceleration you measure is the instantaneous acceleration, and that number can be different from the average acceleration. For example, when you first see red flashing police lights behind you, you may jam on the brakes, which gives you a big acceleration, in the direction opposite to which you're moving (in everyday parlance, you'd say you just *decelerated,* but that term's a no-no in physics circles). But you lighten up a little and coast to a stop, so the acceleration is smaller. The average acceleration, however, is a single value, derived by dividing the overall change in velocity by the overall time.

Acceleration is the rate of change of velocity, not speed. If a velocity's direction changes without a change in speed, this is also a kind of acceleration.

Taking off: Putting the acceleration formula into practice

Here's an acceleration example. As they strap you into the jet on the aircraft carrier deck, the mechanic says you need to take off at a speed of at least 62.0 m/s. You'll be catapulted at an acceleration of 31 m/s^2. Is there going to be enough catapult to do the job? You ask how long the catapult is. "A hundred meters," says the mechanic, finishing strapping you in.

Hmm, you think. Will an acceleration of 31 m/s^2 over a distance of 100 meters do the trick? You take out your clipboard and ask yourself: How far must I be accelerated at 31 m/s^2 to achieve a speed of 62 m/s?

First think of the distance that you need to be accelerated over as the size of the displacement from your initial position. To find this displacement, you can use the equation $s = \bar{v}t$, where s is the displacement, $\bar{v}$ is the average velocity, and t is the time — which means you have to find the time over which you're accelerated. For that, you can use the equation that relates change in velocity, Δv, acceleration a, and change in time, Δt:

$$a = \frac{\Delta v}{\Delta t}$$

Solving for Δt gives you

$$\Delta t = \frac{\Delta v}{a}$$

Plugging in the numbers and solving gives you the change in time:

$$\begin{aligned}\Delta t &= \frac{\Delta v}{a}\\ &= \frac{62 \text{ m/s}}{31 \text{ m/s}^2}\\ &= 2.0 \text{ s}\end{aligned}$$

Okay, so it takes 2.0 seconds for you to reach a speed of 62 m/s if your rate of acceleration is 31 m/s^2. Now you can use this equation to find the total distance you need to travel to get up to this speed; it is the size of the displacement, which is given by $s = \bar{v}t$, where $\bar{v} = (1/2)(v_i + v_f)$, $v_i = 0$ m/s, and $v_f = 62$ m/s. So your equation is

$$s = \frac{1}{2}(v_i + v_f)t$$

Plugging in the numbers gives you

$$\begin{aligned}s &= \frac{1}{2}(v_i + v_f)t\\ &= \frac{1}{2}(0 \text{ m/s} + 62 \text{ m/s})(2.0 \text{ s})\\ &= 62 \text{ m}\end{aligned}$$

So it will take 62 meters of 31 m/s^2 acceleration to get you to takeoff speed — and the catapult is 100 meters long. No problem.

Understanding uniform and nonuniform acceleration

Acceleration can be uniform or nonuniform. Nonuniform acceleration requires a change in acceleration. For example, when you're driving, you encounter stop signs or stop lights often, and when you slow to a stop and then speed up again, you take part in nonuniform acceleration.

Other accelerations are very uniform (in other words, unchanging), such as the acceleration due to gravity near the surface of the Earth. This acceleration is 9.8 meters per second^2 downward, toward the center of the Earth, and it doesn't change (if it did, plenty of people would be pretty startled).

Relating Acceleration, Time, and Displacement

This chapter deals with four quantities of motion: acceleration, velocity, time, and displacement. You work the standard equation relating displacement and time to get velocity:

$$v = \frac{\Delta s}{\Delta t} = \frac{s_f - s_i}{t_f - t_i}$$

And you see the standard equation relating velocity and time to get acceleration:

$$a = \frac{\Delta v}{\Delta t} = \frac{v_f - v_i}{t_f - t_i}$$

But both of these equations only go one level deep, relating velocity to displacement and time and acceleration to velocity and time. What if you want to relate acceleration to displacement and time? This section shows you how you can cut velocity out of the equation.

When you're slinging around algebra, you may find it easier to write single quantities like v (to stand for Δv) rather than $v_f - v_i$. You can usually turn v into $v_f - v_i$ later if necessary.

Not-so-distant relations: Deriving the formula

You relate acceleration, displacement, and time by messing around with the equations until you get what you want. First, note that displacement equals average velocity multiplied by time:

$$s = \bar{v}t$$

You have a starting point. But what's the average velocity? If your acceleration is constant, your velocity increases in a straight line from 0 to its final value, as Figure 3-4 shows.

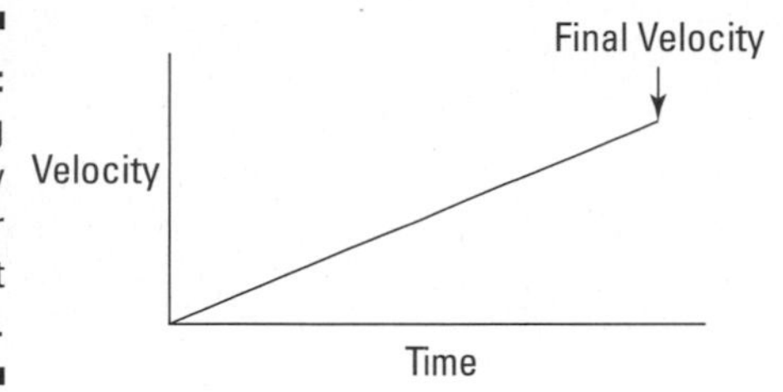

Figure 3-4: Increasing velocity under constant acceleration.

The average velocity is half the final velocity, and you know this because there's constant acceleration. Your final velocity is $v_f = at$, so your average velocity is half this:

$$\bar{v} = \frac{1}{2}at$$

So far, so good. Now you can plug this average velocity into the $s = \bar{v}t$ equation and get

$$s = \bar{v}t = \left(\frac{1}{2}at\right)t$$

And this becomes

$$s = \frac{1}{2}at^2$$

You can also put in $t_f - t_i$ rather than just plain t:

$$s = \frac{1}{2}a(t_f - t_i)^2$$

Congrats! You've worked out one of the most important equations you need to know when you work with physics problems relating acceleration, displacement, time, and velocity.

Notice that when you derived this equation, you had an initial velocity of zero. What if you don't start off at zero velocity, but you still want to relate acceleration, time, and displacement? What if you're initially going 100 miles per hour? That initial velocity would certainly add to the final distance you go. Because distance equals speed multiplied by time, the equation looks like this (don't forget that this assumes the acceleration is constant):

$$s = v_i(t_f - t_i) + \frac{1}{2}a(t_f - t_i)^2$$

You also see this written simply as the following (where t stands for Δt, the time over which the acceleration happened):

$$s = v_i t + \frac{1}{2}at^2$$

Calculating acceleration and distance

With the formula relating distance, acceleration, and time, you can find any of those values, given the other two. If you have an initial velocity, too, finding distance or acceleration isn't any harder. In this section, I work through some physics problems to show you how these formulas work.

Finding acceleration

Given distance and time, you can find acceleration. Say you become a drag racer in order to analyze your acceleration down the dragway. After a test race, you know the distance you went — 402 meters, or about 0.25 miles (the magnitude of your displacement) — and you know the time it took — 5.5 seconds. So what was your acceleration as you blasted down the track?

Well, you know how to relate displacement, acceleration, and time (see the preceding section), and that's what you want — you always work the algebra so that you end up relating all the quantities you know to the one quantity you *don't* know. In this case, you have

$$s = \frac{1}{2}at^2$$

(Keep in mind that in this case, your initial velocity is 0 — you're not allowed to take a running start at the drag race!) You can rearrange this equation with a little algebra to solve for acceleration; just divide both sides by t^2 and multiply by 2 to get

$$a = \frac{2s}{t^2}$$

Great. Plugging in the numbers, you get the following:

$$\begin{aligned} a &= \frac{2s}{t^2} \\ &= \frac{2(402 \text{ m})}{(5.5 \text{ s})^2} \\ &= \frac{804 \text{ m}}{30.25 \text{ s}^2} \\ &\approx 27 \text{ m/s}^2 \end{aligned}$$

Okay, the acceleration is approximately 27 meters per second^2. What's that in more understandable terms? The acceleration due to gravity, *g,* is — 9.8 meters per second^2, so this is about 2.7 g's — you'd feel yourself pushed back into your seat with a force about 2.7 times your own weight.

Figuring out time and distance

Given a constant acceleration and the change in velocity, you can figure out both time and distance. For instance, imagine you're a drag racer. Your acceleration is 26.6 meters per second^2, and your final speed is 146.3 meters per second. Now find the total distance traveled. Got you, huh? "Not at all," you say, supremely confident. "Just let me get my calculator."

You know the acceleration and the final speed, and you want to know the total distance required to get to that speed. This problem looks like a puzzler because the equations in this chapter have involved time up to this point. But if you need the time, you can always solve for it. You know the final speed, v_f, and the initial speed, v_i (which is zero), and you know the acceleration, a. Because $v_f - v_i = at$, you know that

$$\begin{aligned} t &= \frac{v_f - v_i}{a} \\ &= \frac{146.3 \text{ m/s} - 0 \text{ m/s}}{26.6 \text{ m/s}^2} \\ &= 5.50 \text{ s} \end{aligned}$$

Now you have the time. You still need the distance, and you can get it this way:

$$s = v_i t + \frac{1}{2}at^2$$

The second term drops out because $v_i = 0$, so all you have to do is plug in the numbers:

$$\begin{aligned} s &= \frac{1}{2}at^2 \\ &= \frac{1}{2}\left(26.6 \text{ m/s}^2\right)\left(5.50 \text{ s}\right)^2 \\ &\approx 402 \text{ m} \end{aligned}$$

In other words, the total distance traveled is 402 meters, or a quarter mile. Must be a quarter-mile racetrack.

Finding distance with initial velocity

Given initial velocity, time, and acceleration, you can find displacement. Here's an example: There you are, the Tour de France hero, ready to give a demonstration of your bicycling skills. There will be a time trial of 8.0 seconds. Your initial speed is 6.0 meters/second, and when the whistle blows, you accelerate at 2.0 m/s^2 for the 8.0 seconds allowed. At the end of the time trial, how far will you have traveled?

You could use the relation $s = (1/2)at^2$, except you don't start off from zero speed — you're already moving, so you should use the following:

$$s = v_i t + \frac{1}{2}at^2$$

In this case, $a = 2.0$ m/s^2, $t = 8.0$ s, and $v_i = 6.0$ m/s, so you get the following:

$$\begin{aligned} s &= v_i t + \frac{1}{2}at^2 \\ &= \left(6.0 \text{ m/s}\right)\left(8.0 \text{ s}\right) + \frac{1}{2}\left(2.0 \text{ m/s}^2\right)\left(8.0 \text{ s}\right)^2 \\ &= 48.0 \text{ m} + \frac{1}{2}\left(128 \text{ m}\right) \\ &\approx 110 \text{ m} \end{aligned}$$

You write the answer to two significant digits — 110 meters — because you know the time only to two significant digits (see Chapter 2 for info on rounding). In other words, you ride to victory in about 110 meters in 8.0 seconds. The crowd roars.

Linking Velocity, Acceleration, and Displacement

Say you want to relate displacement, acceleration, and velocity without having to know the time. Here's how it works. First, you solve the acceleration formula for the time:

$$t = \frac{v_f - v_i}{a}$$

Because displacement is $s = \bar{v}t$ and average velocity is $\bar{v} = (1/2)(v_i + v_f)$ when the acceleration is constant, you can get the following equation:

$$s = \frac{1}{2}(v_i + v_f)t$$

Substituting for the time, t, you get

$$s = \frac{1}{2}(v_i + v_f)\left(\frac{v_f - v_i}{a}\right)$$

After doing the algebra and simplifying, you get

$$s = \frac{v_f^2 - v_i^2}{2a}$$

Moving the $2a$ to the other side of the equation, you get an important equation of motion:

$$v_f^2 - v_i^2 = 2as$$

Whew. If you can memorize this one, you can relate velocity, acceleration, and displacement. Put this equation to work — you see it often in physics problems.

Finding acceleration

There you are, getting into your Physics racer as the crowd cheers. It's time for some hefty acceleration. You get out your clipboard. What acceleration would you need to end up at 100 miles per hour at the end of a 1.0-mile racetrack?

Okay, you think. You need an equation that relates speed, acceleration, and displacement. It's time for

$$v_f^2 - v_i^2 = 2as$$

In this case, it's even a little easier, because you know that the initial velocity is 0 ($v_i = 0$), so you have

$$v_f^2 = 2as$$

Well, well, it looks like the problem is half-solved. Putting in the numbers gives you

$$(100 \text{ mph})^2 = 2a(1.0 \text{ mile})$$

Now solve for a:

$$\begin{aligned} a &= \frac{(100 \text{ mph})^2}{2(1.0 \text{ mile})} \\ &= 5{,}000 \text{ miles/hour}^2 \end{aligned}$$

Miles per hour^2? What the heck kind of units are those? Change that to something more understandable, such as mph per second. To change one of the per-hour units to per-second, multiply by the conversion factor (see Chapter 2):

$$\begin{aligned} a &= \frac{5{,}000 \text{ miles}}{1 \text{ hour}^2} \\ &= \frac{5{,}000 \text{ miles}}{1 \text{ hour} \times 1 \cancel{\text{hour}}} \times \frac{1 \cancel{\text{hour}}}{60 \cancel{\text{minutes}}} \times \frac{1 \cancel{\text{minute}}}{60 \text{ seconds}} \\ &\approx 1.4 \text{ miles/hour/second} = 1.4 \text{ mph/s} \end{aligned}$$

So your velocity would be increasing by only 1.4 mph every second — that's not too outrageous — you'd feel a mild acceleration, that's all.

Solving for displacement

Now say that you're at the end of the first mile and want to see how far you'd have to go — at the same acceleration — to get to 200 miles per hour. Once again, you need to relate velocity, acceleration, and displacement, so this equation is your baby:

$$v_f^2 = 2as$$

Here, you want to solve for s, the displacement, and you get this:

$$s = \frac{v_f^2 - v_i^2}{2a}$$

Great. Now for some numbers. In this case, $v_f = 200$ mph, $v_i = 100$ mph, and $a = 5{,}000$ miles/hour^2, and you don't know s at this point. To find s, plug your numbers into the equation you found for s to get

$$\begin{aligned} s &= \frac{v_f^2 - v_i^2}{2a} \\ &= \frac{(200 \text{ mph})^2 - (100 \text{ mph})^2}{2(5{,}000 \text{ mph}^2)} \\ &= \frac{40{,}000 \text{ mph}^2 - 10{,}000 \text{ mph}^2}{10{,}000 \text{ mph}^2} \\ &= 3.0 \text{ miles} \end{aligned}$$

So it would take 3.0 additional miles to get you up to 200 mph.

Finding final velocity

Here's one more example. Say you're in your rocket ship, happily speeding along at some 3.25 kilometers per second (about 7,280 miles per hour) when you see a sign: Speed Zone 215 km Ahead — New Speed Limit: 3.0 km/s.

You jam on the brakes (which are a retro rocket in the front of the rocket ship). The retro rocket is capable of accelerating your ship at –10.0 meters/second^2.

It's a tense moment. Will you get your speed down to below 3.0 kilometers per second in 215 kilometers of acceleration? Find out, using your old friend:

$$v_f^2 - v_i^2 = 2as$$

In this case, you want to solve for the final speed, which is

$$v_f^2 = 2as + v_i^2$$

where $a = -10.0 \text{ m/s}^2$, $s = 215 \text{ km} = 215{,}000 \text{ m}$, and $v_i = 3.25 \text{ km/s} = 3{,}250 \text{ m/s}$. Plugging in the data and solving for v_f gives you the following:

$$v_f^2 = 2as + v_i^2$$
$$v_f^2 = 2\left(-10.0 \text{ m/s}^2\right)\left(215{,}000 \text{ m}\right) + \left(3{,}250 \text{ m/s}\right)^2$$
$$v_f^2 = -4{,}300{,}000 \text{ m}^2/\text{s}^2 + 10{,}562{,}500 \text{ m/s}^2$$
$$v_f^2 = 6{,}262{,}500 \text{ m}^2/\text{s}^2$$
$$\sqrt{v_f^2} = \sqrt{6{,}262{,}500 \text{ m}^2/\text{s}^2}$$
$$v_f \approx 2{,}500 \text{ m/s} = 2.50 \text{ km/s}$$

Whew, you think — 2.50 kilometers per second is well under the speed limit of 3.0 kilometes per second. You're safe.

You can now consider yourself a motion master.

Chapter 4

Following Directions: Motion in Two Dimensions

In This Chapter

- Mastering vector addition and subtraction
- Putting vectors into numerical coordinates
- Dividing vectors into components
- Identifying displacement, acceleration, and velocity as vectors
- Completing an exercise in gravity

You aren't limited to moving left and right or forward and backward; you can move in more than one dimension. In the real world, you need to know which way you're going and how far to go. For example, when a person gives you directions, she may point and say something like, "The posse went 15 miles thataway!" When you're helping someone hang a door, the person may say, "Push hard to the left!" And when you swerve to avoid hitting someone in your car, you accelerate in another direction. All these statements involve vectors.

A *vector* is a quantity that has both a size (magnitude) and a direction. Because physics models everyday life, plenty of concepts in physics are vectors, too, including velocity, acceleration, and force. For that reason, you should snuggle up to vectors, because you see them in just about any physics course you take. Vectors are fundamental.

Many people who've had tussles with vectors decide they don't like them, which is a mistake — vectors are easy after you get a handle on them, and you get a handle on them in this chapter. I break down vectors from top to bottom and relate the characteristics of motion (displacement, velocity, and acceleration) to the concept of vectors. Here, balls fly through the air and roll off cliffs, baseball players race to make plays, and you find a great shortcut to the nearest park bench. Read on.

Visualizing Vectors

In one dimension, displacement, velocity, and acceleration are either positive or negative (see Chapter 3). For example, they may be negative if they're to the left and positive to the right. The size of the displacement, velocity, or acceleration is given by the absolute size (regardless of sign) of the number representing it — this is the *magnitude.* The sign of the number indicates direction (left or right).

But what do you do if you have more than one dimension? If the object can move up and down as well as left and right, you can no longer use a single number to represent displacement, velocity, and acceleration. You need vectors. In this section, I represent vectors as arrows and show you what vector addition and subtraction look like.

Asking for directions: Vector basics

When you have a vector, you have to keep in mind two quantities: its direction and its magnitude. Quantities that have only a magnitude are called *scalars*. If you give a scalar magnitude a direction, you create a vector.

Visually, you see vectors drawn as arrows in physics, which is perfect because an arrow has both a clear direction and a clear magnitude (the length of the arrow). Take a look at Figure 4-1. The arrow represents a vector that starts at the arrow's foot (also called the *tail*) and ends at the head.

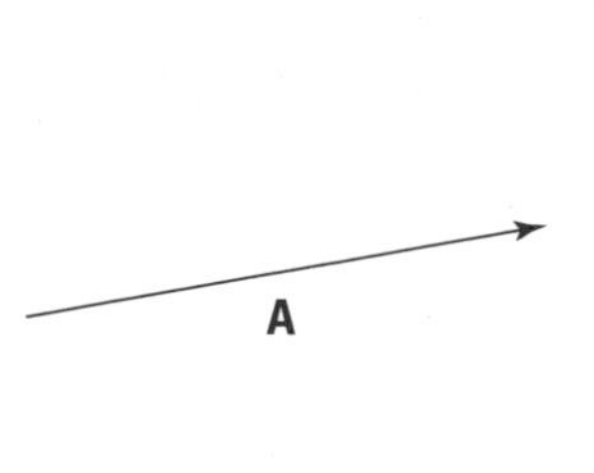

Figure 4-1: A vector, represented by an arrow, has both a direction and a magnitude.

In physics, you use a letter in bold type to represent a vector. This is the notation I use in this book; in some books, however, you see a letter with an arrow on top like this: $\vec{A}$. The arrow means that this is not only a scalar value, which would be represented by *A,* but also something with direction.

Say that you tell some smartypants that you know all about vectors. When he asks you to give him a vector, **A,** you give him not only its magnitude but also its direction, because you need these two bits of info together to define this vector. That impresses him to no end! For example, you may say that **A** is a vector at 15° from the horizontal with a magnitude of 12 meters/second. Smartypants knows all he needs to know, including that **A** is a velocity vector.

Take a look at Figure 4-2, which features two vectors, **A** and **B.** They look pretty much the same — the same length and the same direction. In fact, these vectors are equal. Two vectors are *equal* if they have the same magnitude and direction, and you can write this as **A** = **B**.

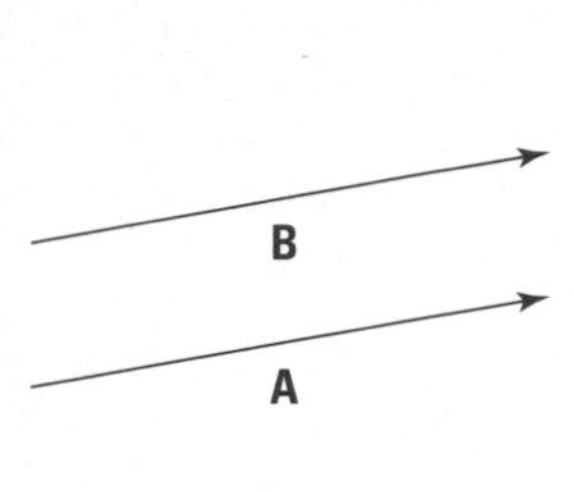

Figure 4-2: Equal vectors have the same length and direction but may have different starting points.

Looking at vector addition from start to finish

Just as you can add two numbers to get a third number, you can add two vectors to get a *resultant* vector. To show that you're adding two vectors, put the arrows together so that one arrow starts where the other arrow ends. The sum is a new arrow that starts at the base of the first arrow and ends at the head (pointy end) of the other.

Consider an example using displacement vectors. A *displacement vector* gives the change in position: the distance from the starting point to the ending point is the magnitude of the displacement vector, and the direction traveled is the direction of the displacement vector.

Assume, for example, that a passerby tells you that to get to your destination, you first have to follow vector **A** and then vector **B.** Just where is that destination? You work this problem just as you find the destination in everyday life. First, you drive to the end of vector **A,** and from that point, you drive to the end of vector **B,** just as you see in Figure 4-3.

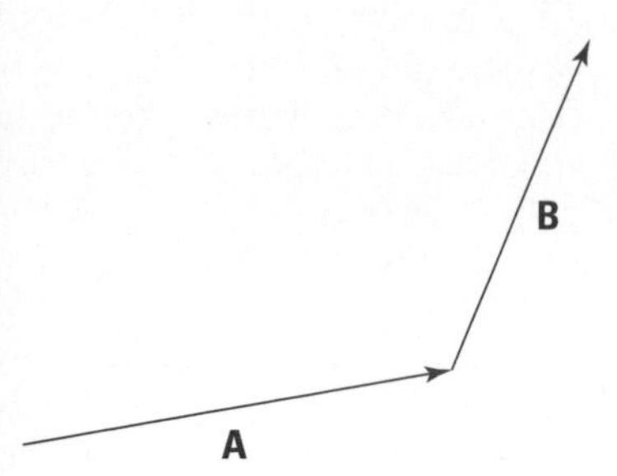

Figure 4-3: Going from the tail of one vector to the head of a second gets you to your destination.

When you get to the end of vector **B,** how far are you from your starting point? To find out, you draw a vector, **C,** from your starting point (foot, or tail, of the first vector) to your ending point (head of the second vector), as you see in Figure 4-4. This new vector represents your complete trip, from start to finish. In other words, **C** = **A** + **B.** The vector **C** is called the *sum,* the *result,* or the *resultant vector.*

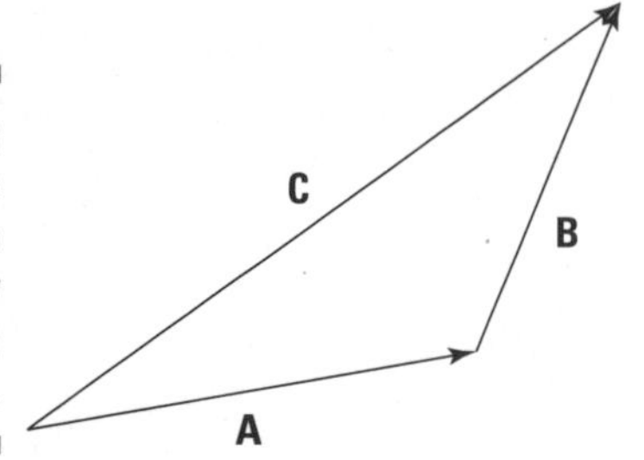

Figure 4-4: Take the sum of two vectors by creating a new vector.

Going head-to-head with vector subtraction

You don't come across vector subtraction very often in physics problems, but it does pop up. To subtract two vectors, you put their feet (the non-pointy parts) together; then draw the resultant vector, which is the difference of the two vectors, from the head of the vector you're subtracting to the head of the vector you're subtracting it from.

To make heads or tails of this, check out Figure 4-5, where you subtract **A** from **C** (in other words, **C** – **A**). As you can see, the result is **B,** because **C** = **A** + **B.**

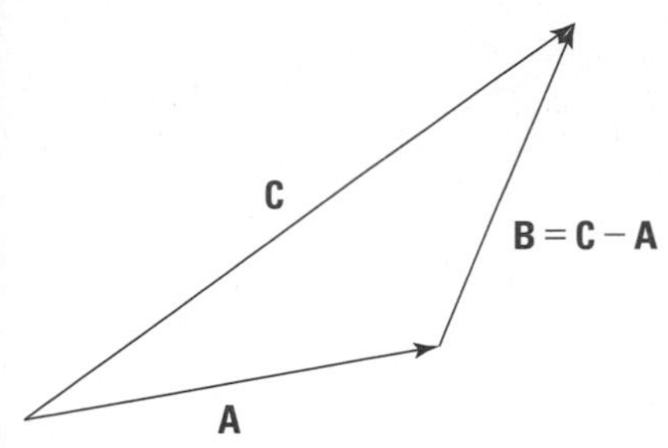

Figure 4-5: Subtracting two vectors by putting their feet together and drawing the result.

Another (and for some people, easier) way to do vector subtraction is to reverse the direction of the second vector (**A** in **C** – **A**) and use vector addition; that is, start with the first vector (**C**), put the reversed vector's foot (**A**) at the first vector's head, and draw the resultant vector.

Putting Vectors on the Grid

Vectors may look good as arrows floating in space, but that's not exactly the most precise way of dealing with them. You can get numerical with vectors, taking them apart as you need them, by putting the arrows in a grid, on the coordinate plane. The coordinate plane allows you to work with vectors using (x, y) coordinates and algebra.

Adding vectors by adding coordinates

In this section, I explain how you can use the components of vectors to add vectors together. Doing so reduces the problem of adding vectors to a simple combination of adding numbers together, which is very useful when you solve problems.

Take a look at the vector addition problem **A** + **B** in Figure 4-6. Now that you have the vectors plotted on a graph, you can see how easy vector addition really is. If the measurements in Figure 4-6 are in meters, that means vector **A** is 5 meters to the right and 1 meter up, and vector **B** is 1 meter to the right and 4 meters up. To add them for the result, vector **C**, you add the horizontal parts together and the vertical parts together.

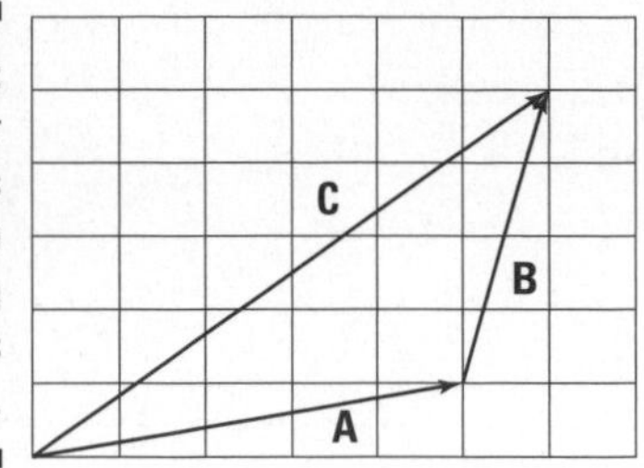

Figure 4-6: Use vector coordinates to make handling vectors easy.

The resultant vector, **C,** ends up being 6 meters to the right and 5 meters up. You can see what that looks like in Figure 4-6: To get the horizontal part of the sum, you add the horizontal part of **A** (5 meters) to the horizontal part of **B** (1 meter). To get the vertical part of the sum, **C,** you just add the vertical part of **A** (1 meter) to the vertical part of **B** (4 meters).

If vector addition still seems cloudy, you can use a notation that was invented for vectors to help physicists and *For Dummies* readers keep it straight. Because **A** is 5 meters to the right (the positive x-axis direction) and 1 up (the positive y-axis direction), you can express it with (x, y) coordinates like this:

$$\mathbf{A} = (5, 1)$$

And because **B** is 1 meter to the right and 4 up, you can express it with (x, y) coordinates like this:

$$\mathbf{B} = (1, 4)$$

Having a notation is great, because it makes vector addition totally simple. To add two vectors together, you just add their x and y parts, respectively, to get the x and y parts of the result:

$$\mathbf{A}\ (5, 1) + \mathbf{B}\ (1, 4) = \mathbf{C}\ (6, 5)$$

The whole secret of vector addition is breaking each vector up into its x and y parts and then adding those separately to get the resultant vector's x and y parts. Nothing to it. Now you can get as numerical as you like, because you're just adding or subtracting numbers. Getting those x and y parts can take a little work, but it's a necessary step. And when you have those parts, you're home free.

Here's a real-world example: Assume you're looking for a hotel that's 20 miles due north and then 20 miles due east. What's the vector that points at the hotel from your starting location? Taking your coordinate info into account, this is an easy problem. Say that the east direction is along the positive x-axis and that north is along the positive y-axis. Step 1 of your travel directions is 20 miles due north, and Step 2 is 20 miles due east. You can write the problem in vector notation like this (east [positive x], north [positive y]):

Step 1: (0, 20)

Step 2: (20, 0)

To add these two vectors together, add the coordinates:

$$(0, 20) + (20, 0) = (20, 20)$$

The resultant vector is (20, 20). It points from your starting point directly to the hotel.

Changing the length: Multiplying a vector by a number

You can perform simple vector multiplication by a scalar (number). For example, say you're driving along at 150 miles per hour eastward on a racetrack and you see a competitor in your rearview mirror. No problem, you think; you'll just double your speed:

$$2(0, 150) = (0, 300)$$

Now you're flying along at 300 miles per hour in the same direction. In this problem, you multiply a vector by a scalar.

A Little Trig: Breaking Up Vectors into Components

Physics problems have a way of not telling you what you want to know directly. As the preceding section explains, a vector can be described by its components, which are enough to uniquely specify a vector. Because a vector, by definition, is a quantity that has both magnitude and direction, another way to specify a vector is to use its magnitude and direction directly. If you know one way of describing the vector, you can work out the other.

These are just two different ways of specifying the same thing, and each has its own use in physics problems. Here's why you may work with vector components:

- **When you have vectors in components, they're easy to add and subtract and manipulate generally.** When a problem gives you vectors in terms of their magnitude and direction (which is often the case), you typically need to calculate their components just so you can work through the problem.
- **Being able to treat the horizontal and vertical directions separately is useful because you can often split one difficult problem into two simple problems. Using components also helps when one direction is more important than the other.** For example, a problem may say that a ball is rolling on a table at angle of 15° with a speed of 7.0 meters/second and ask you how long the ball will take to roll off the table's edge if that edge is 1.0 meter away. In that case, you care only about how quickly the ball is moving horizontally, directly toward the table's edge — the speed in the vertical direction doesn't matter.

After you solve a problem, the answer usually needs to be in terms of the magnitude and direction. So after you find your answer in components, you often have to work out the magnitude and direction again.

This section shows you how you can take the magnitude and direction of a vector and work out its components, as well as how you can take the components of a vector and work out its magnitude and direction.

Finding vector components

When you break a vector into its parts, those parts are called its *components*. For example, in the vector (4, 1), the x-axis (horizontal) component is 4, and the y-axis (vertical) component is 1. Typically, a physics problem gives you an angle and a magnitude to define a vector; you have to find the components yourself using a little trigonometry.

Suppose you know that a ball is rolling on a flat table at 15° from a direction parallel to the bottom edge with a speed of 7.0 meters/second. You may want to find out how long the ball will take to roll off the edge 1.0 meter to the right.

Define your axes so the ball is at the origin initially and the x-axis is parallel to the bottom edge of the table (see Figure 4-7). Therefore, the problem breaks down to finding out how long the ball will take to roll 1.0 meter in the x direction. To find the time, you first need to know how fast the ball is moving in the x direction.

The problem tells you that the ball is rolling at a speed of 7.0 meters/second at 15° to the horizontal (along the positive *x*-axis), which is a vector: 7.0 meters/second at 15° gives you both a magnitude and a direction. What you have here is a velocity — the vector version of speed. The ball's speed is the magnitude of its velocity vector, and when you include a direction to that speed, you get the velocity vector **v.**

To find out how fast the ball is traveling toward the table edge, you need not the ball's total speed but the *x* component of the ball's velocity. The *x* component is a scalar (a number, not a vector), and you write it like this: v_x. The *y* component of the ball's velocity vector is v_y. Therefore, you can say that

$$\mathbf{v} = (v_x, v_y)$$

That's how you express breaking a vector up into its components. So what's v_x here? And for that matter, what's v_y, the *y* component of the velocity? The vector has a length (7.0 meters/second) and a direction (θ = 15° to the horizontal). And you know that the edge of the table is 1.0 meter to the right.

As you can see in Figure 4-7, you have to use some trigonometry to resolve this vector into its components. No sweat. The trig is easy after you get the angles you see in Figure 4-7 down.

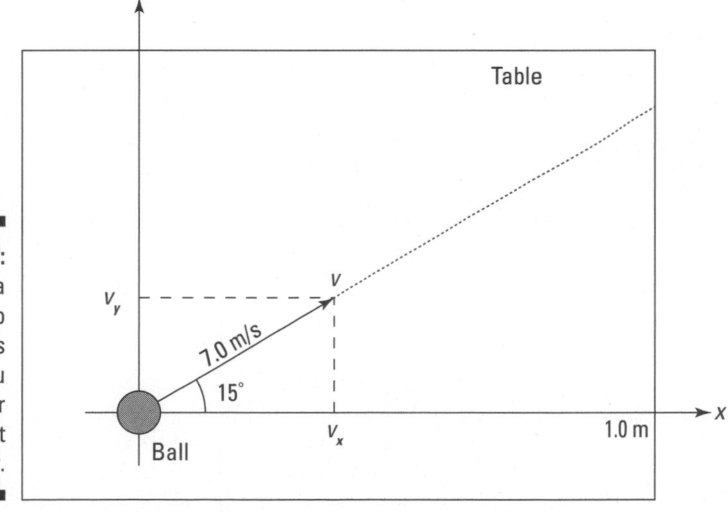

Figure 4-7: Breaking a vector into components allows you to add or subtract them easily.

The magnitude of a vector **v** is expressed as v, and from Figure 4-7, you can see that the following is true:

- **Horizontal component:** $v_x = v \cos \theta$
- **Vertical component:** $v_y = v \sin \theta$

The two vector-component equations are worth knowing because you see them a lot in any beginning physics course. Make sure you know how they work, and always have them at your fingertips.

Of course, if you forget these equations, you can always retrieve them from basic trigonometry. You may remember that the sine and cosine of an angle in a right triangle are defined as the ratio of the opposite side and the adjacent side to the hypotenuse, like so: $\sin \theta = v_y/v$ and $\cos \theta = v_x/v$ (see Chapter 2). By multiplying both sides of these equations by v, you can express the x and y components of the vector as

$$v_x = v \cos \theta$$

$$v_y = v \sin \theta$$

You can go further by relating each side of the triangle to each other side (and if you know that $\tan \theta = \sin \theta / \cos \theta$, you can derive all these from the previous two equations as required; no need to memorize all these):

- $v_x = v \cos\theta = \dfrac{v_y}{\tan\theta}$
- $v_y = v \sin\theta = v_x \tan\theta$
- $v = \dfrac{v_y}{\sin\theta} = \dfrac{v_x}{\cos\theta}$

You know that $v_x = v \cos \theta$, so you can find the x component of the ball's velocity, v_x, this way:

$$v_x = v \cos \theta$$

Plugging in the numbers gives you

$$\begin{aligned} v_x &= v\cos\theta \\ &= (7.0 \text{ m/s})\cos 15° \\ &\approx 6.8 \text{ m/s} \end{aligned}$$

You now know that the ball is traveling at 6.8 meters/second to the right. And because you also know that the table's edge is 1.0 meter away, you can divide distance by speed to get the time:

$$\frac{1.0\text{ m}}{6.8\text{ m/s}} \approx 0.15\text{ s}$$

Because you know how fast the ball is going in the *x* direction, you now know the answer to the problem: The ball will take 0.15 seconds to fall off the edge of the table. What about the *y* component of the velocity? That's easy to find, too:

$$v_y = v\sin\theta = (7.0\text{ m/s})\sin 15° \approx 1.8\text{ m/s}$$

Reassembling a vector from its components

Sometimes you have to find the angle and magnitude of a vector rather than the components. To find the magnitude, you use the Pythagorean theorem. And to find θ, you use the inverse tangent function (or inverse sine or cosine). This section shows you how these formulas work.

For example, assume you're looking for a hotel that's 20 miles due east and then 20 miles due north. From your present location, what is the angle (measured from east) of the direction to the hotel, and how far away is the hotel? You can write this problem in vector notation, like so (see the section "Putting Vectors on the Grid"):

Step 1: (20, 0)

Step 2: (0, 20)

When adding these vectors together, you get this result:

(20, 0) + (0, 20) = (20, 20)

The resultant vector is (20, 20). That's one way of specifying a vector — use its components. But this problem isn't asking for the results in terms of components. The question wants to know the angle and distance to the hotel. In other words, looking at Figure 4-8, the problem asks, "What's *h*, and what's θ?"

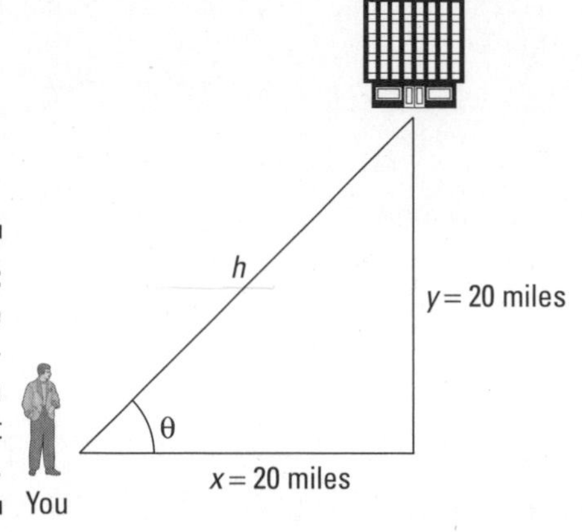

Figure 4-8: Using the angle created by a vector to get to a hotel.

Finding the magnitude

If you know a vector's vertical and horizontal components, finding the vector's magnitude isn't so hard, because you just need to find the hypotenuse of a triangle. You can use the Pythagorean theorem ($x^2 + y^2 = h^2$), solved for h:

$$h = \sqrt{x^2 + y^2}$$

Plugging in the numbers gives you

$$\begin{aligned} h &= \sqrt{x^2 + y^2} \\ &= \sqrt{(20 \text{ mi})^2 + (20 \text{ mi})^2} \\ &= \sqrt{400 \text{ mi}^2 + 400 \text{ mi}^2} \\ &= \sqrt{800 \text{ mi}^2} \\ &\approx 28 \text{ mi} \end{aligned}$$

Finding and checking the angle

When you know the horizontal and vertical components of a vector, you can use the tangent to find the angle because $\tan \theta = y/x$. All you have to do is take the inverse tangent of y/x:

$$\theta = \tan^{-1}\left(\frac{y}{x}\right)$$

Suppose you drive 20 miles east and 20 miles north. Here's how you find θ, the angle between your original position and your final one:

$$\begin{aligned}\theta &= \tan^{-1}\left(\frac{y}{x}\right)\\ &= \tan^{-1}\left(\frac{20\text{ mi}}{20\text{ mi}}\right)\\ &= \tan^{-1}(1)\\ &= 45^\circ\end{aligned}$$

So the hotel is about 28 miles away (as you see from the earlier section "Finding the magnitude") at an angle of 45°.

Be careful when doing calculations with inverse tangents, because angles that differ by 180° have the same tangent. When you take the inverse tangent, you may need to add or subtract 180° to get the actual angle you want. The inverse tangent button on your calculator will always give you an angle between 90° and –90°. If your angle is not in this range, then you have to add or subtract 180°.

For this example, the answer of 45° must be correct. But consider a situation in which you'd need to add or subtract 180°: Suppose that you walk in completely the opposite direction to the hotel. You walk 20 miles west and 20 miles south ($x = -20$ miles, $y = -20$ miles), so if you use the same method to work out the angle, you get the following:

$$\begin{aligned}\theta &= \tan^{-1}\left(\frac{y}{x}\right)\\ &= \tan^{-1}\left(\frac{-20\text{ mi}}{-20\text{ mi}}\right)\\ &= \tan^{-1}(1)\\ &= 45^\circ\end{aligned}$$

You get the same answer for the angle even though you're walking in completely the opposite direction as before! That's because the tangents of angles that differ by 180° are equal. But if you look at the components of the vector ($x = -20$ miles, $y = -20$ miles), they're both negative, so the angle must be between –180° and 0°. If you subtract 180° from your answer of 45°, you get –135°, which is your actual angle.

An alternative method of working out the direction is to find the vector's magnitude (hypotenuse) and then use the components in terms of the sine and cosine of the angle:

- $x = h \cos\theta$
- $y = h \sin\theta$

Then you can write the cosine and sine of the angle as

$$\frac{x}{h} = \cos\theta$$

$$\frac{y}{h} = \sin\theta$$

Now all you have to do is take the inverse cosine or sine:

$$\theta = \cos^{-1}\left(\frac{x}{h}\right)$$

$$\theta = \sin^{-1}\left(\frac{y}{h}\right)$$

Featuring Displacement, Velocity, and Acceleration in 2-D

When an object is moving in only one dimension (as in Chapter 3), you only have to deal with one component, which is just a single number — displacement is just a distance, velocity is just a speed, and acceleration is just speeding up or slowing down. So in one dimension, vectors just look like numbers: The magnitude of the vector is the size of the number, and the direction of the vector is just the sign of the number.

However, displacement, velocity, and acceleration are always vectors. In the real world, an object may be moving in two or more dimensions, so direction is important. In this section, I take another look at the equations for motion, except in more than one dimension so you can see more clearly how the equations are really vector equations.

Displacement: Going the distance in two dimensions

Displacement, which is the change in position (see Chapter 3), has a magnitude and a direction associated with it. When you have a change of position in a particular direction and of a particular distance, then these are given by the magnitude and direction of the displacement vector.

Instead of writing displacement as *s*, you should write it as **s**, a vector (if you're writing on paper, you can put an arrow over the *s* to signify its vector status). When you're talking about displacement in the real world, direction is as important as distance.

For example, say your dreams have come true: You're a big-time baseball or softball hero, slugging another line drive into the outfield. You take off for first base, which is 90 feet away. But 90 feet in which direction? Because you know how vital physics is, you happen to know that first base is 90 feet away at a 45° angle, as you can see in Figure 4-9.

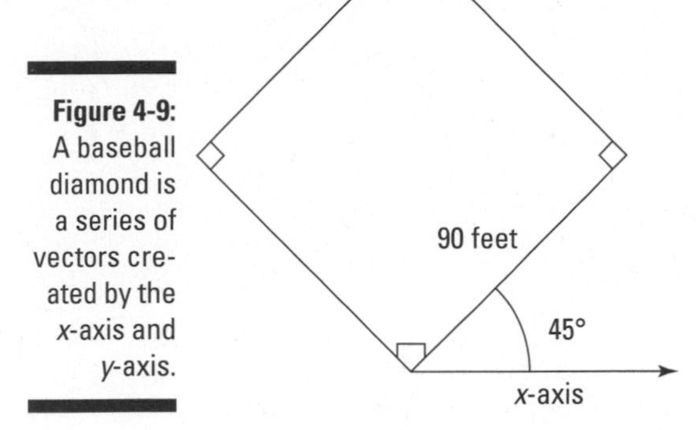

Figure 4-9: A baseball diamond is a series of vectors created by the *x*-axis and *y*-axis.

Now you're set, all because you know that displacement is a vector. In this case, here's the displacement vector:

$$\mathbf{s} = 90 \text{ feet at } 45°$$

What's that in components?

$$\begin{aligned}\mathbf{s} &= (s\cos\theta, s\sin\theta)\\ &= (90 \ \cos 45°, 90 \ \sin 45°) \approx (64 \text{ feet}, 64 \text{ feet})\end{aligned}$$

Sometimes, working with angles and magnitudes isn't as easy as working with *x* and *y* components. For example, say that you're at the park and ask directions to the nearest bench. The person you ask is very precise and deliberate and answers, "Go north 10.0 meters."

"North 10.0 meters," you say. "Thanks."

"Then east 20.0 meters. Then north another 50.0 meters."

"Hmm," you say. "North 10.0 meters, then 20.0 meters east, and then another 50.0 meters east . . . I mean north. Is that right?"

"Then 60.0 meters east."

You look at the person warily. "Is that it?"

"That's it," she says. "Nearest bench."

Okay, time for some physics. The first step is to translate all that north and east business into x and y coordinates like this: (x, y). So assuming that the positive x-axis points east and the positive y-axis points north (as on a map), the first step is 10.0 meters north, which becomes the following (where all measurements are in meters):

$$(0, 10.0)$$

That is, the first step is 10.0 meters north, which translates into 10.0 meters in the positive y direction. Adding the second step, 20.0 meters east (the positive x direction), gives you

$$\begin{aligned}&(0, 10.0)\\+&(20.0, 0)\end{aligned}$$

The third step is 50.0 meters north, and adding that gives you

$$\begin{aligned}&(0, 10.0)\\+&(20.0, 0)\\+&(0, 50.0)\end{aligned}$$

And finally, the fourth step is 60.0 meters east, which gives you

$$\begin{aligned}&(0, 10.0)\\+&(20.0, 0)\\+&(0, 50.0)\\+&(60.0, 0)\end{aligned}$$

Whew. Okay, what's the sum of all these vectors? You just add up the components:

$$\begin{array}{r}(0,\ 10.0)\\+(20.0,\ 0)\\+(0,\ 50.0)\\\underline{+(60.0,\ 0)}\\(80.0,\ 60.0)\end{array}$$

So the resulting vector is (80.0, 60.0). Hmm, that seems a lot easier than the directions you got. Now you know what to do: proceed 80.0 meters east and 60.0 meters north. See how easy adding vectors together is?

You can, if you like, go even further. You have the displacement to the nearest bench in terms of x and y components. But it looks like you'll have to walk 80.0 meters east and then 60.0 meters north to find the bench. Wouldn't it be easier if you just knew the direction to the bench and the total distance? Then you could cut the corner and just walk in a straight line directly to the bench.

This is an example where it's good to know how to convert from the (x, y) coordinate form of a vector into the magnitude-angle form. And you can do it with all the physics knowledge you have. Converting (80.0, 60.0) to the magnitude-angle form allows you to cut the corner when you walk to the bench, saving a few steps.

You know that the x and y components of a vector form a right triangle and that the total magnitude of the vector is equal to the hypotenuse of the right triangle, h. So the magnitude of h is

$$h = \sqrt{x^2 + y^2}$$

Plugging in the numbers gives you the following:

$$\begin{aligned} h &= \sqrt{(80.0 \text{ m})^2 + (60.0 \text{ m})^2} \\ &= \sqrt{6{,}400 \text{ m}^2 + 3{,}600 \text{ m}^2} \\ &= \sqrt{10{,}000 \text{ m}^2} \\ &= 100 \text{ m} \end{aligned}$$

Voilà! The bench is only 100 meters away. So instead of walking 80.0 meters east and then 60.0 meters north, a total distance of 140 meters, you only need to walk 100 meters. Your superior knowledge of vectors has saved you 40 meters.

But in what direction is the bench? You know it's 100 meters away — but 100 meters which way? You find the angle from the x-axis with this trig:

$$\tan\theta = \frac{y}{x}$$

$$\theta = \tan^{-1}\left(\frac{y}{x}\right)$$

So plugging in the numbers, you have

$$\theta = \tan^{-1}\left(\frac{60.0\text{ m}}{80.0\text{ m}}\right)$$

Therefore, the angle θ is the following (using the handy $\tan^{-1}$ button on your calculator):

$$\theta \approx 36.9^\circ$$

And there you have it — the nearest bench is 100 meters away at 36.9° from the *x*-axis. You start off confidently in a straight line at 36.9° from the east, surprising the person who gave you directions, who was expecting you to take off in the goofy zigzag path she'd given you.

Velocity: Speeding in a new direction

Velocity, which is the rate of change of position (or speed in a particular direction), is a vector. Imagine that you just hit a ground ball on the baseball diamond and you're running along the first-base line, or the **s** vector, 90 feet at a 45° angle to the positive *x*-axis. But as you run, it occurs to you to ask, "Will my velocity enable me to evade the first baseman?" A good question, because the ball is on its way from the shortstop. Whipping out your calculator, you figure that you need 3.0 seconds to reach first base from home plate; so what's your velocity? To find your velocity, you quickly divide the **s** vector by the time it takes to reach first base:

$$\frac{\mathbf{s}}{3.0\text{ s}}$$

This expression represents a displacement vector divided by a time, and time is just a scalar. The result must be a vector, too. And it is: velocity, or **v:**

$$\frac{\mathbf{s}}{3.0\text{ s}} = \frac{90\text{ ft at }45^\circ}{3.0\text{ s}} = 30\text{ ft/s at }45^\circ = \mathbf{v}$$

Your velocity is 30 feet/second at 45°, and it's a vector, **v.**

Dividing a vector by a scalar gives you a vector with potentially different units and the same direction.

In this case, you see that dividing a displacement vector, **s,** by a time gives you a velocity vector, **v**. It has the same magnitude as when you divided a distance by a time, but now you see a direction associated with it as well, because the displacement, **s,** is a vector. So you end up with a vector result rather than the scalars you see in the Chapter 3.

Acceleration: Getting a new angle on changes in velocity

What happens when you swerve, whether in a car or on a walk? You accelerate in a particular direction. And just like displacement and velocity, acceleration, **a,** is a vector.

Assume that you've just managed to hit a groundball in a softball game and you're running to first base. You figure you need the *y* component of your velocity to be at least 25.0 feet/second and that you can swerve at 90° to your present path with an acceleration of 60.0 feet/second^2 in an attempt to dodge the first baseman. Is that acceleration going to be enough to change your velocity to what you need it to be in the tenth of a second that you have before the first baseman touches you with the ball? Sure, you're up to the challenge!

Your final time, t_f, minus your initial time, t_i, equals your change in time, Δt. You can find your change in velocity with the following equation:

$$\Delta \mathbf{v} = \mathbf{a}\Delta t$$

Now you can calculate the change in your velocity from your original velocity, as Figure 4-10 shows.

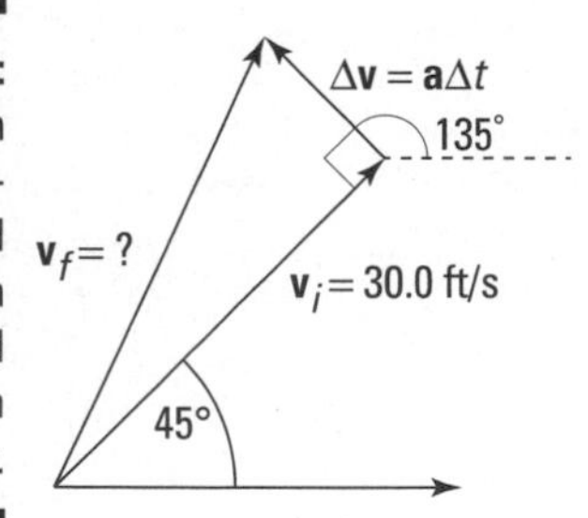

Figure 4-10: You can use acceleration and change in time to find a change in velocity.

Finding your new velocity, $\mathbf{v}_f$, becomes an issue of vector addition. That means you have to break your original velocity, $\mathbf{v}_i$, and your change in velocity, $\Delta\mathbf{v}$, into components. Here's what $\mathbf{v}_i$ equals:

$$\begin{aligned}\mathbf{v}_i &= (\mathbf{v}_i \cos\theta, \mathbf{v}_i \sin\theta)\\ &= ([30.0 \text{ ft/s}]\cos 45°, [30.0 \text{ ft/s}]\sin 45°)\\ &\approx (21.2 \text{ ft/s}, 21.2 \text{ ft/s})\end{aligned}$$

You're halfway there. Now, what about $\Delta\mathbf{v}$, the change in your velocity? You know that $\Delta\mathbf{v} = \mathbf{a}\Delta t$ and that $\mathbf{a}$ = 60.0 feet/second^2 at 90° to your present path, as Figure 4-10 shows. You can find the magnitude of $\Delta\mathbf{v}$, because

$$\Delta\mathbf{v} = \mathbf{a}\Delta t = (60.0 \text{ ft/s}^2)(0.10 \text{ s}) = 6.0 \text{ ft/s}$$

But what about the angle of $\Delta\mathbf{v}$? If you look at Figure 4-10, you can see that $\Delta\mathbf{v}$ is at an angle of 90° to your present path, which is itself at an angle of 45° from the positive x-axis; therefore, $\Delta\mathbf{v}$ is at a total angle of 135° with respect to the positive x-axis. Putting that all together means that you can resolve $\Delta\mathbf{v}$ into its components:

$$\begin{aligned}\Delta\mathbf{v} &= (6.0 \text{ ft/s} \cos 135°, 6.0 \text{ ft/s} \sin 135°)\\ &\approx (-4.2 \text{ ft/s}, 4.2 \text{ ft/s})\end{aligned}$$

You now have all you need to perform the vector addition to find your final velocity:

$$\begin{aligned}\mathbf{v}_f &= \mathbf{v}_i + \Delta\mathbf{v}\\ &= (21.2 \text{ ft/s}, 21.2 \text{ ft/s}) + (-4.2 \text{ ft/s}, 4.2 \text{ ft/s})\\ &= (17.0 \text{ ft/s}, 25.4 \text{ ft/s})\end{aligned}$$

You've done it: $\mathbf{v}_f$ = (17.0, 25.4). The y component of your final velocity is more than you need, which is 25.0 feet/second. Having completed your calculation, you put your calculator away and swerve as planned. And to everyone's amazement, it works — you evade the startled first baseman and make it to first base safely without going out of the baseline (some tight swerving on your part!). The crowd roars, and you tip your helmet, knowing that it's all due to your superior knowledge of physics. After the roar dies down, you take a shrewd look at second base. Can you steal it at the next pitch? It's time to calculate the vectors, so you get out your calculator again (not as pleasing to the crowd).

Notice that total displacement is a combination of where your initial velocity takes you in the given time, added to the displacement you get from constant acceleration.

Accelerating Downward: Motion under the Influence of Gravity

Gravity problems present good examples of working with vectors in two dimensions. Because the acceleration due to gravity is only vertical, it's especially useful to treat the horizontal and vertical components separately. Because there's no acceleration in the horizontal direction, the horizontal component of motion is just uniform. The vertical component undergoes a constant acceleration of magnitude g, directed straight down. You can use this idea to make the solutions to trajectory problems really easy.

The golf-ball-off-the-cliff exercise

Here's an example of the motion of an object accelerating under the influence of gravity. Treating the horizontal and vertical components separately is natural to the problem and can really help you solve it. In this example, the horizontal motion is uniform (as always in gravitational trajectories near the surface of the Earth), and the vertical component of the motion is just the same as that of an object dropping from a height.

Imagine that a golf ball traveling horizontally at 1.0 meter/second is about to hurtle off a 5.0-meter cliff, as Figure 4-11 shows. The question: Where will the ball hit the ground, and what will be its total speed immediately before landing? First you must find the amount of time the golf ball will be flying through the air before it lands.

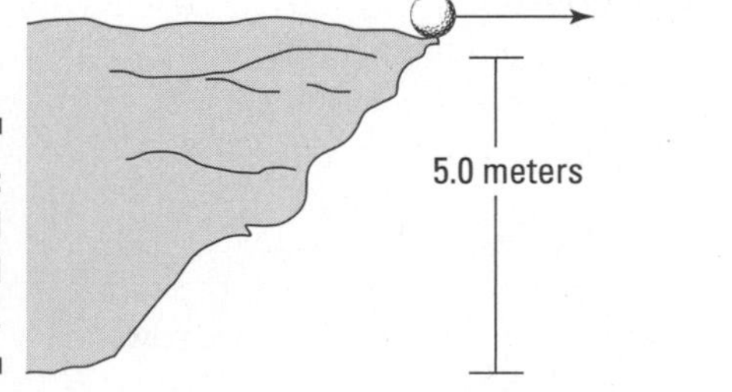

Figure 4-11: A golf ball about to roll off a cliff.

Time to gather the facts. You know that the golf ball has the velocity vector (1.0, 0) and that it flies off the cliff from a position that is 5.0 meters above the ground. When it falls, it comes down with a constant acceleration, g, the acceleration due to gravity, and that's 9.8 meters/second^2 directed straight down.

So how can you find out where the golf ball will hit the ground? One way to solve this problem is to determine how much time the ball will have before it hits. Because the golf ball accelerates only in the y direction (straight down), the x component of its velocity, v_x, doesn't change, which means that the horizontal distance at which it hits will be $v_x t$, where t is the time the golf ball is in the air. Gravity is accelerating the ball as it falls, so the following equation, which relates displacement, acceleration, and time, is a good one to use:

$$\mathbf{s} = \mathbf{v}_i t + \frac{1}{2}\mathbf{a}t^2$$

Here, **s** is the displacement of the ball, $\mathbf{v}_i$ is the initial velocity of the ball, and the acceleration **a** is equal to the acceleration due to gravity, **g.** Write down the components of these vectors.

First consider the displacement, **s.** You know that the ball starts at the top of the cliff and falls to the bottom, so the vertical component of the displacement is –5.0 meters. The vertical displacement has a magnitude of 5.0 meters, equal to the height of the cliff. Displacement is negative because the ball falls in the negative direction. You don't know the horizontal displacement of the ball yet, so write it as s_x. So you can write the displacement vector as

$$\mathbf{s} = (s_x, -5.0 \text{ m})$$

Second, write down the initial velocity, $\mathbf{v}_i$, of the ball. You know that the ball is initially rolling along the horizontal top of the cliff with a speed of $v_x = 1.0$ m/s, so the initial velocity of the ball is

$$\mathbf{v}_i = (1.0 \text{ m/s}, 0 \text{ m/s})$$

Finally, you know that the acceleration is just equal to the acceleration due to gravity, g, directed straight down, and it's constant. So the ball's acceleration, **a,** is

$$\begin{aligned}\mathbf{a} &= (0, -g)\\ &= (0, -9.8 \text{ m/s}^2)\end{aligned}$$

Now you have all you need to know to work out the horizontal displacement s_x. Take each component of the preceding equation for displacement under constant acceleration separately.

First write the vertical component of the equation by putting in the vertical components of displacement, initial velocity, and acceleration:

$$\begin{aligned}s_y &= v_y t + \frac{1}{2}(-g)t^2\\ -5.0 \text{ m} &= 0t - \frac{1}{2}(9.8 \text{ m/s}^2)t^2\end{aligned}$$

You can simplify and rearrange this equation to find t, the time that the ball is falling:

$$\begin{aligned}t &= \sqrt{\frac{2(5.0 \text{ m})}{(9.8 \text{ m/s}^2)}}\\ &\approx 1.0 \text{ s}\end{aligned}$$

So you know now that the ball falls for 1.0 second. Great! Now use this to look at the horizontal component. If you write out the horizontal component of the displacement equation, you have:

$$s_x = v_x t + \frac{1}{2}(0)(t^2)$$
$$= (1.0 \text{ m/s})\, t$$

And now that you know that $t = 1.0$ s, you can work out how far the ball moves horizontally as it falls from the top to the bottom of the cliff:

$$s_x = (1.0 \text{ m/s})(1.0 \text{ s}) = 1.0 \text{ m}$$

So there you have it — the ball will land 1.0 meter to the right.

Time to figure out what the speed of the golf ball will be when it hits. You already know half the answer, because the *x* component of its velocity, v_x, isn't affected by gravity, so it doesn't change. Gravity is pulling on the golf ball in the *y* direction, not the *x,* which means that the final velocity of the golf ball will look like this: (1.0, ?). So you have to figure out the *y* component of the velocity, or the *?* business in the (1.0, ?) vector. For that, you can use the following equation:

$$\mathbf{v}_f - \mathbf{v}_i = \mathbf{a}t$$

In this case, $\mathbf{v}_i = 0$, the acceleration is $-g$, and you want the final velocity of the golf ball in the *y* direction, so the equation looks like this:

$$v_y = -gt$$

The acceleration due to gravity, *g,* is also a vector, **g.** That makes sense because **g** is an acceleration. This vector happens to point to the center of the Earth — that is, in the negative *y* direction — and on the surface of the Earth, its value is 9.8 meters/second^2.

The negative sign here indicates that **g** is pointing downward, toward negative *y*. So the real result is

$$v_y = -gt$$
$$= (-9.8 \text{ m/s}^2)(1.0 \text{ s})$$
$$= -9.8 \text{ m/s}$$

The final velocity vector of the golf ball when it hits the ground is (1.0, –9.8) meters/second. You still need to find the golf ball's speed when it hits, which is the magnitude of its velocity. You can figure that out easily enough:

$$v_f = \sqrt{(1.0 \text{ m/s})^2 + (-9.8 \text{ m/s})^2} \approx 9.9 \text{ m/s}$$

You've triumphed! The golf ball will hit 1.0 meter to the right, and its speed at that time will be 9.9 meters/second.

Not bad, but if you're still not satisfied, you can work out the angle at which the ball strikes the ground, too. You can simply use the components of the final velocity vector to work out the angle as usual, using the inverse tangent:

$$\tan\theta = \frac{v_y}{v_x}$$

$$\begin{aligned}\theta &= \tan^{-1}\left(\frac{v_y}{v_x}\right)\\ &= \tan^{-1}\left(\frac{-9.8 \text{ m/s}}{1.0 \text{ m/s}}\right)\\ &\approx -84^\circ\end{aligned}$$

If the ball were traveling straight down, the angle would be –90°, so the ball is only 6° from traveling straight down.

The how-far-can-you-kick-the-ball exercise

This example uses the same principles and strategies as the one in the preceding section, except that this time the trajectory is not quite as simple. In this example, the object is projected up at an angle before it falls back down again. With your new projectile skills gleaned from the preceding section, you can determine how far the object will go.

You're at the tryouts for your favorite soccer team, with World Cup dreams on your mind. The only thing left is to prove that you can kick the ball far enough. The situation is as you see in Figure 4-12. You kick the ball at an angle θ with a certain speed, and you want to know how far the ball will go before hitting the ground.

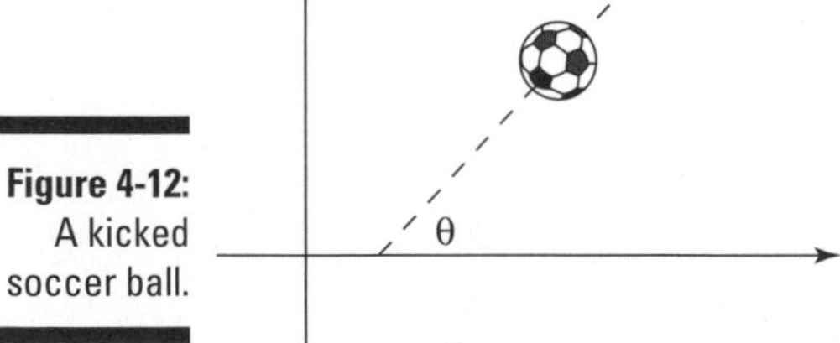

Figure 4-12: A kicked soccer ball.

Say that $\theta = 45^\circ$ and that the initial speed of the ball is 50.0 meters/second. How far will it travel in the x direction before it hits the ground?

Most people would be lost here, but you have your knowledge of physics to guide you. You consider the problem carefully — you know that the horizontal distance the ball travels is equal to

$$x = v_x t$$

where v_x is the speed of the ball in the *x* direction. But what's *t?*

The variable *t* is the time the ball takes to leave your foot, travel through the air, and then hit the ground again. How on Earth can you calculate that time?

During the time *t,* the ball leaves your foot, travels upward, then goes downward, and then hits the ground. Here's where you can be clever. The vertical speed of the ball is

$$v_y = v_{yi} + at$$

where v_{yi} is the ball's original speed vertically, *a* is the acceleration of the ball, and *t* is time.

So how does that help? It helps because you know the vertical speed of the ball at the top of its flight — and that's zero. Think about it — the ball starts by flying upward, then it stops rising, and then it starts going down. So at a particular time, at the top of its flight, the ball has zero speed in the vertical direction for just an instant. That happens exactly halfway through the ball's flight. So if you can solve for the time at which the ball has zero vertical speed and then double that time, you'll have the total time the ball is in the air.

Looked at purely in the vertical direction, the ball starts off at its maximum vertical speed, then reaches the top of its flight. The ball stops traveling vertically for an instant and then falls, hitting the ground with the same maximum speed (only in the opposite direction — down, not up). So if you can find the time at which the ball instantaneously has zero speed vertically and then double that time, you'll have the total time of the ball's flight.

To find the time at which the ball has zero vertical speed temporarily, turn to the equation for its vertical speed:

$$v_y = v_{yi} + at$$

The vertical component of acceleration, *a,* is equal to –*g* (it's negative because it's in the downward direction). That means you have:

$$v_y = v_{yi} - gt$$

Halfway through the flight, at time = $t_{1/2}$, $v_y = 0$, so you have

$$0 = v_{yi} - gt_{1/2}$$

$$v_{yi} = gt_{1/2}$$

$$t_{1/2} = \frac{v_{yi}}{g}$$

Okay, so what is v_{yi}, the original speed in the vertical direction? You know that $\theta = 45°$ and that the speed of the ball is $v_i = 50.0$ m/s. The vertical component of this speed is

$$v_{yi} = v_i \sin\theta$$

And plugging in the numbers, you have

$$\begin{aligned} v_{yi} &= (50.0 \text{ m/s}) \sin 45° \\ &\approx 35.4 \text{ m/s} \end{aligned}$$

Great! Now recalling that $t_{1/2} = v_{yi}/g$ and that $g = 9.8 \text{ m/s}^2$, you have the following:

$$\begin{aligned} t_{1/2} &= \frac{35.4 \text{ m/s}}{9.8 \text{ m/s}^2} \\ t_{1/2} &\approx 3.6 \text{ s} \end{aligned}$$

Because $t_{1/2}$ is the time halfway through the flight, the time the full flight takes, t, must be twice that:

$$t = 2t_{1/2} = 2(3.6 \text{ s}) = 7.2 \text{ s}$$

So how far does the ball go before it hits the ground? The horizontal distance is

$$x = v_x t$$

where v_x is the speed of the ball in the x direction (which doesn't change throughout the whole flight). Taking the horizontal component of the ball's velocity vector gives you

$$\begin{aligned} v_x &= v_i \cos\theta \\ &= (50.0 \text{ m/s}) \cos 45° \\ &\approx 35.4 \text{ m/s} \end{aligned}$$

Because $x = v_x t$, you can plug in the numbers and find out how far the ball sailed downfield:

$$x = v_x t = (35.4 \text{ m/s})(7.2 \text{ s}) \approx 255 \text{ m}$$

Wow — 255 meters. That's a hefty kick. You not only made the team but most certainly set a world record for distance in the process!

Part II

May the Forces of Physics Be with You

In this part . . .

Part II gives you the lowdown on famous laws related to forces, such as "For every action, there is an equal and opposite reaction." The subject of forces is where Isaac Newton gets to shine. His laws of motion and the equations in this part allow you to predict what will happen when you apply a force to an object or even to fluids. Mass, acceleration, friction — all the key topics having to do with forces are here.

Chapter 5

When Push Comes to Shove: Force

In This Chapter

- Discovering Newton's three takes on force
- Utilizing force vectors with Newton's laws

You can't get away from forces in your everyday world; you use force to open doors, type at a keyboard, steer a car, drive a bulldozer through a wall, climb the stairs of the Statue of Liberty (not everyone, necessarily), take your wallet out of your pocket — even to breathe or talk. You unknowingly take force into account when you cross bridges, walk on ice, lift a hot dog to your mouth, unscrew a jar's cap, or flutter your eyelashes at your sweetie. Force is integrally connected to making objects move, and physics takes a big interest in understanding how it works.

Force is fun stuff. Like other physics topics, you may assume it's difficult, but that's before you get into it. Like your old buddies displacement, velocity, and acceleration (see Chapters 3 and 4), force is a vector, meaning it has a magnitude and a direction (unlike, say, speed, which just has a magnitude).

This chapter is where you find Newton's famous three laws of motion. You've heard these laws before in various forms, such as "For every action, there's an equal and opposite reaction." That's not quite right; it's more like "For every force, there's an equal and opposite force," and this chapter is here to set the record straight. In this chapter, I use Newton's laws as a vehicle to focus on force and how it affects the world.

Newton, Einstein, and the laws of physics

In the 17th century, Sir Isaac Newton was the first to put the relationship among force, mass, and acceleration into equation form. (He's also famous for watching apples drop off trees and developing the consequent mathematical expression of gravity.)

As with other advances in physics, Newton made observations first, modeled them mentally, and then expressed those models in mathematical terms. Newton expressed his model by using three assertions, which have come to be known as Newton's laws. But don't forget that physics just models the world, and as such, it's all subject to later revision.

Newton's laws have been heavily revised by the likes of Albert Einstein and his theory of relativity. Newton's laws are based on ideas of space and time and mass that make sense to most people in everyday terms: Everyone agrees when two events are simultaneous, mass is a constant that doesn't depend on speed, and so on. But Einstein's theory of relativity takes the speed of light as a constant for all observers however they're moving, and this leads to some very different ideas of space and time, which in turn brings about very different laws of motion. However, Einstein's theory becomes important only for motion close to the speed of light. At speeds that you see around you every day, Newton's laws of motion are extremely accurate and, therefore, are still very important to understand.

Newton's First Law: Resisting with Inertia

Newton's laws explain what happens with forces and motion, and his first law states, "An object continues in a state of rest, or in a state of motion at a constant velocity along a straight line, unless compelled to change that state by a net force." Translation? If you don't apply a *net,* or unbalanced, force to an object at rest or in motion, it will stay at rest or in that same motion along a straight line. Forever.

For example, when scoring a hockey goal, the hockey puck slides toward the open goal in a straight line because the ice it slides on is nearly frictionless. If you're lucky, the puck won't come into contact with the opposing goalie's stick, which would cause it to change its motion.

Newton's first law may not seem very intuitive because most things don't seem to continue moving in straight lines forever. Left to themselves, most moving things come to a halt. The idea that the natural tendency of an object in motion is to come to a halt was Aristotle's, and it was accepted wisdom for 2,000 years. It took the tremendous insight of Newton to see that the natural state of motion is actually to continue in a straight line at constant velocity. Only when acted on by a force does the motion change.

In everyday life, objects don't coast around in straight lines at constant velocity. This is because most objects around you are subject to friction forces. So, for example, when you slide a coffee mug across your desk, it slows and comes to a stop (or spills over). That's not to say Newton's first law is invalid, just that friction provides a force to change the mug's motion to stop it.

Saying that if you don't apply a force to an object in motion, it will stay in motion at constant velocity forever sounds an awful lot like a *perpetual-motion machine,* a theoretical machine that would run indefinitely without the input of any energy. Interestingly, such a machine is perfectly possible according to Newton's laws. In practice, you just can't get away from forces that will ultimately affect an object in motion. Even in the farthest reaches of space, the rest of the mass in the universe pulls at you, if only very slightly. And that means your motion is affected. So much for perpetual motion!

What Newton's first law really says is that the only way to get something to change its motion is to use force. It also says that an object in motion tends to stay in motion, which introduces the idea of inertia.

Resisting change: Inertia and mass

Inertia is the natural tendency of an object to resist any change in its motion, which means that it tends to stay at rest or in constant motion along a straight line. Inertia is a quality of mass, and the mass of an object is really just a measurement of its inertia. To get a stationary object to move — that is, to change its current state of motion — you have to apply a force to overcome its inertia.

Be careful to distinguish between mass and weight. The *weight* of an object is the force of gravity on it, so weight depends on where the mass is. For example, a 1-kilogram object would have a different weight on the moon than it does on Earth, but the mass would be the same. Even in space, with no significant gravitational field and therefore no weight, the mass would still be 1 kilogram. If you tried to push this object in space, you'd feel a resistance to the acceleration, which is inertia. The larger the mass of the object, the more resistance you would feel.

Say, for example, you're at your summer vacation house, taking a look at the two boats at your dock: a dinghy and an oil tanker. If you apply the same net force to each with your foot, the boats respond in different ways. The dinghy scoots away and glides across the water. The oil tanker moves away more slowly (what a strong leg you have!). That's because they both have different masses and, therefore, different amounts of inertia. When responding to the same net force, an object with little mass — and a small amount of inertia — will have greater acceleration than an object with large mass, which has a large amount of inertia.

Inertia, the resistance of an object to changes in its velocity, can be a problem at times. Refrigerated meat trucks, for example, have large amounts of frozen meat hanging from their ceilings, and when the drivers of the trucks begin turning corners, they create a pendulum motion they can't stop from the driver's seat. Trucks with inexperienced drivers can end up tipping over because of the inertia of the swinging frozen load in the back.

Because objects have inertia, they resist changing their motion, which is why you have to start applying forces to get changes in velocity and therefore acceleration. Mass ties force and acceleration together.

Measuring mass

The units of mass (and, therefore, inertia) depend on your measuring system. In the meter-kilogram-second (MKS) system, or the International System of Units (SI), mass is measured in kilograms (under the influence of the Earth's gravity, 1 kilogram of mass weighs about 2.205 pounds). What's the unit of mass in the foot-pound-second system? Brace yourself: It's the *slug*. Under the influence of the Earth's gravity, a slug has a weight of about 32 pounds. If you need to convert between the slug and the kilogram, then you'll be pleased to know that one slug is equal to about 14.59 kilograms.

Mass isn't the same as weight. Mass is a measure of inertia; when you put that mass into a gravitational field, you get weight. So, for example, a slug is a certain amount of mass. When you subject that slug to the gravitational pull on the surface of the Earth, it has weight. And that weight is about 32 pounds. If you took that same slug of mass to the moon, which doesn't have as much gravitational pull as Earth, the slug would weigh only around 5.3 pounds, which is about 1/6 of its weight on Earth.

Newton's Second Law: Relating Force, Mass, and Acceleration

Newton's first law says that an object remains in uniform motion unless acted on by a net force. When a net force is applied, the object accelerates. Newton's second law details the relationship among net force, the mass, and the acceleration:

- **The acceleration of an object is in the direction of the net force.** If you push or pull an object in a particular direction, it accelerates in that direction.
- **The acceleration has a magnitude proportional to the magnitude of the net force.** If you push twice as hard (and no other forces are present), the acceleration is twice as big.
- **The magnitude of the acceleration is inversely proportional to the mass of the object.** That is, the larger the mass, the smaller the acceleration for a given force (which is just as you'd expect from inertia).

All these features of the relation among net force (ΣF), acceleration (a), and mass (m) are contained in the following equation:

$$\Sigma F = ma$$

Note that you use the term ΣF to describe the net force because the Greek letter sigma, Σ, stands for "sum"; therefore, ΣF means the sum of all the separate forces acting on the object. If this is not zero, then there's a net force.

Relating the formula to the real world

You can see that the equation $\Sigma F = ma$ is consistent with Newton's first law of motion (which deals with inertia), because if there's no net force (ΣF) acting on a mass m, then the left-hand side of this equation is zero; therefore, the acceleration must also be zero — just as you'd expect from the first law.

If you rearrange the net-force equation to solve for acceleration, you can see that if the size of the net force doubles, then so does the size of the acceleration (if you push twice as hard, the object accelerates twice as much), and if the mass doubles, then the acceleration halves (if the mass is twice as big, it accelerates half as much — inertia):

$$a = \frac{\Sigma F}{m}$$

Take a look at the hockey puck in Figure 5-1 and imagine it's sitting there all lonely in front of a net. These two should meet.

In a totally hip move, you decide to apply your knowledge of physics to this one. You figure that if you apply the force of your stick to the puck for a tenth of a second, you can accelerate it in the appropriate direction. You try the experiment, and sure enough, the puck flies into the net. Score! Figure 5-1 shows how you made the goal. You applied a net force to the puck, which has a certain mass, and off it went — accelerating in the direction you pushed it.

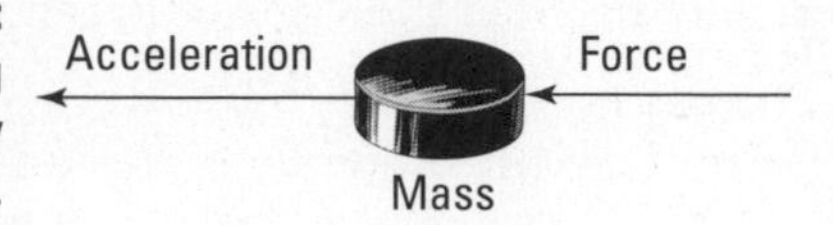

Figure 5-1: Accelerating a hockey puck.

What's its acceleration? That depends on the force you apply (along with any other forces that may be acting on the puck), because $\Sigma F = ma$.

Naming units of force

So what are the units of force? Well, $\Sigma F = ma$, so in the MKS or SI system, force must have these units:

kilogram-meters/second^2

This is a *derived* unit because you reach it by using a formula. Because most people think this unit line looks a little awkward, the MKS units are given a special name: *newtons* (named after guess who). Newtons are often abbreviated as simply N. Table 5-1 shows unit names for force in the MKS and foot-pound-second systems of measurement.

Table 5-1 Units of Force

System of Measurement	*Derived Unit*	*Special Unit Name*
Meter-kilogram-second (MKS) or SI	kilogram-meters/second^2 ($kg \cdot m/s^2$)	newton (N)
Foot-pound-second	slug-feet/second^2 ($slug \cdot ft/s^2$)	pound (lb)

So how do these units relate to each other? Well, 1.0 pound is about 4.448 newtons.

Vector addition: Gathering net forces

Most books shorten $\Sigma F = ma$ to simply $F = ma$, which is what I do, too, but I must note that F stands for *net force.* An object you apply force to responds to the net force — that is, the vector sum of all the forces acting on it.

Take a look, for example, at all the forces (represented by arrows) acting on the ball in Figure 5-2. Which way will the golf ball be accelerated?

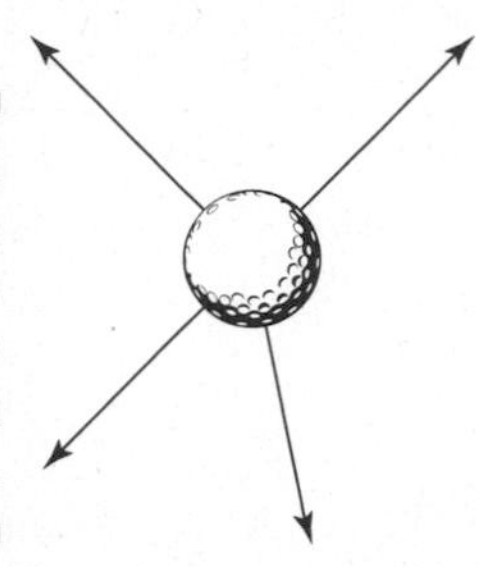

Figure 5-2: A ball in flight may face many forces that act on it.

Because Newton's second law talks about net force, the problem becomes easier. All you have to do is add the various forces together as vectors to get the resultant, or net, force vector, ΣF, as Figure 5-3 shows. When you want to know how the ball will accelerate, you can apply the equation $\Sigma F = ma$.

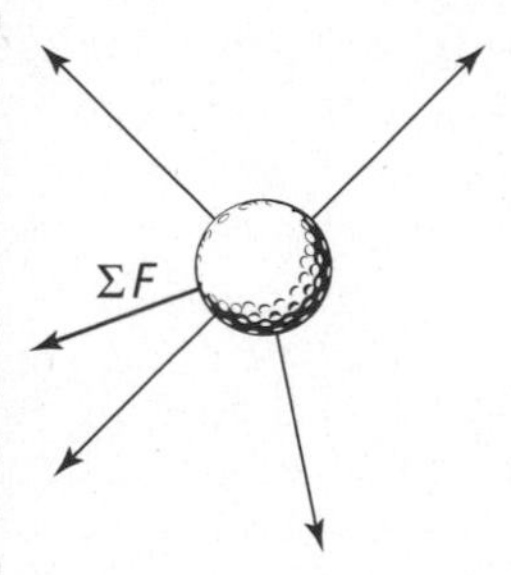

Figure 5-3: The net force vector factors in all forces to determine the ball's acceleration.

Calculating displacement given a time and acceleration

Assume that you're on your traditional weekend physics data-gathering expedition, and you happen upon a football game. Very interesting, you think. In a certain situation, you observe that the football, although it starts from rest, has three players subjecting forces on it, as you see in Figure 5-4. This figure shows a free-body diagram.

A *free-body diagram* shows all the forces acting on an object, making it easier to determine their components and find the net force.

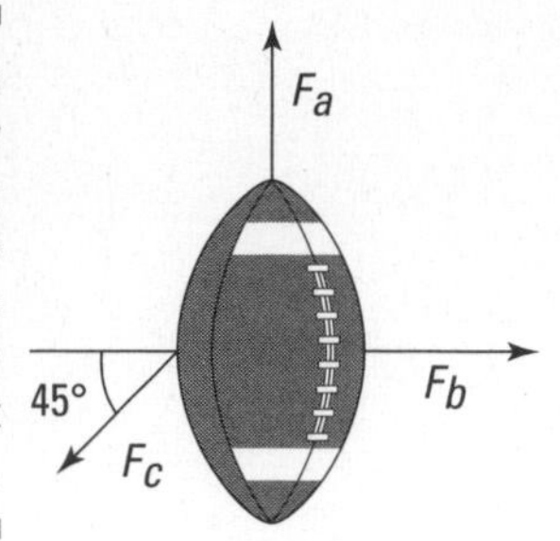

Figure 5-4: A free-body diagram of all the forces acting on a football at one time.

Slipping intrepidly into the mass of moving players, risking injury in the name of science, you measure the magnitude of these forces and mark them down on your clipboard:

F_a = 15.0 N

F_b = 12.5 N

F_c = 16.5 N

You measure the mass of the football as 0.40 kilograms (I don't include the force of gravity). Now you wonder where the football will be in 1.0 second, assuming the forces shown act on the ball continuously during that second. Follow these steps to calculate the displacement of an object in a given time with a given constant acceleration:

1. **Find the net force, ΣF, by using vector addition to add all the forces acting on the object (see Chapter 4 for info on vector addition).**
2. **Use $\Sigma F = ma$ to determine the acceleration vector.**
3. **Use $s = v_i t + (1/2)at^2$ to get the distance traveled in the specified time.**

   Refer to Chapter 3 to find this original equation.

Step 1: Finding net force

Time to get out your calculator. Because you want to relate net force, mass, and acceleration, the first order of business is to find the net force on the mass. To do that, you need to break up the force vectors you see in Figure 5-4 into their components and then add those components together to get the net force (see Chapter 4 for more info on breaking up vectors in components).

Determining F_a and F_b is easy because F_a is straight up — along the positive *y*-axis — and F_b is to the right — along the positive *x*-axis. That means

F_a = (0 N, 15.0 N)

F_b = (12.5 N, 0 N)

Finding the components of F_c is a little trickier. You need the x and y components of this force this way:

$$F_c = (F_{cx}, F_{cy})$$

F_c is along an angle 45° with respect to the negative x-axis, as you see in Figure 5-4. If you measure all the way from the positive x-axis, you get an angle of $180° + 45° = 225°$. This is the way you break up F_c:

$$F_c = (F_{cx}, F_{cy}) = (F_c \cos \theta, F_c \sin \theta)$$

Plugging in the numbers gives you

$$\begin{aligned} F_c &= (16.5 \text{ N} \cos 225°, 16.5 \text{ N} \sin 225°) \\ &\approx (-11.7 \text{ N}, -11.7 \text{ N}) \end{aligned}$$

Look at the signs here — both components of F_c are negative. You may not follow that business about the angle of F_c being $180° + 45° = 225°$ without some extra thought, but you can always make a quick check of the signs of your vector components. F_c points downward and to the left, along the negative x- and negative y-axis. That means that both components of this vector, F_{cx} and F_{cy}, have to be negative. I've seen many people get stuck with the wrong signs for vector components because they didn't think to make sure their numbers matched the reality.

Always compare the signs of your vector components with their actual directions along the axes. It's a quick check, and it saves you plenty of problems later.

Now you know the components of the three forces on the football:

$$F_a = (0 \text{ N}, 15.0 \text{ N})$$

$$F_b = (12.5 \text{ N}, 0 \text{ N})$$

$$F_c = (-11.7 \text{ N}, -11.7 \text{ N})$$

You're ready for some vector addition:

$$\begin{aligned} F_a &= (0,\ 15.0 \text{ N}) \\ +F_b &= (12.5 \text{ N},\ 0) \\ +F_c &= (-11.7 \text{ N},\ -11.7 \text{ N}) \\ \hline \Sigma F &= (0.8 \text{ N},\ 3.3 \text{ N}) \end{aligned}$$

You calculate that the net force, ΣF, is (0.8 N, 3.3 N). That also gives you the direction in which the football will move, assuming it was at rest when you made the force measurements.

Step 2: Finding acceleration

The next step is to find the acceleration of the football. From Newton, you know that $\Sigma F = (0.8 \text{ N}, 3.3 \text{ N}) = ma$, which means the following:

$$a = \frac{\Sigma F}{m} = \frac{(0.8 \text{ N}, 3.3 \text{ N})}{m}$$

Because the mass of the football is 0.40 kilograms, the problem works out like this:

$$\begin{aligned} a &= \frac{\Sigma F}{m} \\ &= \frac{(0.8 \text{ N}, 3.3 \text{ N})}{0.40 \text{ kg}} \\ &\approx (2.0 \text{ m/s}^2, 8.3 \text{ m/s}^2) \end{aligned}$$

You're making good progress; you now know the acceleration of the football.

Step 3: Finding displacement

To find out where the football will be in 1.0 second, you can apply the following equation (found in Chapter 3), where s is the distance and the acceleration is assumed to be for one full second due to the forces continuously being applied:

$$s = v_i t + \frac{1}{2} a t^2$$

Plugging in the numbers gives you the following (note that the football's initial velocity is 0 meters/second, so the first term drops out):

$$\begin{aligned} s &= v_i t + \frac{1}{2} a t^2 \\ &= (0 \text{ m/s})(1.0 \text{ s}) + \frac{1}{2}(2.0 \text{ m/s}^2, 8.3 \text{ m/s}^2)(1.0 \text{ s})^2 \\ &\approx (1.0 \text{ m}, 4.2 \text{ m}) \end{aligned}$$

Well, well, well. At the end of 1.0 second, the football will be 1.0 meter along the positive x-axis and 4.2 meters along the positive y-axis. You get your stopwatch out of your lab-coat pocket and measure off 1.0 second. Sure enough, you're right. The football moves 1.0 meter toward the sideline and 4.2 meters toward the goal line. Satisfied, you put your stopwatch back into your pocket and put a checkmark on the clipboard. Another successful physics experiment.

Calculating net force given a time and velocity

What if you want to find how much net force is necessary in a specific time to produce a particular velocity? Say, for example, that you want to accelerate your car from 0 to 60.0 miles per hour in 10.0 seconds; how much net force is necessary? You start by converting 60.0 miles/hour to feet/second. First, convert to miles/second:

$$\frac{60.0\text{ miles}}{1\ \cancel{\text{hour}}}\times\frac{1\ \cancel{\text{hour}}}{60\ \cancel{\text{minutes}}}\times\frac{1\ \cancel{\text{minute}}}{60\ \text{seconds}}\approx 1.67\times10^{-2}\text{ miles/second}$$

Notice that the *hours* and *minutes* cancel out to leave you with *miles* and *seconds* for the units. Next, you take the result to feet/second:

$$\frac{1.67\times10^{-2}\ \cancel{\text{miles}}}{1\text{ second}}\times\frac{5{,}280\text{ feet}}{1\ \cancel{\text{mile}}}\approx 88\text{ feet/second}$$

You want to get to 88 feet/second in 10.0 seconds. If the car weighs 3,000 pounds, how much net force do you need? First you find the acceleration with the following equation from Chapter 3:

$$a=\frac{\Delta v}{\Delta t}$$

Plugging in some numbers, you get

$$a=\frac{v}{t}=\frac{88\text{ ft/s}}{10.0\text{ s}}$$
$$=8.8\text{ ft/s}^2$$

You calculate that 8.8 feet/second^2 is the acceleration you need.

From Newton's second law, you know that $\Sigma F = ma$, and you know that the weight of the car is 3,000 pounds. What's the car's mass in the foot-pound-second system of units, or slugs? In this system of units, you can find an object's mass given its weight by dividing by the acceleration due to gravity — 32 feet/second^2 (converted from 9.8 meters/second^2 — the number given to you in most physics problems):

$$m=\frac{3{,}000\ \cancel{\text{pounds}}}{32\ \cancel{\text{feet/second}^2}}\times\frac{1\text{ slug-}\cancel{\text{foot/second}^2}}{1\ \cancel{\text{pound}}}\approx 94\text{ slugs}$$

You have all you need to know. You have to accelerate 94 slugs of mass by 8.8 feet per second^2; so, what net force do you need? Just multiply to get your answer:

$$\Sigma F = ma = (94\text{ slugs})(8.8\text{ ft/s}^2)\approx 830\text{ pounds}$$

There needs to be a net force of about 830 pounds on the car for those 10.0 seconds to accelerate you to the speed you want: 60.0 miles per hour.

Note that this solution ignores pesky little issues like friction and upward grade on the road; you get to those issues in Chapter 6. Even on a flat surface, friction would be large in this example, so you'd need to maybe double the magnitude of force in real life.

Newton's Third Law: Looking at Equal and Opposite Forces

Newton's third law of motion is famous, especially in wrestling and drivers' ed circles, but you may not recognize it in all its physics glory: "Whenever one body exerts a force on a second body, the second body exerts an oppositely directed force of equal magnitude on the first body."

The more popular version of this, which I'm sure you've heard many times, is "For every action, there's an equal and opposite reaction." But for physics, it's better to express the originally intended version, in terms of forces, not actions (which, from what I've seen, can apparently mean everything from voting trends to temperature forecasts!).

Seeing Newton's third law in action

Here's a real-world example to show you how Newton's third law of motion works. Say that you're in your car, speeding up with constant acceleration. To do this, your car has to exert a force against the road; otherwise, the car wouldn't be accelerating. And the road has to exert the same force on your car. You can see what this looks like, tire-wise, in Figure 5-5.

The two forces in the Figure 5-5 are equal in magnitude but opposite in direction. However, they do not cancel out because the two forces are acting on different bodies — one on the car and the other on the road. The force that the car exerts on the road is equal and opposite to the force the road exerts on the car. The force on the car accelerates it.

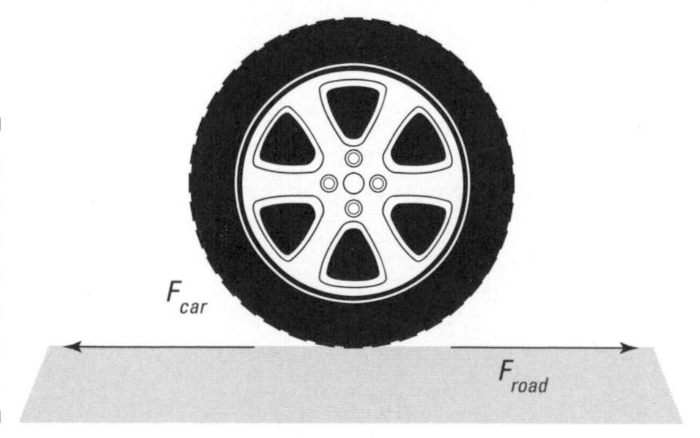

Figure 5-5: Equal forces acting on a car tire and the road during acceleration.

So why doesn't the road accelerate? The car accelerates, so shouldn't the road accelerate in the opposite direction? Believe it or not, it does; Newton's law is in full effect. Your car pushes the Earth, affecting the motion of the

Earth in just the tiniest amount. Given the fact that the Earth is about 6,000,000,000,000,000,000,000 times as massive as your car, however, any effects aren't too noticeable!

Similarly, when a hockey player slaps a puck, the puck accelerates away from the spot of contact, and so does the hockey player. If hockey pucks weighed 1,000 pounds — with a mass of about 31 slugs, or 450 kilograms — you'd notice this effect much more; in fact, the puck wouldn't move much at all, but the player would hurtle off in the opposite direction after striking it. (More on what happens in this case in Part III of this book.)

Pulling hard enough to overcome friction

Because of Newton's third law, whenever you apply a force to an object, say, by pulling it, then the object applies an equal and opposite force on you. Here's an example that lets you work out how much force you're subject to when you drag something along. For fantasy physics purposes, say that a hockey game ends, and you get the job of dragging a 31-slug hockey puck off the rink. You use a rope to do the trick, as shown in Figure 5-6.

Physics problems are very fond of using ropes, including ropes with pulleys, because with ropes, the force you apply at one end is the same as the force that the rope exerts on what you tie it to at the other end.

In this case, the massive hockey puck will have some friction that resists you — not a terrific amount, given that it slides on top of ice, but still, some. Therefore, the net force on the puck is

$$\Sigma F = F_{rope} - F_{friction}$$

Figure 5-6: Pulling a heavy puck with a rope to exert equal force on both ends.

Because F_{rope} is greater than $F_{friction}$, the puck will accelerate and start to move. In fact, if you pull on the rope with a constant force, the puck will accelerate at a constant rate, which obeys the equation

$$\Sigma F = F_{rope} - F_{friction} = ma$$

Because some of the force you exert on the puck goes into accelerating it and some goes into overcoming the force of friction, the force you exert on the puck is the same as the force it exerts on you (but in the opposite direction), as Newton's third law predicts:

$$F_{rope} = F_{friction} + ma$$

Pulleys: Supporting double the force

No force can be exerted without an equal and opposite force (even if some of that opposing force comes from making an object accelerate). A rope and pulley can act together to change the direction of the force you apply, but not for free. In order to change the direction of your force from $-F$ (that is, downward) to $+F$ (upward on the mass), the pulley's support has to respond with a force of $2F$.

Here's how this works: When you pull a rope in a pulley system to lift a stationary object, you lift the mass if you exert enough force to overcome its weight, mg, where g is the acceleration due to gravity at the surface of the Earth, 9.8 meters/second^2. Take a look at Figure 5-7, in which a rope goes over a pulley and down to a mass m.

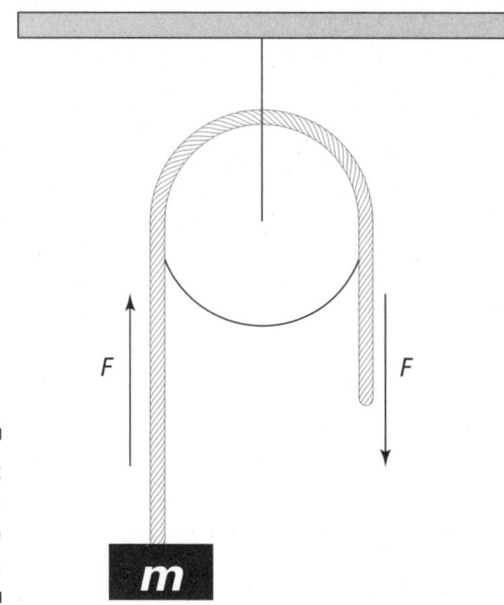

Figure 5-7: Using a pulley to exert force.

The rope and pulley together function not only to transmit the force, F, you exert but also to change the direction of that force, as you see in the figure. The force you exert downward is exerted upward on the mass, because the rope, going over the pulley, changes the force's direction. In this case, if F is greater than mg, you can lift the mass. If you apply no force on the object, then the only force acting on it is gravity, $F_{gravity}$, and so the object accelerates at a rate of $-mg$ (the negative sign indicates that the acceleration is downward) because

$$F_{gravity} = -mg$$

If you apply a force on the rope of magnitude F, then it is transmitted by the rope and pulley to the object as an upwardly directed force of the same magnitude. Therefore, the total force on the object is given by the sum of these two forces, $F_{gravity} + F$. The force F, acting alone without gravity, would accelerate the object upward at a rate that you can call a:

$$F = ma$$

When the two forces act together, you have the following sum:

$$\begin{aligned} F_{gravity} + F &= -mg + ma \\ &= m(a - g) \end{aligned}$$

So you can see that if F is greater than mg, then a is greater than g and the object accelerates upward.

But this force-changing use of a rope and pulley comes at a cost, because you can't cheat Newton's third law. Assume that you lift the mass and it hangs in the air. In this case, F must equal mg to hold the mass stationary. The direction of your force is being changed from downward to upward. How does that happen?

To figure this out, consider the force that the pulley's support exerts on the ceiling. What's that force? Because the pulley isn't accelerating in any direction, you know that $\Sigma F = 0$ on the pulley. That means that all the forces on the pulley, when added up, give you zero.

From the pulley's point of view, two forces pull downward: the force F you pull with and the force mg that the mass exerts on you (because nothing is moving at the moment). That's $2F$ downward. To balance all the forces and get 0 total, the pulley's support must exert a force of $2F$ upward.

Analyzing angles and force in Newton's third law

To take angles into account when measuring force, you need to do a little vector addition. Take a look at Figure 5-8. Here, the mass m isn't moving, and you're applying a force F to hold it stationary. Here's the question: What force is the pulley's support exerting, and in which direction, to keep the pulley where it is?

You're sitting pretty here. Because the pulley isn't moving, you know that $\Sigma F = 0$ on the pulley. So what are the forces on the pulley? You can account for the force due to the mass's weight, which has magnitude *mg* and is directed straight down. Putting that in terms of vector components (see Chapter 4), it looks like this (keep in mind that the *y* component of F_{mass} has to be negative, because it points downward, which is along the negative *y*-axis):

$$F_{mass} = (0, -mg)$$

You also have to account for the force of the rope on the pulley, which, because you're holding the mass stationary and the rope transmits the force you're applying, must be of magnitude *mg* and directed to the right — along the positive *x*-axis. That force looks like this:

$$F_{rope} = (mg, 0)$$

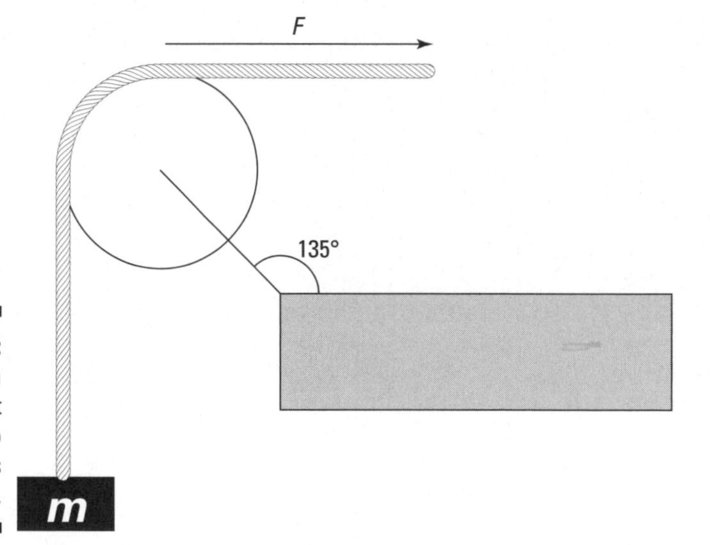

Figure 5-8: Using a pulley at an angle to keep a mass stationary.

You can find the force exerted on the pulley by the rope and the mass by adding the vectors F_{mass} and F_{rope}:

$$\begin{aligned} F_{mass + rope} &= F_{mass} + F_{rope} \\ &= (0, -mg) + (mg, 0) \\ &= (mg, -mg) \end{aligned}$$

The force exerted by both the mass and the rope, $F_{mass + rope}$, is $(mg, -mg)$. You know that the total force on the pulley is zero (because it is not accelerating):

$\Sigma F = 0$. Two forces are acting on the pulley, $F_{mass + rope}$ and $F_{support}$, so the sum of these two must be zero:

$$F_{mass + rope} + F_{support} = 0$$

This means that

$$F_{support} = -F_{mass + rope}$$

Therefore, $F_{support}$ must equal

$$\begin{aligned} F_{support} &= -F_{mass + rope} \\ &= -(mg, -mg) \\ &= (-mg, mg) \end{aligned}$$

As you can see by checking Figure 5-8, the directions of this vector make sense — the pulley's support must exert a force to the left ($-mg$) and upward ($+mg$) to hold the pulley where it is.

You can also convert $F_{support}$ to magnitude and direction form (see Chapter 4), which gives you the full magnitude of the force. The magnitude is equal to

$$F_{support} = \sqrt{(-mg)^2 + (mg)^2} \approx mg\sqrt{2}$$

Note that this magnitude is greater than the force you exert or the force the mass exerts on the pulley, because the pulley support has to change the direction of those forces.

Now find the direction of the force $F_{support}$. You can find the angle it makes with the horizontal axis, θ, using the components of the force. You know from basic trigonometry that the components can be expressed in terms of θ, like so:

$$F_{support, x} = F_{support} \cos \theta$$

$$F_{support, y} = F_{support} \sin \theta$$

where $F_{support}$ indicates the magnitude of the force in these equations. This relates the components of the vector to its magnitude and direction; you can use this to isolate the direction in terms of the components in the following way: If you divide the y component by the x component in the preceding form, you find the tangent of the angle:

$$\begin{aligned} \tan\theta &= \frac{F_{support,y}}{F_{support,x}} \\ &= \frac{mg}{-mg} \\ &= 1 \end{aligned}$$

Now if you take the inverse tangent, you get an answer for θ:

$$\tan^{-1}(1) = 45^\circ$$

However, this answer can't be right, because this angle would mean that the force pointed to the right and up. But you may remember that angles that differ by a multiple of 180° give the same tangent, so you can subtract the preceding answer from 180° to get

$$\theta = 135^\circ$$

This direction is to the left and upward and has the correct tangent, so this is the direction of the force. See Chapter 4 for more info on trig.

If you get confused about the signs when doing this kind of work, check your answers against the directions you know the force vectors actually go in. A picture's worth a thousand words, even in physics!

Finding equilibrium

In physics, an object is in *equilibrium* when it has zero acceleration — when the net force acting on it is zero. The object doesn't actually have to be at rest — it can be going 1,000 miles per hour as long as the net force on it is zero and it isn't accelerating. Forces may be acting on the object, but they all add up, as vectors, to zero.

For example, take a look at Figure 5-9, where you've started your own grocery store and bought a wire rated at 15 newtons to hang the sign with.

The sign weighs only 8.0 newtons, so hanging it should be no problem, right? Obviously, you can tell from my phrasing that you have a problem here. Coolly, you get out your calculator to figure out what force the wire, F_1 in the diagram, has to exert on the sign to support it. You want the sign to be at equilibrium, which means that the net force on it is zero. Therefore, the entire weight of the sign, *mg*, has to be balanced out by the upward force exerted on it.

In this case, the only upward force acting on the sign is the *y* component of F_1, where F_1 is the tension in the wire, as you can see in Figure 5-9. Force exerted by the horizontal brace, F_2, is only horizontal, so it can't do anything for you in the vertical direction. Using your knowledge of trigonometry (see Chapter 4), you can determine from the figure that the *y* component of F_1 is

$$F_{1y} = F_1 \sin 30^\circ$$

To hold up the sign, F_{1y} must equal the weight of the sign, *mg*:

$$F_{1y} = F_1 \sin 30^\circ = mg$$

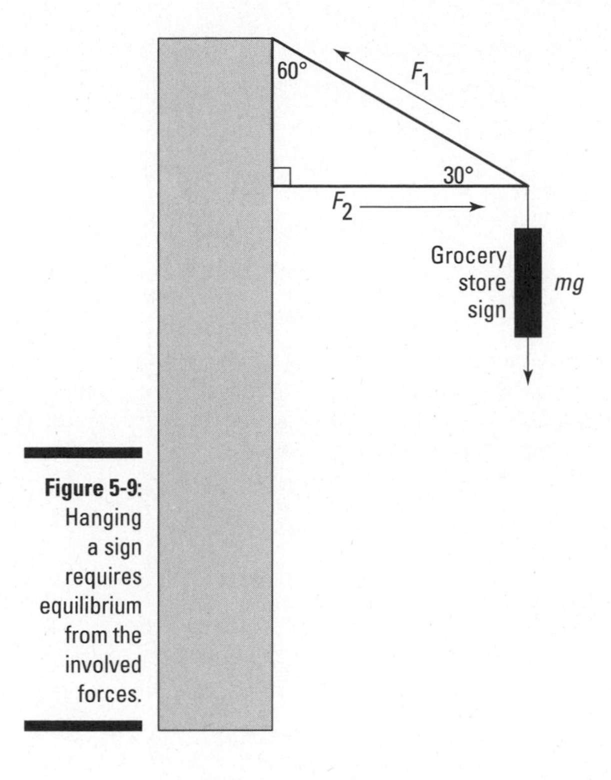

Figure 5-9: Hanging a sign requires equilibrium from the involved forces.

This tells you that the tension in the wire, F_1, must be

$$F_1 = \frac{mg}{\sin 30°}$$

You know that the weight of the sign is 8.0 newtons, so

$$F_1 = \frac{8.0 \text{ N}}{\sin 30°} = 16 \text{ N}$$

Uh oh. Looks like the wire will have to be able to withstand a force of 16 newtons, not just the 15 newtons it's rated for. You need to get a stronger wire.

Assume that you get a stronger wire. Now you may be worried about the brace that provides the horizontal force, F_2, you see diagrammed in Figure 5-9. What force does that brace have to be capable of providing? Well, you know that the figure has only two horizontal forces: F_{brace} and the x component of F_1. And you already know that F_1 = 16 N. You have all you need to figure F_{brace}. To start, you need to determine what the x component of F_1 is. Looking at Figure 5-9 and using a little trig, you can see that

$$F_{1x} = F_1 \cos 30°$$

This is the force whose magnitude must be equal to F_{brace}:

$$F_{brace} = F_1 \cos 30°$$

This tells you that

$$F_{brace} = (16\text{ N}) \cos 30° \approx 14\text{ N}$$

The brace you use has to be able to exert a force of about 14 newtons.

To support a sign of just 8 newtons, you need a wire that supports at least 16 newtons and a brace that can provide a force of 14 newtons. Look at the configuration here — the y component of the tension in the wire has to support all the weight of the sign, and because the wire is at a pretty small angle, you need a lot of tension in the wire to get the force you need. And to be able to handle that tension, you need a pretty strong brace.

Chapter 6

Getting Down with Gravity, Inclined Planes, and Friction

In This Chapter

- Jumping into gravity
- Examining angles on an inclined plane
- Adjusting for the forces of friction
- Measuring flight paths

Gravity, one of the fundamental forces of the universe, is a very big part of our everyday lives. Any object that has mass exerts an attractive force on any other object that has mass. All objects on the surface of the Earth are subject to significant gravitational forces, and gravity plays an important role throughout the whole universe. For these reasons, an understanding of gravity is a vital part of physics.

In this chapter, you find out how to handle gravity along ramps and how to work friction into your calculations. You also see how gravity affects the trajectory of objects flying through the air.

This discussion sticks pretty close to the ground, er, Earth, where the acceleration due to gravity is constant. But Chapter 7 takes off into orbit, looking at gravity from the moon's point of view. The farther you get away from the Earth, the less its gravity affects you.

Acceleration Due to Gravity: One of Life's Little Constants

When you're on or near the surface of the Earth, the pull of gravity is constant. It's a constant force directed straight down with magnitude equal to mg, where m is the mass of the object being pulled by gravity and g is the magnitude of the acceleration due to gravity:

$$g = 9.8 \text{ meters/second}^2 = 32.2 \text{ feet/second}^2$$

Acceleration is a vector, meaning it has a direction and a magnitude (see Chapter 4), so this equation really boils down to **g**, an acceleration straight down toward the center of the Earth. The fact that $\mathbf{F}_{gravity} = m\mathbf{g}$ is important because it says that the acceleration of a falling body doesn't depend on its mass:

$$\mathbf{F}_{gravity} = m\mathbf{a} = m\mathbf{g}$$

In other words, $m\mathbf{a} = m\mathbf{g}$.

Because $\mathbf{a} = \mathbf{g}$, a heavier object doesn't fall faster than a lighter one. Gravity gives any freely falling body the same acceleration downward (g near the surface of Earth), assuming that no other forces, such as air resistance, are present.

Finding a New Angle on Gravity with Inclined Planes

Plenty of gravity-oriented problems in introductory physics involve inclined planes, or ramps. Gravity accelerates objects down ramps — but not the full force of gravity; only the component of gravity acting along the ramp accelerates the object. That's why an object rolling down a steep ramp rolls quickly: The ramp slopes sharply downward, close to the direction of gravity, so most of the force of gravity can act along the ramp.

To find out how much of the force of gravity accelerates an object on a ramp, you have to break the gravity vector into its components along and perpendicular to the ramp.

Check out Figure 6-1. Here, a cart is about to roll down a ramp. The cart travels not only vertically but also horizontally along the ramp, which is inclined at an angle θ. Say that $\theta = 30°$ and that the length of the ramp is 5.0 meters. How fast will the cart be going at the bottom of the ramp?

You know the length of the ramp (the cart's displacement) and the cart's mass, so if you can find the cart's acceleration along the ramp, you can calculate the cart's final velocity.

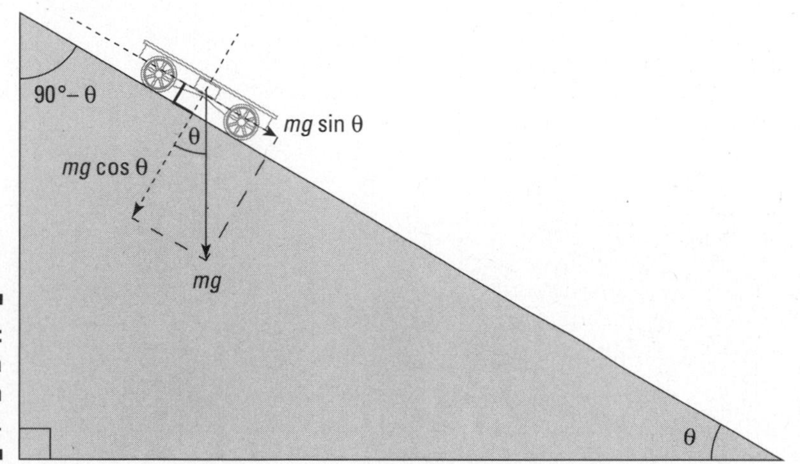

Figure 6-1: Racing a cart down a ramp.

Finding the force of gravity along a ramp

You can break the weight of the cart down into components that are parallel to and perpendicular to the ramp. The component perpendicular to the ramp presses the cart into the surface of the ramp. The component of the weight that acts along the ramp accelerates the cart down the ramp. In this section, you find the component of gravity acting along the ramp when the vertical force due to gravity is F_g.

Figuring out the angle

To work out the components of the weight parallel to and perpendicular to the ramp, you need to know the relationship between the direction of the total weight and the direction of the ramp. The simplest way to determine this is to work out the angle between the weight and a line perpendicular to the ramp. This angle is labeled in Figure 6-1 as θ, which is equal to the angle of the ramp.

There are various ways you can use geometry to show that θ is equal to the angle of the ramp. For example, you may note that the angle between the weight and the line perpendicular to the ramp must be complementary to the angle at the top of the ramp, which is $90° - \theta$ (two angles are *complementary* if they add up to 90°).

Look at Figure 6-2. The angle of the ramp is given by the angle *ABC*. The angle at the top of the ramp is the complement of this because the angles of a triangle add up to 180°, so the angle $BDE = 90° - \theta$. The angle BCA must be equal to the angle BDE because the triangles EBD and ABC are similar, so you can say that the angle $BCA = 90° - \theta$. Finally, the angle BCA must be complementary to the angle ACF because they clearly add up to 90° (along with right angle FCD, they form a straight line), so you finally have your answer: $ACF = \theta$.

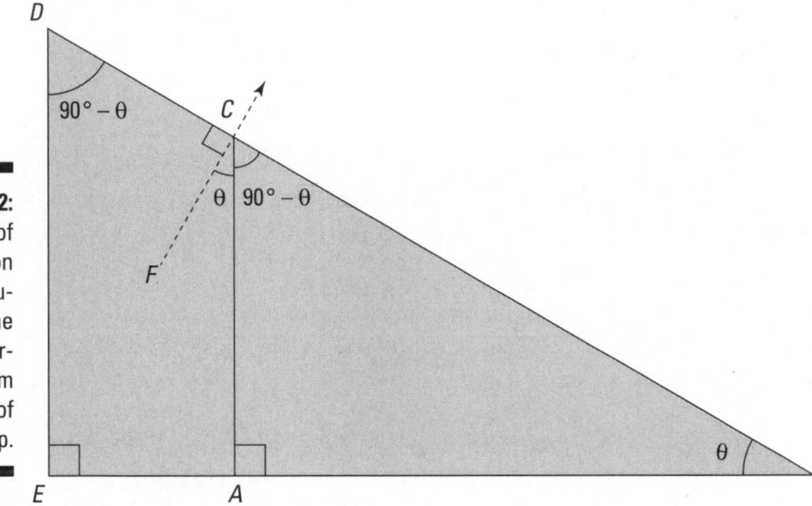

Figure 6-2: The angle of the direction perpendicular to the ramp surface from the angle of the ramp.

Finding the component of the weight along a ramp

If you use trigonometry to project the weight vector onto the lines perpendicular to and parallel to the ramp (refer to Figure 6-1 and rotate the book by 30° if doing so helps you see what's going on), you obtain the expression for the component of the weight perpendicular to the ramp as this:

$$-mg\ \cos\theta$$

And the component of the weight that's along the ramp is this:

$$mg\ \sin\theta$$

Because you know the force, you can use Newton's second law to work out the acceleration:

$$a = \frac{mg\ \sin\theta}{m} = g\sin\theta$$

At this point, you know that the acceleration of the cart along the ramp is given by $a = g \sin\theta$. This equation holds for any object that gravity accelerates down a ramp, as long as friction doesn't apply.

Figuring the speed along a ramp

All you speed freaks may be wondering, "What's the speed of the cart at the bottom of the ramp?" This looks like a job for the following equation (presented in Chapter 3):

$$v_f^2 - v_i^2 = 2as$$

The initial velocity along the ramp, v_i, is 0 meters/second; the displacement of the cart along the ramp, *s*, is 5.0 meters; and the acceleration along the ramp is $g \sin \theta$, so you get the following:

$$v_f^2 = 2as$$

$$v_f^2 = 2(9.8 \text{ m/s}^2 \sin 30°)(5.0 \text{ m})$$

$$v_f^2 = 49 \text{ m}^2/\text{s}^2$$

$$v_f = 7.0 \text{ m/s}$$

This works out to v_f = 7.0 meters/second, or a little under 16 miles/hour. That doesn't sound too fast until you try to stop an 800-kilogram automobile at that speed — don't try it at home! (Actually, this example is a little simplified, because some of the motion goes into the angular velocity of the wheels and such. More on this topic in Chapter 11.)

Quick: How fast would an ice cube on the ramp from Figures 6-1 and 6-2 go at the bottom of the ramp if friction weren't an issue? Answer: the same speed you just figured, 7.0 meters/second. The acceleration of an object along a ramp that's at an angle θ with respect to the ground is $g \sin \theta$. The mass of the object doesn't matter — this simply takes into consideration the component of the acceleration due to gravity that acts along the ramp. And after you know the acceleration along the ramp's surface, which has a length equal to *s*, you can use this equation:

$$v_f^2 = 2as$$

Mass doesn't enter into it.

Getting Sticky with Friction

You know all about friction. It's the force that holds objects in motion back — or so it may seem. Actually, friction is essential for everyday living. Imagine a world without friction: no way to drive a car on the road, no way to walk on pavement, no way to pick up that tasty sandwich. Friction may seem like an enemy to the hearty physics follower, but it's also your friend.

Friction comes from the interaction of surface irregularities. If you introduce two surfaces that have plenty of microscopic pits and projections, you produce friction. And the harder you press those two surfaces together, the more friction you create as the irregularities interlock more and more.

Physics has plenty to say about how friction works. For example, imagine that you decide to put all your wealth into a huge gold ingot (a bar of gold), only to have someone steal your fortune. The thief applies a force to the ingot to accelerate it away as the police start after him. Thankfully, the force of friction comes to your rescue, because the thief can't accelerate away nearly as fast as he thought — all that gold drags heavily along the ground. See Figure 6-3, which shows the forces on the gold ingot.

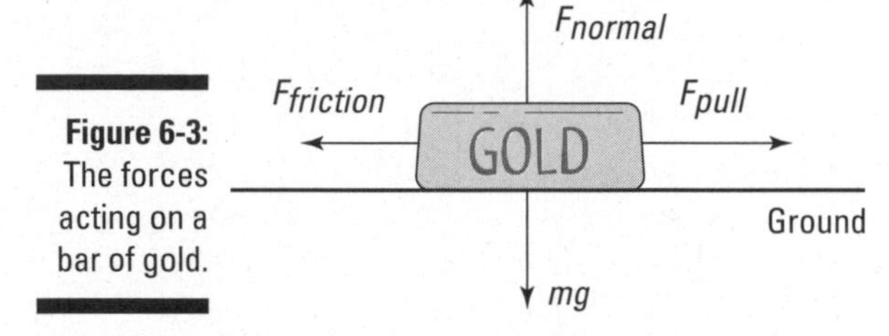

Figure 6-3: The forces acting on a bar of gold.

So if you want to get quantitative here, what would you do? You'd say that the pulling force, F_{pull}, minus the force due to friction, $F_{friction}$, is equal to the net force in the *x*-axis direction, which gives you the acceleration in that direction:

$$F_{pull} - F_{friction} = ma$$

That looks straightforward enough. But how do you calculate $F_{friction}$? You start by calculating the normal force.

Calculating friction and the normal force

The force of friction, $F_{friction}$, always acts to oppose the force you apply when you try to move an object. Friction is proportional to the force with which an object pushes against the surface you're trying to slide it along.

As you can see in Figure 6-3, the force with which the gold ingot presses against the ground in this situation is just its weight, or *mg*. The ground presses back with the same force in accordance with Newton's third law. The force that pushes up against the ingot, perpendicular to the surface, is called the *normal force,* and its symbol is *N*. The normal force isn't necessarily equal to the force due to gravity; it's the force perpendicular to the surface an object is sliding on. In other words, the normal force is the force pushing the two surfaces together, and the stronger the normal force, the stronger the force due to friction.

In the case of Figure 6-3, because the ingot slides along the horizontal ground, the normal force has the same magnitude as the weight of the ingot, so $F_{normal} = mg$. You have the normal force, which is the force pressing the ingot and the ground together. But where do you go from there? You find the force of friction.

Conquering the coefficient of friction

The force of friction comes from the surface characteristics of the materials that come into contact. How can physics predict those characteristics theoretically? It doesn't. Detailed knowledge of the surfaces that come into contact is something people have to measure themselves (or they can check a table of information after someone else has done all the work).

What you measure is how the normal force (a force perpendicular to the surface an object is sliding on) relates to the friction force. It turns out that to a good degree of accuracy, the two forces are proportional, and you can use a constant, μ, to relate the two:

$$F_{friction} = \mu F_{normal}$$

Usually, you see this equation written in the following terms:

$$F_F = \mu F_N$$

This equation tells you that when you have the normal force, F_N, all you have to do is multiply it by a constant to get the friction force, F_F. This constant, μ, is called the *coefficient of friction,* and it's something you measure for contact between two particular surfaces. (***Note:*** Coefficients are simply numbers; they don't have units.)

Here are a couple of things to remember:

- **The equation $F_F = \mu F_N$ relates the magnitude of the force of friction to the magnitude of the normal force.** The normal force is always directed perpendicular to the surface, and the friction force is always directed parallel to the surface. F_F and F_N are perpendicular to each other.
- **The force due to friction is generally independent of the contact area between the two surfaces.** This means that even if you have an ingot that's twice as long and half as high, you still get the same frictional force when dragging it over the ground. This makes sense, because if the area of contact doubles, you may think that you should get twice as much friction. But because you've spread out the gold into a longer ingot, you halve the force on each square centimeter, because less weight is above it to push down.

On the move: Understanding static and kinetic friction

Okay, are you ready to get out your lab coat and start calculating the forces due to friction? Not so fast — you need to know whether the objects in contact with each other are moving. You have two different coefficients of friction for each pair of surfaces because two different physical processes are involved:

- **Static:** When two surfaces aren't moving but are pressing together, they have the chance to interlock on the microscopic level. That's static friction. The coefficient of static friction is μ_s.
- **Kinetic:** When the surfaces are sliding, the microscopic irregularities don't have the same chance to connect, and you get kinetic friction. Kinetic friction is weaker than static friction; however, for most hard, smooth surfaces, these two coefficients are quite similar. The coefficient of kinetic friction is μ_k.

Therefore, you must account for two different coefficients of friction for each pair of surfaces: a static coefficient of friction, μ_s, and a kinetic coefficient of friction, μ_k.

You can notice yourself that static friction is stronger than kinetic friction. Imagine that a box you're unloading onto a ramp starts to slide. To make it stop, you can put your foot in its way, and after you stop it, the box is more likely to stay put and not start sliding again. That's because static friction, which happens when the box is at rest, is greater than kinetic friction, which happens when the box is sliding.

Starting motion with static friction

You experience static friction when you push something that starts at rest. This is the friction that you have to overcome to get something to slide.

For example, say that the static coefficient of friction between the ingot from Figure 6-3 and the ground is 0.30, and the ingot has a mass of 1,000 kilograms (quite a fortune in gold). What's the horizontal force that a thief has to exert to get the ingot moving? You know that the magnitude of the force of friction is related to the magnitude of the normal force by

$$F_F = \mu_s F_N$$

And because the surface is flat, the normal force — the force that presses the two surfaces together — is in the opposite direction of the ingot's weight and has the same magnitude. That means that

$$F_F = \mu_s mg$$

where m is the mass of the ingot and g is the acceleration due to gravity near the surface of the Earth. Plugging in the numbers gives you

$$F_F = \mu_s mg$$
$$= (0.30)(1{,}000\ \text{kg})(9.8\ \text{m/s}^2)$$
$$\approx 2{,}900\ \text{N}$$

The thief needs about 2,900 newtons of force just to get the ingot started. There are 4.448 newtons to a pound, so that translates to about 650 pounds of force. Pretty respectable force for any thief. What happens after the burly thief gets the ingot going? How much force does he need to keep it moving? He needs to look at kinetic friction.

Sustaining motion with kinetic friction

The force due to kinetic friction, which occurs when two surfaces are already sliding, isn't quite as strong as static friction, but that doesn't mean you can predict what the coefficient of kinetic friction is going to be, even if you know the coefficient of static friction — someone has to measure both forces.

Say that the gold ingot from Figure 6-3, which has a mass of 1,000 kilograms, has a coefficient of kinetic friction, μ_k, of 0.18. How much force does the thief need to pull the ingot along at a constant speed during his robbery? You have all you need — the magnitude of the kinetic coefficient of friction is related to the magnitude of the normal force by:

$$F_F = \mu_k F_N = \mu_k mg$$

Putting in the numbers gives you

$$F_F = \mu_k mg$$
$$= (0.18)(1{,}000\ \text{kg})(9.8\ \text{m/s}^2)$$
$$\approx 1{,}800\ \text{N}$$

The thief needs approximately 1,800 newtons of force to keep your gold ingot sliding while evading the police. That converts to about 400 pounds of force (4.448 newtons to a pound) — not exactly the kind of force you can keep going while trying to run at top speed, unless you have some friends helping you. Lucky you! Physics states that the police are able to recover your gold ingot. The cops know all about friction — taking one look at the prize, they say, "We got it back. You drag it home."

A not-so-slippery slope: Handling uphill and downhill friction

Frictional forces depend on the normal force acting. However, when frictional forces are acting on a ramp, the angle of the ramp tilts the normal force at an angle. When you work out the frictional forces, you need to take this into account.

What if you have to drag a heavy object up a ramp? Say, for example, you have to move a refrigerator. You want to go camping, and because you expect to catch plenty of fish, you decide to take your 100-kilogram refrigerator with you. The only catch is getting the refrigerator into your vehicle (see Figure 6-4). The refrigerator has to go up a 30° ramp that happens to have a static coefficient of friction with the refrigerator of 0.20 and a kinetic coefficient of friction of 0.15 (see the earlier section "On the move: Understanding static and kinetic friction"). The good news is that you have two friends to help you move the fridge. The bad news is that you can supply only 350 newtons of force each, so your friends panic.

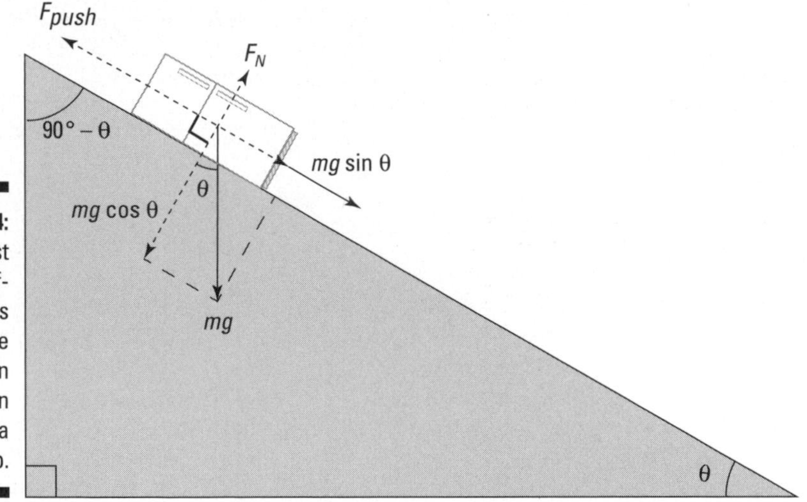

Figure 6-4: You must battle different types of force and friction to push an object up a ramp.

The minimum force needed to push that refrigerator up the ramp has a magnitude F_{push}, and it has to counter the component of the weight of the refrigerator acting along the ramp and the force due to friction. I tackle these components one at a time in the following sections.

Figuring out the weight components parallel and perpendicular to the ramp

The first step in this problem is to resolve the weight of the refrigerator into components parallel and perpendicular to the ramp. Take a look at Figure 6-4, which shows the refrigerator and the forces acting on it. I show you how to resolve the components of the weight vector on a ramp in the earlier section "Finding the component of the weight along a ramp." The component of the weight of the refrigerator along the ramp is $mg \sin \theta$, and the component of the refrigerator's weight perpendicular to the ramp is $-mg \cos \theta$.

When you know the component of the weight along the ramp, you can work out the minimum force required to push the refrigerator up the ramp. The minimum force has to overcome the static force of friction acting down the ramp and the component of the refrigerator's weight acting down the ramp, so the minimum force is

$$F_{push} = mg \sin \theta + F_F$$

Determining the force of friction

The next question is "What's the force of friction, F_F?" Should you use the static coefficient of friction or the kinetic coefficient of friction? Because the static coefficient of friction is greater than the kinetic coefficient of friction, the static coefficient is your best choice. After you and your friends get the refrigerator to start moving, you can keep it moving with less force. Because you're going to use the static coefficient of friction, you can get F_F this way:

$$F_F = \mu_s F_N$$

You also need the normal force, F_N, to continue (see the section "Calculating friction and the normal force" earlier in this chapter). F_N is equal and opposite to the component of the refrigerator's weight acting perpendicularly to the ramp. The component of the refrigerator's weight acting perpendiculary to the ramp is $-mg \cos \theta$ (see the preceding section), so you can say that the normal force acting on the refrigerator is

$$F_N = mg \cos \theta$$

You can verify this by letting θ go to zero, which means that F_N becomes mg, as it should.

The static force of friction, F_F, is then given by $F_F = \mu_s mg \cos \theta$. So the minimum force required to overcome the component of the weight acting along the ramp and the static force of friction is given by

$$F_{pull} = mg \sin \theta + \mu_s mg \cos \theta$$

Now just plug in the numbers:

$$
\begin{aligned}
F_{pull} &= mg \sin \theta + \mu_s mg \cos \theta \\
&= (100 \text{ kg})(9.8 \text{ m/s}^2)(\sin 30°) + (0.20)(100 \text{ kg})(9.8 \text{ m/s}^2)(\cos 30°) \\
&\approx 490 \text{ N} + 170 \text{ N} \\
&= 660 \text{ N}
\end{aligned}
$$

You need 660 newtons of force to push the refrigerator up the ramp. In other words, your two friends, who can exert 350 newtons each, are enough for the job. "Get started," you say, pointing confidently at the refrigerator. Unfortunately, just as they get to the top of the ramp, one of them stumbles. The refrigerator begins to slide down the ramp, and they jump off, abandoning it to its fate.

Object on the loose: Calculating how far an object will slide

Assuming that the ramp and the ground both have the same kinetic coefficient of friction and that the refrigerator starts to slide from the top of the ramp, how far will the refrigerator that your friends drop (in the preceding section) slide? Take a look at Figure 6-5, which shows the refrigerator as it slides down the 3.0-meter ramp. As you watch with dismay, it picks up speed. A car is parked behind the ramp, only 7.2 meters away. Will the errant refrigerator smash it?

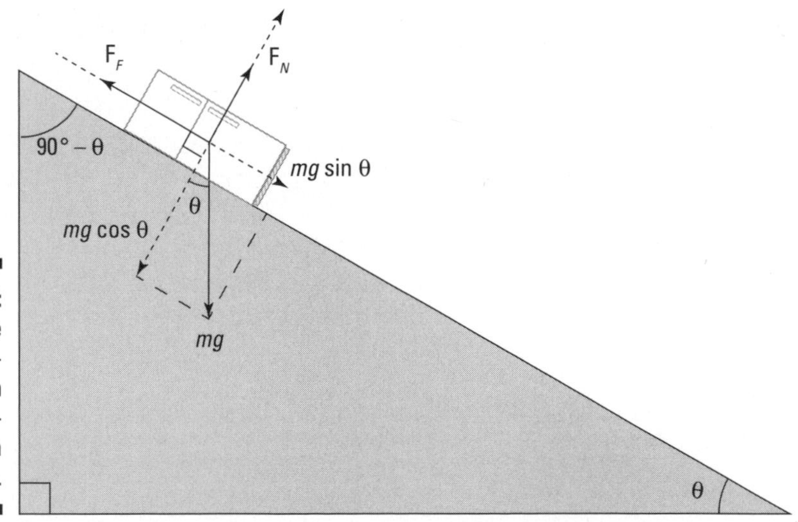

Figure 6-5: All the forces acting on an object sliding down a ramp.

Figuring the acceleration and final velocity at the end of the ramp

When an object slides downward, the forces acting on it change (see Figure 6-5). With the fridge, there's no more F_{pull} force to push it up the ramp. Instead, the component of the refrigerator's weight acting along the ramp pulls the refrigerator downward. And while the fridge slides down, friction opposes that force. So what force accelerates the refrigerator downward? The weight acting along the ramp is $mg \sin \theta$ and the normal force is $mg \cos \theta$, which means that the kinetic force of friction is

$$F_F = \mu_k F_N = \mu_k mg \cos \theta$$

The net force accelerating the refrigerator down the ramp, $F_{accleration}$, is the the difference between the component of the refrigerator's weight along the ramp and the frictional force opposing it:

$$\begin{aligned} F_{acceleration} &= mg \sin \theta - F_F \\ &= mg \sin \theta - \mu_k \, mg \cos \theta \end{aligned}$$

Note that you subtract F_F, the force due to friction, because that force always acts to oppose the force causing the object to move. Plugging in the numbers gives you

$$\begin{aligned} F_{acceleration} &= (100 \text{ kg})(9.8 \text{ m/s}^2)\sin 30° - (0.15)(100 \text{ kg})(9.8 \text{ m/s}^2)\cos 30° \\ &\approx 360 \text{ N} \end{aligned}$$

The force pulling the refrigerator down the ramp is 360 newtons. Because the refrigerator is 100 kilograms, you have an acceleration of 360 N/100 kg = 3.6 m/s^2, which acts along the entire 3.0-meter ramp. You can calculate the final speed of the refrigerator at the bottom of the ramp this way:

$$v_f^2 = 2as$$

Plugging in the numbers, you get

$$v_f^2 = 2(3.6 \text{ m/s}^2)(3.0 \text{ m})$$

$$v_f^2 = 21.6 \text{ m}^2/\text{s}^2$$

$$v_f \approx 4.6 \text{ m/s}$$

The final speed of the refrigerator when it starts traveling along the street toward the parked car is about 4.6 meters per second.

Figuring the distance traveled

With your calculations from the preceding section, do you know how far the refrigerator will travel after your friends let go of it on a ramp?

You have a refrigerator heading down the street at 4.6 meters per second, and you need to calculate how far it's going to go. Because it's traveling along the pavement now, you need to factor in the force due to friction. Gravity will no longer accelerate the object, because the street is flat. Sooner or later, the refrigerator will come to a stop. But how close will it come to a car that's parked in the street 7.2 meters away? As usual, your first calculation is the force acting on the object. In this case, you figure the magnitude of the force due to friction:

$$F_F = \mu_k F_N$$

Because the refrigerator is moving along a horizontal surface, the normal force, F_N, is simply the weight of the refrigerator, *mg,* which means the force of friction is

$$F_F = \mu_k F_N = \mu_k mg$$

Plugging in the data gives you

$$F_F = \mu_k mg = (0.15)(100 \text{ kg})(9.8 \text{ m/s}^2) \approx 150 \text{ N}$$

A force of 150 newtons acts to stop the sliding refrigerator that's now terrorizing the neighborhood. So how far will it travel before it comes to rest? If you take the refrigerator to be moving horizontally in the positive direction, then because the force is acting in the opposite direction, its horizontal component is negative. Because of Newton's second law, the acceleration is also negative and is given by

$$a = \frac{F_F}{m} = \frac{-150 \text{ N}}{100 \text{ kg}} = -1.5 \text{ m/s}^2$$

You can find the distance through the equation $v_f^2 - v_i^2 = 2as$. The distance the refrigerator slides is

$$s = \frac{v_f^2 - v_i^2}{2a}$$

In this case, you want the final velocity, v_f, to be zero, because you need to know where the refrigerator will stop. Therefore, this equation breaks down to

$$s = \frac{v_{fi}^2 - v^2}{2a} = \frac{0^2 - (4.6 \text{ m/s})^2}{2(-1.5 \text{ m/s}^2)} \approx 7.1 \text{ meters}$$

Whew! The refrigerator slides only 7.1 meters, and the car is 7.2 meters away. With the pressure off, you watch the show as your panic-stricken friends hurtle after the refrigerator, only to see it come to a stop right before hitting the car — just as you expected.

Let's Get Fired Up! Sending Objects Airborne

This section is all about how what goes up must come down — the behavior of objects under the influence of constant gravitational attraction. With Newton's second law, you can relate the acceleration of a body to the net force acting on it. You know that gravity exerts a force on a mass, called its *weight,* which has the magnitude *mg.* So you can work out the constant *g,* the acceleration of a mass under the sole influence of gravity. When you know how constant acceleration relates to velocity and displacement, you can work out the motion of a projectile.

In this section, you sling projectiles around and let gravity do its work on shaping their trajectories. You'll see that because the force of gravity only acts downward — that is, in the vertical direction — you can treat the vertical and horizontal components separately. I start with just vertical motion before going on to look at trajectories with both horizontal and vertical components to them. Armed with this information, you can calculate things like the time for a projectile to strike the ground or reach the top of its trajectory and the distance that a projectile will travel.

Shooting an object straight up

To start simply, figure out how far a projectile can travel straight up in the air. Say, for example, that on your birthday, your friends give you just what you've always wanted: a cannon. It has a muzzle velocity of 860 meters/second, and it shoots 10-kilogram cannonballs. Anxious to show you how it works, your friends shoot it off. The only problem: The cannon is pointing straight up. How long do you have to get out of the way?

Going up: Maximum height

Wow, you think, watching the cannonball. You wonder how high it will go, so everyone starts to guess. Because you know your physics, you can figure this one out exactly.

You know the initial vertical velocity, v_i, of the cannonball, and you know that gravity will accelerate it downward. How can you determine how high the ball will go? At the cannonball's maximum height, its vertical velocity will be zero, and then it will head down to Earth again. Therefore, you can use the following equation at the cannonball's highest point, where its vertical velocity will be zero:

$$v_f^2 - v_i^2 = 2as$$

You want to know the cannonball's displacement from its initial position, so solve for s. This gives you

$$s = \frac{v_f^2 - v_i^2}{2a}$$

Plugging in what you know — v_f is 0 meters/second, v_i is 860 meters/second, and the acceleration is g downward (g being 9.8 meters/second^2, the acceleration due to gravity on the surface of the Earth), or $-g$. You get this:

$$s = \frac{v_{fi}^2 - v^2}{2a} = \frac{(0 \text{ m/s})^2 - (860 \text{ m/s})^2}{2(-9.8 \text{ m/s}^2)} \approx 3.8 \times 10^4 \text{ meters}$$

Whoa! The ball will go up 38 kilometers, or nearly 24 miles. Not bad for a birthday present.

Floating on air: Hang time

How long would it take a cannonball shot 24 miles straight up (see the preceding section) to reach its maximum height?

You know that the vertical velocity of the cannonball at its maximum height is 0 meters/second, so you can use the following equation to find the time the cannonball will take to reach its maximum height:

$$v_f = v_i + at$$

Because $v_f = 0$ meters/second and $a = -g = -9.8$ meters/seconds^2, it works out to this:

$$0 = v_i - gt$$

Solving for time, you get the following:

$$t = \frac{v_i}{g}$$

You enter the numbers into your calculator as follows:

$$t = \frac{v_i}{g} = \frac{860 \text{ m/s}}{9.8 \text{ m/s}^2} \approx 88 \text{ s}$$

It takes about 88 seconds for the cannonball to reach its maximum height.

Note: This equation is one way to come to the solution, but you have all kinds of ways to solve a problem like this. You look at a somewhat similar problem in Chapter 4, where a golf ball falls off a cliff; there, you use the equation $s = \frac{1}{2}at^2$ to determine how long the ball is in the air, given the height of the cliff.

Going down: Factoring the total time

How long would it take a cannonball shot 24 miles straight into the air to complete its entire trip — up and then down, from muzzle to lawn — half of which takes 88 seconds (to reach its maximum height)? Flights like the one taken by the cannonball are symmetrical; the trip up is a mirror of the trip down. The velocity at any point on the way up has exactly the same magnitude as on the way down, but on the way down, the velocity is in the opposite direction. Ignoring air resistance, this means that the total flight time is double the time it takes the cannonball to reach its highest point, or

$$t_{total} = 2(88\text{ s}) = 176\text{ s}$$

You have 176 seconds, or 2 minutes and 56 seconds, until the cannonball hits the ground.

Projectile motion: Firing an object at an angle

Firing projectiles at an angle introduces a horizontal component to the motion. However, the force of gravity acts only in the vertical direction, so the horizontal component of the trajectory is uniform. You can tackle this kind of problem by separating out the horizontal and vertical components of the motion.

Here's an example: Imagine that one of your devious friends decides to fire a cannonball at an angle, as Figure 6-6 shows. The following sections cover the cannonball's motion when you shoot at an angle.

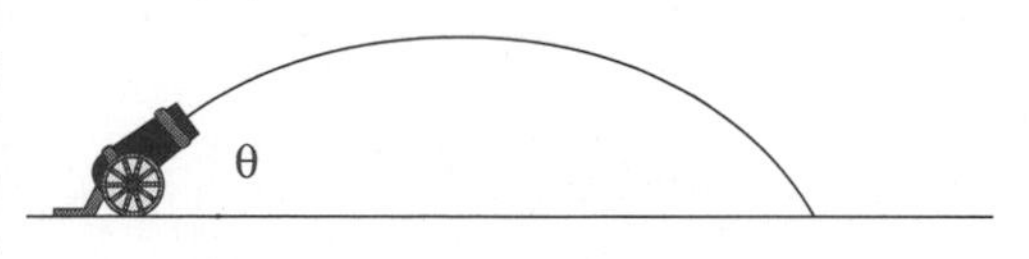

Figure 6-6: Shooting a cannon at a particular angle with respect to the ground.

Breaking down a cannonball's motion into its components

How do you handle the motion of an object shot up at an angle? Because you can always break motion in two dimensions up into *x* and *y* components, and because gravity acts only in the *y* component, your job is easy. All you have to do is break the initial velocity into *x* and *y* components (see Chapter 4 for the basics of this task):

$v_x = v_i \cos\theta$

$v_y = v_i \sin\theta$

These velocity components are independent, and gravity acts only in the y direction, which means that v_x is constant; only v_y changes with time, as follows:

$v_y = v_i \sin\theta - gt$

If you want to know the x and y positions of the cannonball at any time, you can easily find them. You know that x is simply

$x = v_x t = (v_i \cos\theta)t$

And because gravity accelerates the cannonball vertically, here's what y looks like (the t^2 here is what gives the cannonball's trajectory in Figure 6-6 its parabolic shape):

$$y = v_y t - \frac{1}{2}gt^2$$

You figure out in previous sections the time it takes a cannonball to hit the ground when shot straight up: $t = 2v_y / g$. Knowing the time allows you to find the range of the cannon in the x direction:

$$s = v_x t = \frac{2v_x v_y}{g} = \frac{2v_0^2 \sin\theta\cos\theta}{g}$$

So there you have it — now you can figure out the range of the cannon given the speed of the cannonball and the angle at which it was shot.

Discovering the cannon's maximum range

What's the range for your new cannon if you aim it at 45°, which gives you your maximum range? If the cannonball has an initial velocity of 860 meters/second, the equation you use looks like this:

$$s = v_x t = \frac{2v_i^2 \sin\theta\cos\theta}{g} = \frac{2(860 \text{ m/s})^2 \sin 45°\cos 45°}{9.8 \text{ m/s}^2} \approx 75{,}000 \text{ m}$$

Your range is 75 kilometers, or nearly 47 miles. Not bad.

Chapter 7

Circling around Rotational Motion and Orbits

In This Chapter

- Working with centripetal acceleration
- Feeling the pull of centripetal force
- Incorporating angular displacement, velocity, and acceleration
- Orbiting with Newton's laws and gravity
- Staying in the loop with vertical circular motion

Circular motion can include rockets' moving around planets, race cars' whizzing around a track, or bees' buzzing around a hive. In this chapter, you look at the velocity and acceleration of objects that are moving in circles. This discussion leads to more general forms of rotational motion, where it's useful to talk about motion in angular terms.

Angular equivalents exist for displacement, velocity, and acceleration. Instead of dealing with linear displacement as a distance, you deal with angular displacement as an angle. Angular velocity indicates what angle you sweep through in so many seconds, and angular acceleration gives you the rate of change in the angular velocity. All you have to do is take linear equations and substitute the angular equivalents: angular displacement for displacement, angular velocity for velocity, and angular acceleration for acceleration.

Centripetal Acceleration: Changing Direction to Move in a Circle

In order to keep an object moving in circular motion, its velocity constantly changes direction. Because velocity changes, you have acceleration. Specifically, you have *centripetal acceleration* — the acceleration needed to keep an object moving in circular motion. At any point, the velocity of the object is perpendicular to the radius of the circle.

If the string holding the ball in Figure 7-1 breaks at the top, bottom, left, or right moment you see in the illustration, which way would the ball go? If the velocity points to the left, the ball would fly off to the left. If the velocity points to the right, the ball would fly off to the right. And so on. That's not intuitive for many people, but it's the kind of physics question that may come up in introductory courses.

The velocity of an object in circular motion is always at right angles to the radius of the object's path. At any one moment, the velocity points along the tiny section of the circle's circumference where the object is, so the velocity is tangential to the circle.

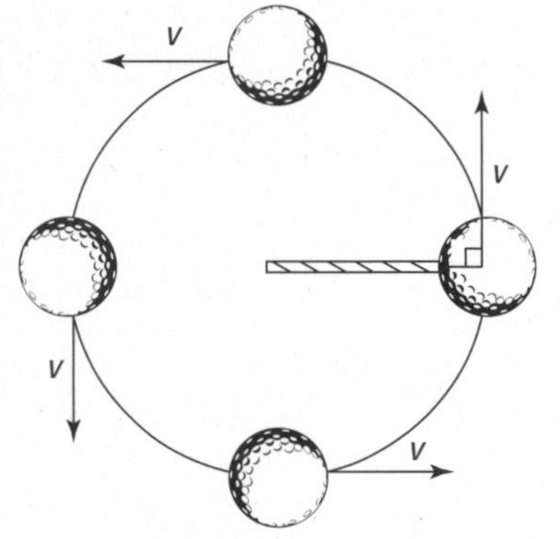

Figure 7-1: Velocity constantly changes direction when an object is in circular motion.

Keeping a constant speed with uniform circular motion

An object with *uniform circular motion* travels in a circle with a constant speed. Practical examples may be hard to come by, unless you see a race car driver on a perfectly circular track with his accelerator stuck, a clock with a seconds hand that's in constant motion, or the moon orbiting the Earth.

Take a look at Figure 7-2, where a golf ball tied to a string is whipping around in circles. The golf ball is traveling at a uniform speed as it moves around in a circle, so you can say it's traveling in uniform circular motion.

An object in uniform circular motion does not travel with a uniform velocity, because its direction changes all the time.

Describing the period

Any object that travels in uniform circular motion always takes the same amount of time to move completely around the circle. That time is called its *period*, designated by T.

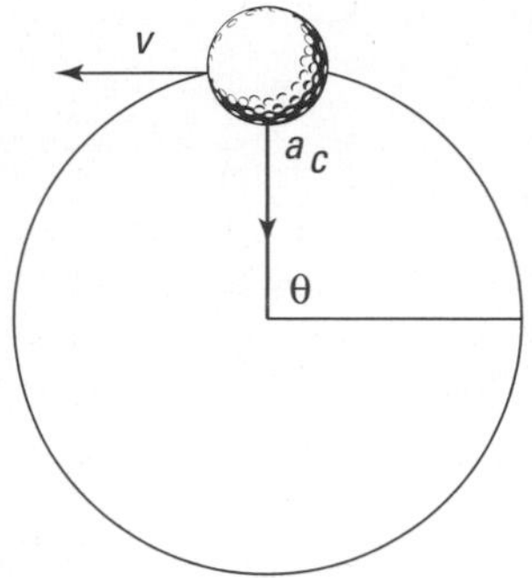

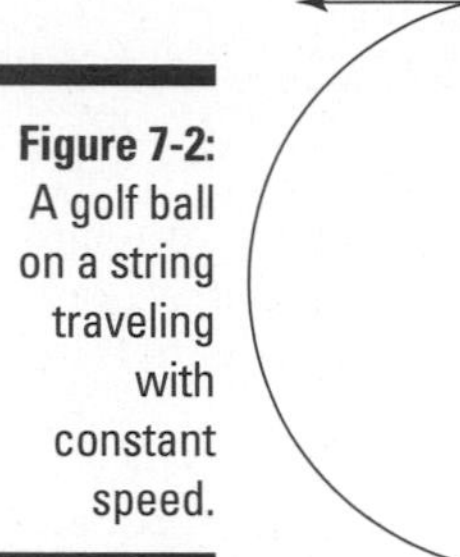
Figure 7-2: A golf ball on a string traveling with constant speed.

If you're swinging a golf ball around on a string at a constant speed, you can easily relate the ball's speed to its period. You know that the distance the ball must travel each time around the circle equals the circumference of the circle, which is $2\pi r$ (where r is the radius of the circle), so you can get the equation for finding an object's period by first finding its speed:

$$v = \frac{\text{circumference}}{\text{period}} = \frac{2\pi r}{T}$$

If you solve for T, you get the equation for the period:

$$T = \frac{2\pi r}{v}$$

Say that you're spinning a golf ball in a circle at the end of a 1.0-meter string every half-second. How fast is the ball moving? Time to plug in the numbers:

$$\begin{aligned} v &= \frac{2\pi r}{T} \\ &= \frac{2\pi(1.0\text{ m})}{0.50\text{ s}} \approx 12.6\text{ m/s} \end{aligned}$$

The ball moves at a speed of 12.6 meters/second. Just make sure you have a strong string!

Accelerating toward the center

When an object travels in uniform circular motion, its speed is constant, which means that the magnitude of the object's velocity doesn't change. Therefore, acceleration can have no component in the same direction as the velocity; if it did, the velocity's magnitude would change.

However, the velocity's direction is constantly changing — it always bends so that the object maintains movement in a constant circle. To make that happen, the object's centripetal acceleration is always concentrated toward the center of the circle, perpendicular to the object's velocity at any one time. The acceleration changes the direction of the object's velocity while keeping the magnitude of the velocity constant.

In the ball's case (refer to Figures 7-1 and 7-2), the string exerts a force on the ball to keep it going in a circle — a force that provides the ball's centripetal acceleration. In order to provide that force, you have to constantly pull on the ball toward the center of the circle. (Picture what it feels like, force-wise, to whip an object around on a string.) You can see the centripetal acceleration vector, a_c, in Figure 7-2.

If you accelerate the ball toward the center of the circle to provide the centripetal acceleration, why doesn't it hit your hand? The answer is that the ball is already moving at a high speed. The force, and therefore the acceleration, that you provide always acts at right angles to the velocity.

Finding the magnitude of the centripetal acceleration

You always have to accelerate an object toward the center of the circle to keep it moving in circular motion. So can you find the magnitude of the acceleration you create? No doubt. If an object is moving in uniform circular motion at speed v and radius r, you can find the magnitude of the centripetal acceleration with the following equation:

$$a_c = \frac{v^2}{r}$$

For a practical example, imagine you're driving around curves at a high speed. For any constant speed, you can see from the equation $a_c = v^2/r$ that the centripetal acceleration is inversely proportional to the radius of the curve. In other words, on tighter curves (as the radius decreases), your car needs to provide a greater centripetal acceleration (the acceleration increases).

Seeking the Center: Centripetal Force

When you're driving a car around a bend, you create centripetal acceleration by the friction of your tires on the road. How do you know what force you need to create to turn the car at a given speed and turning radius? That depends on the *centripetal force* — the center-seeking, inward force needed to keep an object moving in uniform circular motion.

In this section, you discover how the centripetal force keeps the object moving in a circle and how the details of the circular motion such as radius and velocity depend upon the centripetal force.

Looking at the force you need

Centripetal force isn't some new force that appears out of nowhere when an object travels in a circle; it's the force the object *needs* to keep traveling in that circle.

As you know from Newton's first law (see Chapter 5), if there's no net force on a moving object, the object will continue to move uniformly in a straight line. If a force (or a component of a force) acts in the same direction as the object's velocity, then the object begins to speed up, and if the force acts in the opposite direction to the velocity, then the object slows down. However, if the force always acts perpendicularly to the velocity while remaining of constant magnitude, then the magnitude of the velocity (the speed) does not change; only its direction does — the object moves in a circle. In this case, the force is called *centripetal force.*

If you're spinning a ball on a string, then the centripetal force comes from the tension in the string. When the moon orbits the Earth, the centripetal force comes from gravity. And when you drive your car in a circle, the centripetal force comes from the friction of the tires against the road. The origin of the force is not important, only that it remains of constant magnitude and always acts perpendicularly to the velocity, toward the center of the circle.

The fictitious centrifugal force

You've probably heard of centrifugal force and most likely have felt it when a car you were in turned a corner. However, centrifugal force is not really a force as defined in Newton's laws. It only *appears* to be a force. When you're in a car turning a corner, your body has inertia and is naturally inclined to move at uniform speed in a straight line. But because the car is turning, it feels as though your body is being thrown outward, toward the car door.

Seeing how the mass, velocity, and radius affect centripetal force

Because force equals mass times acceleration, **F** = *m***a,** and because centripetal acceleration is equal to v^2/r (see the earlier section "Finding the magnitude of the centripetal acceleration"), you can determine the magnitude of the centripetal force needed to keep an object moving in uniform circular motion with the following equation:

$$F_c = \frac{mv^2}{r}$$

This equation tells you the magnitude of the force that you need to move an object of a given mass, *m,* in a circle at a given radius, *r,* and speed, *v.* (Remember that the direction of the force is always toward the center of the circle.)

Think about how force is affected if you change one of the other variables. The equation shows that if you increase mass or speed, you'll need a larger force; if you decrease the radius, you're dividing by a smaller number, so you'll also need a larger force. Here's how these ideas play out in the real world:

- **Increasing mass:** You may have an easy time swinging a golf ball on a string in a circle, but if you replace the golf ball with a cannonball, watch out. You may now have to whip 10 kilograms around on the end of a 1.0-meter string every half-second. As you can tell, you need a heck of a lot more force.
- **Increasing speed:** Not interested in spinning cannonballs? Then imagine you're driving your car around in a circle. If you're going quite slowly around the circle, your tires have no problem generating enough frictional force to keep you going in the circle. But if you go too fast, then your tires can no longer generate the frictional force acting toward the center of the circle, so you start to skid.
- **Decreasing the radius:** You can see the effect of the radius in your car going around in a circle. If you drive your car at a fixed speed in a circle of smaller and smaller radius, eventually your tires won't be able to supply enough centripetal force from the friction, and you'll skid off the circular path.

Try plugging some numbers into the formula. The ball from Figure 7-2 is moving at 12.6 meters/second on a 1.0-meter string. How much force do you need to make a 10.0-kilogram cannonball move in the same circle at the same speed? Here's what the equation looks like:

$$\begin{aligned} F_c &= \frac{mv^2}{r} \\ &= \frac{(10.0 \text{ kg})(12.6 \text{ m/s})^2}{1.0 \text{ m}} \approx 1{,}590 \text{ N} \end{aligned}$$

You need about 1,590 newtons, or about 357 pounds of force (4.448 newtons are in a pound; see Chapter 5). Pretty hefty, if you ask me; I just hope your arms can take it.

Negotiating flat curves and banked turns

Imagine that you're driving a car and you come to a curve. On a flat road, the centripetal force you need to negotiate the curve comes from the friction of the tires against the ground. If the surface is covered with a substance such as ice, you have less friction, and you can't turn as safely at high speeds.

To make turns safer, engineers design roads so that curves are banked. With the road at an angle, there's a component of the normal force of the road against your car, toward the center of the circle. This means that you don't require as much friction from your tires to make the turn.

Relying on friction to turn on a flat road

When you're driving on a flat road, friction provides the centripetal force — toward the center of the circle — that allows you to make a turn.

Say you're sitting in the passenger seat of the car, approaching a turn with a 200.0-meter radius (with a level, non-banked road surface). You know that the coefficient of static friction is 0.8 on this road (you use the coefficient of static friction because the tires aren't slipping on the road's surface) and that the car has a mass of about 1,000 kilograms. What's the maximum speed the driver can go and still keep you safe? You get out your calculator as the driver shoots you a look with raised eyebrows. The frictional force needs to supply the centripetal force, so you come up with the following:

$$F_c = \frac{mv^2}{r} = \mu_s mg$$

where m is the mass of the car, v is the velocity, r is the radius, μ_s is the coefficient of static friction, and g is the acceleration due to gravity, 9.8 meters/second^2. Solving for the speed on one side of the equation gives you

$$v = \sqrt{\mu_s gr}$$

This looks simple enough — you just plug in the numbers to get

$$\begin{aligned} v &= \sqrt{\mu_s gr} \\ &= \sqrt{(0.8)(9.8\ \text{m/s}^2)(200.0\ \text{m})} \approx 40\ \text{m/s} \end{aligned}$$

You calculate 40 meters/second, or about 87 miles/hour. You look at the speedometer and see a speed of 70 miles/hour. You can negotiate the turn safely at your present speed.

Depending on the normal force to make a banked turn

If a curve is banked, then a component of the normal force of the road against the car contributes to the centripetal force, and so you can go around the curve much faster. Because you don't have to rely on friction to supply the centripetal force, the question of whether you can safely make the turn no longer depends on road conditions.

Take a look at Figure 7-3, which shows a car banking around a turn. The engineers can make the driving experience enjoyable if they bank the turn so that drivers garner the centripetal force needed to go around the turn entirely by the component of the normal force of the road against the car acting toward the center of the turn's circle. That component is $F_N \sin\theta$ (F_N is the normal force, the upward force perpendicular to the road; see Chapter 6), so

$$F_c = F_N \sin\theta = \frac{mv^2}{r}$$

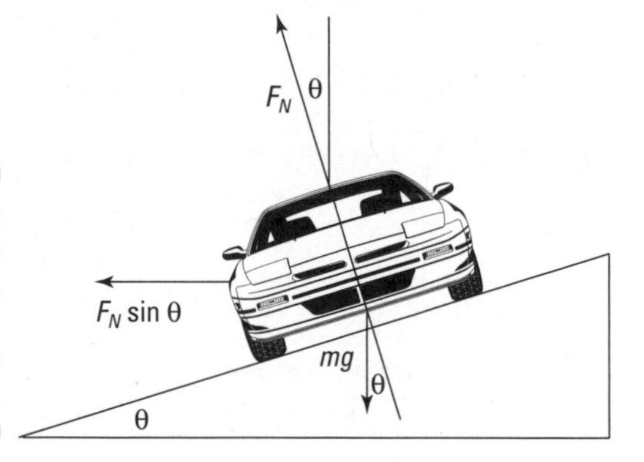

Figure 7-3: The forces acting on a car banking around a turn.

To find the centripetal force, you need the normal force, F_N. If you look at Figure 7-3, you can see that F_N comes from a combination of the centripetal force due to the car's banking around the turn and the car's weight. The purely vertical component of F_N must equal *mg,* because no other forces are operating vertically, so

$$F_N \cos\theta = mg$$

$$F_N = \frac{mg}{\cos\theta}$$

Plugging this result into the equation for centripetal force gives you

$$F_c = F_N \sin\theta$$

$$\frac{mv^2}{r} = \left(\frac{mg}{\cos\theta}\right)\sin\theta$$

Because $\sin\theta / \cos\theta = \tan\theta$, you can also write this as

$$\frac{mv^2}{r} = mg\tan\theta$$

$$\frac{mv^2}{mg\,r} = \tan\theta$$

Solve for θ to find the angle of the road. The equation finally breaks down to

$$\theta = \tan^{-1}\left(\frac{v^2}{gr}\right)$$

You don't have to memorize this result, in case you're panicking — this is the kind of equation used by highway engineers when they have to bank curves (notice that the mass of the car cancels out, meaning that it holds for vehicles regardless of weight). You can always derive this equation from your knowledge of Newton's laws and circular motion.

What should the angle θ be if drivers go around a 200-meter-radius turn at 60 miles/hour? Plug in the numbers; 60 miles/hour is about 27 meters/second and the radius of the turn is 200 meters, so

$$\theta = \tan^{-1}\left(\frac{v^2}{gr}\right)$$

$$= \tan^{-1}\frac{(27\text{ m/s})^2}{(9.8\text{ m/s}^2)(200\text{ m})} \approx 20°$$

The designers should bank the turn at about 20° to give drivers a smooth experience. Remember though, that you made this calculation such that all the centripetal force comes from the normal force of the road against the car. You could go around the corner faster than this if you have some friction from your tires — but not too fast, or you'll be skidding off into the verge!

Getting Angular with Displacement, Velocity, and Acceleration

For objects moving in a circle, you can work with acceleration and velocity using the horizontal and vertical components, just as in previous chapters on motion. But when objects are undergoing rotational motion, using angular variables instead makes a lot of sense. With these variables, instead of specifying the horizontal and vertical components, you specify the radius and the angle of rotation.

In this section, you discover the angular equivalents of displacement, velocity, and acceleration. You can apply these variables to rotating objects and objects moving in a circle.

Measuring angles in radians

The natural unit of measurement of angles is the radian, not the degree. A full circle is made up of 2π radians, which is also 360°, so 360° = 2π radians. If you travel in a full circle, you go 360°, or 2π radians. (If an object rotates one revolution, then the angle has magnitude of 2π radians. Therefore, sometimes instead of *radians per second,* you see *revolutions per second.*) A half-circle is π radians, and a quarter-circle is $\pi/2$ radians.

The radian is a natural measure of angle because a circular arc that has a length of one radius extends an angle of 1 radian (see Figure 7-4). So if you know the radius and the angle that an object has moved through in radians, you can easily find the distance that the object has moved in proportion to the radius. If the object moves θ radians in a circle of radius r, then the object travels a distance of θr along the circle.

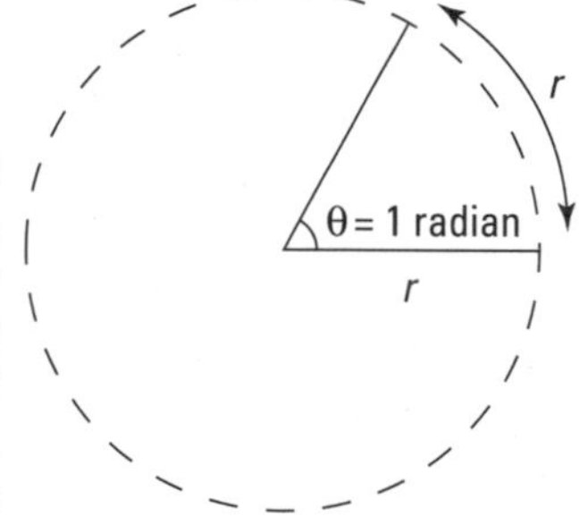

Figure 7-4: A circular arc extends an angle of one radian.

This idea is useful in relating the angular velocity to the speed of an object moving in a circle. In addition, you can see why a full circle has an angle of 2π radians: You know that the circumference of a circle is $2\pi r$ and that to go the whole way around the 360° of a circle, you need to travel 2π times the radius. Therefore, there are 2π radians to 360°.

How do you convert from degrees to radians and back again? Because 360° = 2π radians (or 2 multiplied by 3.14, the rounded version of pi), you have an easy calculation. If you have 45° and want to know how many radians that translates to, just use this conversion factor:

$$45^\circ\left(\frac{2\pi \text{ radians}}{360^\circ}\right) = \frac{\pi}{4} \text{ radians}$$

You find out that 45° = $\pi/4$ radians. If you have, say, $\pi/2$ radians and want to know how many degrees that converts to, you do this conversion:

$$\left(\frac{\pi \not{\text{radians}}}{2}\right)\left(\frac{360°}{2\pi \not{\text{radians}}}\right) = 90°$$

You calculate that $\pi/2$ radians = 90°.

Relating linear and angular motion

The fact that you can think of the angle, θ, in rotational motion just as you think of the displacement, s, in linear motion is great, because it means you have an angular counterpart for many of the linear motion equations (see Chapter 3). Here are the variable substitutions you make to get the angular motion formulas:

- **Displacement:** Instead of s, which you use in linear travel, use θ, the angular displacement; θ is measured in radians.
- **Velocity:** In place of the velocity, v, use the angular velocity, ω; angular velocity is the number of radians covered per second.
- **Acceleration:** Instead of acceleration, a, use the *angular acceleration*, α; the unit for angular acceleration is radians per second^2.

Table 7-1 compares the formulas for both linear and angular motion.

Table 7-1 Linear and Angular Motion Formulas

Type of Formula	*Linear*	*Angular*
Velocity	$v = \frac{\Delta s}{\Delta t}$	$\omega = \frac{\Delta \theta}{\Delta t}$
Acceleration	$a = \frac{\Delta v}{\Delta t}$	$\alpha = \frac{\Delta \omega}{\Delta t}$
Displacement	$s = v_i t + \frac{1}{2} a t^2$	$\theta = \omega_i t + \frac{1}{2} \alpha t^2$
Motion with time canceled out	$v_f^2 - v_i^2 = 2as$	$\omega_f^2 - \omega_i^2 = 2\alpha\theta$

Say, for example, that you have a ball tied to a string. What's the angular velocity of the ball if you whirl it around? It makes a complete circle, 2π radians, in 0.5 seconds, so its angular velocity is

$$\omega = \frac{\Delta \theta}{\Delta t} = \frac{2\pi \text{ rad}}{0.5 \text{ s}} = 4\pi \text{ rad/s}$$

Another demonstration of the usefulness of radians in measuring angles is that the linear speed can easily be related to the angular speed. If you take the equation

$$\omega = \frac{\Delta\theta}{\Delta t}$$

And multiply both sides by the radius, r, you get

$$r\omega = \frac{r\Delta\theta}{\Delta t}$$

The term $r\Delta\theta$ is simply the distance traveled by an object moving in a circle of radius r, so this equation becomes

$$r\omega = \frac{\Delta s}{\Delta t}$$

You may recognize the right side of this equation as the equation for speed. So you can see that linear speed and angular speed are related by $r\omega = v$.

If the ball speeds up from 4π radians per second to 8π radians per second in 2 seconds, what would its average angular acceleration be? Work it out by plugging in the numbers:

$$\alpha = \frac{\Delta\omega}{\Delta t} = \frac{8\pi \text{ rad} - 4\pi \text{ rad}}{2 \text{ s}} = \frac{4\pi \text{ rad}}{2 \text{ s}} = 2\pi \text{ rad/s}$$

To find out more about angular displacement, angular velocity, and angular acceleration, see the discussion on angular momentum and torque in Chapter 11. Keep in mind, however, that these angular variables, like their linear counterparts, are actually vector quantities. What you've seen so far are simply components of vectors in one dimension. Because they only have one component, the sign of the component gives the direction in the single dimension (for example, positive indicates movement to the right, and negative indicates movement to the left). In Chapter 11, you see more about the direction and the vector nature of these variables.

Letting Gravity Supply Centripetal Force

You don't have to tie objects to strings to observe travel in circular motion; larger bodies such as planets move in circular motion, too. Gravity provides the necessary centripetal force.

In this section, you discover Newton's take on the gravitational force between two objects, and I show you how his theory relates to 9.8 meters/second^2, the value experimenters identified as the acceleration due to gravity near the surface of the Earth. Then you put Newton's formula to use in looking at the orbits of satellites.

Using Newton's law of universal gravitation

Sir Isaac Newton came up with one of the heavyweight laws in physics for you: the *law of universal gravitation*. This law says that every mass exerts an attractive force on every other mass. If the two masses are m_1 and m_2 and the distance between them is r, the magnitude of the force is

$$F = \frac{Gm_1m_2}{r^2}$$

where G is a constant equal to 6.67×10^{-11} N·m²/kg².

This equation allows you to figure the gravitational force between any two masses. What, for example, is the pull between the sun and the Earth? The sun has a mass of about 1.99×10^{30} kilograms, and the Earth has a mass of about 5.98×10^{24} kilograms. A distance of about 1.50×10^{11} meters separates the two bodies. Plugging the numbers into Newton's equation gives you

$$\begin{aligned} F &= \frac{Gm_1m_2}{r^2} \\ &= \frac{\left(6.67\times10^{-11}\ \text{N}\cdot\text{m}^2/\text{kg}^2\right)\left(1.99\times10^{30}\text{kg}\right)\left(5.98\times10^{24}\text{kg}\right)}{\left(1.50\times10^{11}\text{m}\right)^2} \\ &\approx 3.52\times10^{22}\text{N} \end{aligned}$$

Your answer of 3.52×10^{22} newtons converts to about 8.0×10^{20} pounds of force (4.448 newtons are in a pound).

On the land-based end of the spectrum, say that you're out for your daily physics observations when you notice two people on a park bench, looking at each other and smiling. As time goes on, you notice that they seem to be sitting closer and closer to each other each time you take a glance. In fact, after a while, they're sitting right next to each other. What could be causing this attraction? If the two lovebirds have masses of about 75 kilograms each, what's the force of gravity pulling them together, assuming they started out 0.50 meters away? Your calculation looks like this:

$$\begin{aligned} F &= \frac{Gm_1m_2}{r^2} \\ &= \frac{\left(6.67\times10^{-11}\ \text{N}\cdot\text{m}^2/\text{kg}^2\right)\left(75\ \text{kg}\right)\left(75\ \text{kg}\right)}{\left(0.50\ \text{m}\right)^2} \\ &\approx 1.5\times10^{-6}\text{N} \end{aligned}$$

The force of attraction is roughly 5 millionths of an ounce. Maybe not enough to shake the surface of the Earth, but that's okay. The Earth's surface has its own forces to deal with.

Deriving the force of gravity on the Earth's surface

The equation for the force of gravity — $F = (Gm_1m_2)/r^2$ — holds true no matter how far apart two masses are. But you also come across a special gravitational case (which most of the work on gravity in this book is about): the force of gravity near the surface of the Earth.

The gravitational force between a mass and the Earth is the object's *weight*. *Mass* is considered a measure of an object's inertia, and its weight is the force exerted on the object in a gravitational field. On the surface of the Earth, the two forces are related by the acceleration due to gravity: $F_g = mg$. Kilograms and slugs are units of mass; newtons and pounds are units of weight.

You can use Newton's law of gravitation to get the acceleration due to gravity, *g*, on the surface of the Earth just by knowing the gravitational constant *G*, the radius of the Earth, and the mass of the Earth. The force on an object of mass m_1 near the surface of the Earth is

$$F = m_1g$$

This force is provided by gravity between the object and the Earth, according to Newton's gravity formula, and so you can write

$$m_1g = \frac{Gm_1m_2}{r_e^2}$$

The radius of the Earth, r_e, is about 6.38×10^6 meters, and the mass of the Earth is 5.98×10^{24} kilograms. Putting in the numbers, you have

$$m_1g = \frac{\left(6.67 \times 10^{-11}\ \text{N}\cdot\text{m}^2/\text{kg}^2\right)m_1\left(5.98 \times 10^{24}\,\text{kg}\right)}{\left(6.38 \times 10^6\ \text{m}\right)^2}$$

Dividing both sides by m_1 gives you the acceleration due to gravity:

$$g = \frac{\left(6.67 \times 10^{-11}\ \text{N}\cdot\text{m}^2/\text{kg}^2\right)\left(5.98 \times 10^{24}\,\text{kg}\right)}{\left(6.38 \times 10^6\ \text{m}\right)^2}$$

$$\approx 9.8\ \text{m/s}^2$$

Newton's law of gravitation gives you the acceleration due to gravity near the surface of the Earth: 9.8 meters/second^2.

Of course, you can measure g by letting an apple drop and timing it, but what fun is that when you can calculate it in a roundabout way that requires you to first measure the mass of the Earth?

Using the law of gravitation to examine circular orbits

In space, bodies are constantly orbiting other bodies due to gravity. Satellites (including the moon) orbit the Earth, the Earth orbits the sun, the sun orbits around the center of the Milky Way, the Milky Way orbits around the center of its local group of galaxies. This is big-time stuff. In the case of orbital motion, gravity supplies the centripetal force that causes the orbits.

The force of gravity between orbiting bodies is quite a bit different from small-time orbital motion — such as when you have a ball on a string — because for a given distance and two masses, the gravitational force is always going to be the same. You can't increase the force to increase the speed of an orbiting planet as you can with a ball. The following sections examine the speed and period of orbiting bodies in space.

Calculating a satellite's speed

A particular satellite can have only one speed when in orbit around a particular body at a given distance because the force of gravity doesn't change. So what's that speed? You can calculate it with the equations for centripetal force and gravitational force. You know that for a satellite of a particular mass, m_1, to orbit, you need a corresponding centripetal force (see the section "Seeking the Center: Centripetal Force"):

$$F_c = \frac{m_1 v^2}{r}$$

This centripetal force has to come from the force of gravity, so

$$\frac{G m_1 m_2}{r^2} = \frac{m_1 v^2}{r}$$

You can rearrange this equation to get the speed:

$$v = \sqrt{\frac{G m_2}{r}}$$

This equation represents the speed that a satellite at a given radius must have in order to orbit if the orbit is due to gravity. The speed can't vary as long as the satellite has a constant orbital radius — that is, as long as it's going around in circles. This equation holds for any orbiting object where the attraction is the force of gravity, whether it's a human-made satellite orbiting the Earth or the Earth orbiting the sun. If you want to find the speed for satellites that orbit the Earth, for example, you use the mass of the Earth in the equation:

$$v = \sqrt{\frac{Gm_E}{r}}$$

Here are a few details you should note on reviewing the orbiting speed equation:

- **You have to use the distance from the *center* of the Earth, not the distance above Earth's surface, as the radius.** Therefore, the distance you use in the equation is the distance between the two orbiting bodies. In this case, you add the distance from the center of the Earth to the surface of the Earth, 6.38×10^6 meters, to the satellite's height above the Earth.
- **The equation assumes that the satellite is high enough off the ground that it orbits out of the atmosphere.** That assumption isn't really true for artificial satellites; even at 400 miles above the surface of the Earth, satellites do feel air friction. Gradually, the drag of friction brings them lower and lower, and when they hit the atmosphere, they burn up on reentry. When a satellite is less than 100 miles above the surface, its orbit decays appreciably each time it circles the Earth. (Look out below!)
- **The equation is independent of mass.** If the moon rather than the artificial satellite orbited at 400 miles and you could ignore air friction and collisions with the Earth, it would have to go at the same speed as the satellite in order to preserve its close orbit (which would make for some pretty spectacular moonrises).

Human-made satellites typically orbit at heights of 400 miles from the surface of the Earth (about 640 kilometers, or 6.4×10^5 meters). What's the speed of such a satellite? All you have to do is put in the numbers:

$$v = \sqrt{\frac{Gm_E}{r}} = \sqrt{\frac{\left(6.67\times10^{-11}\ \text{N}\cdot\text{m}^2/\text{kg}^2\right)\left(5.98\times10^{24}\text{kg}\right)}{\left(6.38\times10^{6}\text{m}\right)+\left(6.40\times10^{5}\text{m}\right)}} \approx 7.54\times10^{3}\ \text{m/s}$$

This converts to about 16,800 miles per hour.

You can think of a satellite in motion around the Earth as always falling. The only thing that keeps it from striking the Earth is that its velocity points over the horizon. The satellite *is* falling, but its velocity takes it over the horizon — that is, over the curve of the world as it falls — so it doesn't get any closer to the

Earth. (The same is true of the astronauts inside. They only have the appearance of being weightless, but they're continuously falling, too.)

Calculating the period of a satellite

Sometimes it's more important to know the period of an orbit rather than the speed, such as when you're counting on a satellite to come over the horizon before communication can take place. The *period* of a satellite is the time it takes it to make one full orbit around an object. The period of the Earth as it travels around the sun is one year.

If you know the satellite's speed and the radius at which it orbits (see the preceding section), you can figure out its period. The satellite travels around the entire circumference of the circle — which is $2\pi r$ if r is the radius of the orbit — in the period, T. This means the orbital speed must be $2\pi r/T$, giving you

$$\sqrt{\frac{Gm_E}{r}} = \frac{2\pi r}{T}$$

If you solve this for the period of the satellite, you get

$$T = 2\pi\sqrt{\frac{r^3}{Gm_E}}$$

You, the intuitive physicist, may be wondering: What if you want to examine a satellite that simply stays stationary over the same place on the Earth at all times? In other words, a satellite whose period is the same as the Earth's 24-hour period? Can you do it? Such satellites do exist. They're very popular for communications, because they're always orbiting in the same spot relative to the Earth; they don't disappear over the horizon and then reappear later. They also allow for the satellite-based global positioning system, or GPS, to work.

In cases of stationary satellites, the period, T, is 24 hours, or about 86,400 seconds. Can you find the radius a stationary satellite needs to have? Using the equation for periods, you see that

$$r^3 = \frac{T^2 Gm_E}{4\pi^2}$$

Plugging in the numbers, you get

$$\begin{aligned} r^3 &= \frac{T^2 Gm_E}{4\pi^2} \\ &= \frac{\left(8.64\times10^4\,\text{s}\right)^2\left(6.67\times10^{-11}\ \text{N}\cdot\text{m}^2/\text{kg}^2\right)\left(5.98\times10^{24}\,\text{kg}\right)}{4\pi^2} \\ &\approx 7.542\times10^{22}\ \text{m}^3 \end{aligned}$$

Understanding Kepler's laws of orbiting bodies

Johannes Kepler (1571–1630), a German national born in the Holy Roman Empire, came up with three laws that helped explain a great deal about orbits before Newton came up with his law of universal gravitation. Here are Kepler's laws:

- **Law 1:** Planets orbit in ellipses. An ellipse is a shape like a squashed circle, and the degree of squashedness is called the *eccentricity* of the ellipse. The orbits allowed can have any degree of eccentricity. When the eccentricity is zero, you have a circular orbit.
- **Law 2:** Planets move so that a line between the sun and the planet sweeps out the same area in the same time, independent of where they are in their orbits. This means that when the planet is in that part of its orbit where it is close to the sun, it has to travel faster to sweep out the same area that it does when it's farther away.
- **Law 3:** The square of a planet's orbital period (the time it takes the planet to make one complete orbit) is proportional to its average distance from the sun cubed.

You can see how the third law can be derived from Newton's laws by looking at the section "Calculating the period of a satellite." It takes the form of this equation: $r^3 = (T^2 G m_E)/4\pi^2$. Although Kepler's third law says that T^2 is proportional to r^3, you can get the exact constant relating these quantities by using Newton's law of gravitation.

If you take the cube root of this, you get a radius of 4.23×10^7 meters. Subtracting the Earth's radius of 6.38×10^6 meters, you get 3.59×10^7 meters, which converts to about 22,300 miles. This is the distance from the Earth geosynchronous satellites need to orbit. At this distance, they orbit the Earth at the same rate the Earth is turning, which means that they stay put over the same piece of real estate.

In practice, it's very hard to get the speed just right, which is why geosynchronous satellites have either gas boosters that can be used for fine-tuning or magnetic coils that allow them to move by pushing against the Earth's magnetic field.

Looping the Loop: Vertical Circular Motion

Maybe you've watched extreme sports on television and wondered how bikers or skateboarders can ride into a loop on a track and go upside down without falling to the ground. Shouldn't gravity bring them down? How fast do they have to go? The answers to these vertical circular-motion questions lie in centripetal force and the force of gravity.

Take a look at Figure 7-5, where a ball is looping around a circular track. A question you may come across in introductory physics classes asks, "What speed is necessary so that the ball makes the loop safely?" The crucial point is at the very top of the track — if the ball is going to peel away from its circular track, the top is where it'll fall. To answer the crucial question, you must know what criterion the ball must meet to hold on. Ask yourself, "What's the constraint that the ball must meet?"

To travel in a loop, an object must have a net force acting on it that equals the centripetal force it needs to keep traveling in a circle of the given radius and speed. At the top of its path, as you can see in Figure 7-5, the ball barely stays in contact with the track. Other points along the track provide normal force (see Chapter 6) because of the speed and the fact that the track is curved. If you want to find out what minimum speed an object needs to have to stay on a loop, you need to look at where the object is just barely in contact with the track — in other words, on the verge of falling out of its circular path.

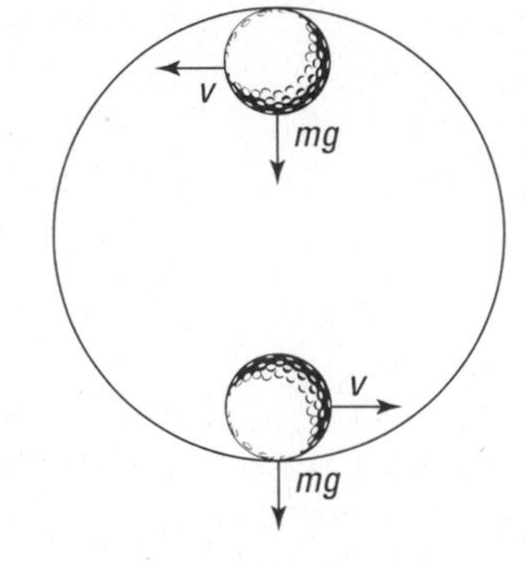

Figure 7-5: The force and velocity of a ball on a circular track.

The normal force the track applies to an object at the top is just about zero. The only force keeping the object on its circular track is the force of gravity, which means that at the apex, the speed of the object has to be such that the centripetal force equals the object's weight to keep it going in a circle whose radius is the same as the radius of the loop. That means that if this is the force needed

$$F_c = \frac{mv^2}{r}$$

then the force of gravity at the top of the loop is

$$F_g = mg$$

And because F_g must equal F_c, you can write

$$\frac{mv^2}{r} = mg$$

You can simplify this equation into the following form:

$$v = \sqrt{rg}$$

The mass of any object traveling around a circular track, such as a motorcycle or a race car, drops out of the equation.

The square root of r times g is the minimum speed an object needs at the top of the loop in order to keep going in a circle. Any object with a slower speed will peel off the track at the top of the loop (it may drop back into the loop, but it won't be following the circular track at that point). For a practical example, if the loop from Figure 7-5 has a radius of 20.0 meters, how fast does the ball have to travel at the top of the loop in order to stay in contact with the track? Just put in the numbers:

$$v = \sqrt{rg} = \sqrt{(20.0\text{ m})(9.8\text{ m/s}^2)} \approx 14.0\text{ m/s}$$

The golf ball has to travel 14.0 meters per second at the top of the track, which is about 31 miles per hour.

What if you want to do the same trick on a flaming circular loop on a motorcycle to impress your pals? The same speed applies — you need to be going about 31 miles per hour minimum at the top of the track, which has a radius of 20 meters. If you want to try this at home, don't forget that this is the speed you need at the top of the track — you have to go faster at the bottom of the track in order to travel at 31 miles per hour at the top, simply because you're twice the radius, or 40 meters, higher up in the air, much like having coasted to the top of a 40-meter hill.

So how much faster do you need to go at the bottom of the track? How about $\sqrt{5}$ times faster? Check out Chapter 9, where kinetic energy (the kind of energy moving motorcycles have) is turned into potential energy (the kind of energy that motorcycles have when they're high up in the air against the force of gravity).

Chapter 8

Go with the Flow: Looking at Pressure in Fluids

In This Chapter

- Examining mass density
- Understanding pressure in liquids and gases
- Floating with Archimedes's principle
- Looking at fluids in motion

On a hot summer day, nothing is better than taking a dip in the neighbor's pool. As you execute the perfect swan dive and slip gracefully to the depths, you notice a curious sensation: water pressure. It increases with every foot you're below the surface. You note the rapid increase of that pressure as you go deeper, and you wonder what being miles under the surface of the ocean would feel like. "Hmm," you think. "Just how does water pressure increase with depth?"

This chapter is all about pressures in fluids, and I cover a lot more than pounds per square inch. You also encounter info on Archimedes's principle (which is all about floating and buoyancy), hydraulic machines, fluids moving in pipes, streamlines, and lots more. With all that coming up, it's time to get your feet wet in the physics of fluids.

Both liquids and gases are considered fluids. A *fluid* is defined as any continuous distribution of matter that cannot support a shear stress without moving. If, on the other hand, you *shear* a solid piece of material by applying different forces to different parts of it, then the solid deforms to some degree but eventually finds a balance. For example, if you hold a piece of rubber in one hand and then push the top of it with the other, it bends over, supporting the shear stress you apply to it.

Mass Density: Getting Some Inside Information

Density is the ratio of mass to volume. Any solid object that's less dense than water floats. Density is an important property of a fluid because mass is continuously distributed throughout a fluid; the static forces and motions within the fluid depend on the concentration of mass (density) rather than the fluid's overall mass.

Calculating density

Density (ρ) is mass (m) divided by volume (V), so here's the formula for density:

$$\rho = \frac{m}{V}$$

In the MKS system, the units are kilograms per cubic meter, or kg/m^3.

Say you have a whopper diamond with a volume of 0.0500 cubic meters (that's a cube that's about 1 foot on each side, so it's truly a whopper). You measure its mass as 176.0 kilograms. So what's its density?

Plugging in the numbers and doing the calculations gives you your answer:

$$\begin{aligned}\rho &= \frac{m}{V}\\ &= \frac{176.0 \text{ kg}}{0.0500 \text{ m}^3}\\ &= 3{,}520 \text{ kg/m}^3\end{aligned}$$

So the density of diamond is 3,520 kg/m^3. That's pretty dense.

You can see a sample of the densities of common materials in Table 8-1. Note that ice is less dense than water, so ice floats. Generally, solids and gases expand with temperature and therefore become less dense (you can find out more about the expansion of solids in Chapter 14 and the expansion of gases in Chapter 16). This table includes the density of water at 4°C as a reference point because the density of water varies with temperature. The densities of the gases generally have a stronger dependence on temperature than the solids do, though.

Table 8-1 Densities of Common Materials

Substance	*Density (kg/m³)*
Gold (near room temperature)	19,300
Mercury (near room temperature)	13,600
Silver (near room temperature)	10,500
Copper (near room temperature)	8,890
Diamond (near room temperature)	3,520
Aluminum (near room temperature)	2,700
Blood (near body temperature)	1,060
Water (4°C)	1,000
Ice (0°C)	917
Oxygen (at 0°C, 101.325 kPa)	1.43
Helium (at 0°C, 101.325 kPa)	0.179

Comparing densities with specific gravity

A substance's *specific gravity* is the ratio of that substance's density to the density of water at 4°C. Because the density of water at 4°C is 1,000 kg/m³, that ratio is easy to find. For example, the density of gold is 19,300 kg/m³, so its specific gravity is the following:

$$\text{specific gravity}_{\text{gold}} = \frac{19{,}300 \text{ kg/m}^3}{1{,}000 \text{ kg/m}^3} = 19.3$$

Specific gravity has no units, because it's a ratio of density divided by density, so all units cancel out. Therefore, the specific gravity of gold is simply 19.3.

Anything with a specific gravity greater than 1,000 sinks in pure water at 4°C, and anything with a specific gravity less than 1,000 floats. As you'd expect, gold, with a specific gravity of 19,300, sinks. Ice, on the other hand, with a specific gravity of 917, floats. So how can a ship, which is made of metal with a specific gravity very much greater than water, float? The ship floats because of the shape of its hull. The ship is mostly hollow and displaces water weighing more than the weight of the ship. Averaged throughout, the ship is less dense than the water overall, so the effective specific gravity of the ship is less than that of water.

Applying Pressure

Everyone who's ever talked about car or bicycle tires or blown up a balloon knows about air pressure. And if you've gone swimming underwater, you know about water pressure. When you push on something, people say you exert pressure on it.

In physics terms, *pressure* is force per area — a fact you may already know if you've filled a tire to a certain number of pounds per square inch. The equation for pressure, *P,* is the following:

$$P = \frac{F}{A}$$

where F is force and A is area. Note that pressure is not a vector — it's a scalar (that is, just a number without a direction).

In this section, you look at the units of pressure, see how pressure changes along with depth or altitude, and discover how hydraulic machines work.

Looking at units of pressure

Because pressure is force divided by area, its MKS units are newtons per square meter, or N/m^2. In the foot-pound-second (FPS) system, the units are pounds per square inch, or psi.

The unit *newtons per square meter* is so common in physics that it has a special name: the *pascal,* which equals 1 newton per square meter. The pascal is abbreviated as Pa.

You don't have to be underwater to experience pressure from a fluid. Air exerts pressure, too, due to the weight of the air above you. Here's how much pressure the air exerts on you at sea level:

$$\text{air pressure}_{\text{sea level}} = 1.013 \times 10^5 \text{ Pa}$$

The air pressure at sea level is a standard pressure that people refer to as 1 *atmosphere* (abbreviated atm):

$$\text{air pressure}_{\text{sea level}} = 1.013 \times 10^5 \text{ Pa} = 1 \text{ atm}$$

If you convert an atmosphere to pounds per square inch, it's about 14.7 psi. That means that 14.7 pounds of force are pressing in on every square inch of your body at sea level.

Your body pushes back with 14.7 psi, so you don't feel any pressure on you at all. But if you suddenly got transported to outer space, the inward pressure of the air pushing on you would be gone, and all that would remain would be the 14.7 pounds per square inch your body exerted outward. You wouldn't explode, but your lungs could burst if you tried to hold your breath. The change in pressure could also cause the nitrogen in your blood to form bubbles and give you the bends!

Here's a pressure example problem using water pressure. Say you're in your neighbor's pool, waiting near the bottom until your neighbors give up trying to chase you off and go back into the house. You're near the deep end of the pool, and using the handy pressure gauge you always carry, you measure the pressure on the back of your hand as 1.2×10^5 pascals. What force does the water exert on the back of your hand? The back of your hand has an area of about 8.4×10^{-3} square meters. You reason that if $P = F/A$, then the following is true:

$$F = PA$$

Plugging in the numbers and solving gives you the answer:

$$\begin{aligned} F &= PA \\ &= (1.2 \times 10^5 \text{ Pa})(8.4 \times 10^{-3} \text{ m}^2) \\ &= (1.2 \times 10^5 \text{ N/m}^2)(8.4 \times 10^{-3} \text{ m}^2) \\ &\approx 1.0 \times 10^3 \text{ N} \end{aligned}$$

Yikes. A thousand newtons! You whip out your underwater calculator to find that's about 230 pounds. Forces add up quickly when you're underwater because water is a heavy liquid. The force you feel is the weight of the water above you.

Connecting pressure to changes in depth

You know that pressure increases the farther you go underwater, but by how much? As a physicist, you can put some numbers in and get numerical results out. Just what pressure would you expect for a given depth?

Say that you're underwater and you're considering the imaginary cube of water you see in Figure 8-1. At the top of the cube, the water pressure is P_1. At the bottom of the cube, it's P_2. The cube has horizontal faces of area A and a height h. First find the forces on the top and bottom of the cube.

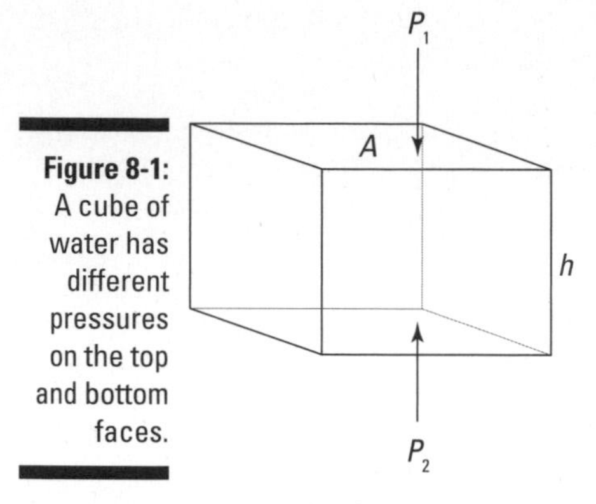

Figure 8-1: A cube of water has different pressures on the top and bottom faces.

The sum of the forces is the difference between the force on the bottom face of the cube, F_2, and the force on the top face of the cube, F_1:

$$\Sigma F = F_2 - F_1$$

You can say the force pushing down on the top face is $F_1 = P_1A$ and that the force pushing on the bottom face is $F_2 = P_2A$. Therefore, in terms of pressure, the sum of forces is the following:

$$\Sigma F = P_2A - P_1A$$

So what's the net force upward on the cube of water? The upward force must be equal to the weight of the water, mg, where m is the mass of the water and g is the gravitational constant (9.8 meters/second^2). So you have the following equation:

$$P_2A - P_1A = mg$$

Hmm. You don't know m, the mass of the water. Can you get the weight of the water in terms of A, the area of the top and bottom faces of the cube? The mass of the water is the density of water, ρ, multiplied by the volume of the cube, which is Ah. So you can replace m with ρAh, which gives you the following equation:

$$P_2A - P_1A = \rho gAh$$

Now you're talking. Dividing everything by A gives you the difference in pressures:

$$P_2 - P_1 = \rho gh$$

If you call the difference in the pressures ΔP, you get the following equation:

$$\Delta P = \rho g h$$

The preceding equation is an important, general result that holds for any fluid: water, air, gasoline, and so on. This equation says that the difference in pressure between two points in a fluid is equal to the fluid's density multiplied by g (the acceleration due to gravity) multiplied by the difference in height between the two points.

The next few sections provide some example problems so you can see what the pressure formula looks like in practice.

Diving down

How much does the pressure increase for every meter you go underwater? You know that $\Delta P = \rho g h$, so plug in the numbers and do the math:

$$\Delta P = \rho g h = (1{,}000\ \text{kg/m}^3)(9.8\ \text{m/s}^2)(1.0\ \text{m}) = 9{,}800\ \text{Pa}$$

That works out to be about 1.4 pounds per square inch added pressure for every meter you go down.

If you were wondering how the units work out, rearrange the units from the first equation:

$$\left(\frac{\text{kg}}{\text{m}^3}\right)\left(\frac{\text{m}}{\text{s}^2}\right)(\text{m}) = \left(\frac{\text{kg}\cdot\text{m}}{\text{s}^2}\right)\left(\frac{1}{\text{m}^2}\right)$$

A kg·m/s^2 is just a newton, and a N/m^2 is a pascal, so the units boil down to pascals:

$$\begin{aligned}\left(\frac{\text{kg}\cdot\text{m}}{\text{s}^2}\right)\left(\frac{1}{\text{m}^2}\right) &= \frac{\text{N}}{\text{m}^2}\\ &= \text{Pa}\end{aligned}$$

That's a fair bit of added pressure. But what if you decided to take a dip in a pool of mercury instead (don't try this at home)? Mercury has a density of 13,600 kg/m^3, as opposed to water's density at 1,000 kg/m^3. In this case, the added pressure for every meter would be

$$\Delta P = \rho g h = (13{,}600\ \text{kg/m}^3)(9.8\ \text{m/s}^2)(1.0\ \text{m}) \approx 133{,}000\ \text{Pa}$$

That's an increase of about 19 pounds per square inch for every meter you go down — and that's a lot of pressure.

So does that mean that the pressure 1 meter under the surface of a pool of mercury is about 19 pounds per square inch? No, because you have to add to that pressure the pressure of the air on top of it, so you have the following:

$$P_t = P_m + P_a$$

where P_t is the total pressure, P_m is the pressure due to the mercury, and P_a is the pressure due to the air.

To find the total pressure on something submerged in a liquid, you have to add the pressure due to the liquid to the atmospheric pressure, which is about 14.7 pounds per square inch, or 1.013×10^5 pascals.

Varying blood pressure

Say that your head is 1.5 meters above your feet. What's the difference in blood pressure between your head and your feet (neglecting the action of the heart) when you're lying down and when you're standing? You can use the following equation to settle these questions:

$$\Delta P = \rho g h$$

The calculation for the case where you're lying down is simple because *h,* the vertical distance between your heart and your feet, is 0:

$$\Delta P = \rho g h = \rho g(0) = 0$$

Therefore, you see no difference in pressure between your heart and feet when you're lying down (neglecting heart action). What about when you're standing up? In that case, $h = 1.5$ m:

$$\Delta P = \rho g h = \rho g(1.5 \text{ m})$$

As you can see in Table 8-1, the density, ρ, of blood is 1,060 kg/m^3. Putting in the numbers and doing the math gives you the following difference in pressure:

$$\Delta P = \rho g h = (1{,}060 \text{ kg/m}^3)(9.8 \text{ m/s}^2)(1.5 \text{ m}) \approx 1.6 \times 10^4 \text{ Pa}$$

That pressure works out to be slightly less than 2.0 pounds per square inch.

Pumping water upward

Suppose that a water slide park has been drilling a well to get water. The water in the well is 20 meters down, and the park owners hire you to find out how powerful of a pump they need to get a satisfactory water flow. Hmm, you think — a well that's 20 meters deep, with a water pump on top. Will that even work?

How much pressure can the pump exert on the water at the bottom of the well? The pump is pulling up air in the pipe, creating a vacuum, which the water will follow. But the amount of suction you can create with a pump sucking air is limited. You can create the most pressure with a complete vacuum, $P = 0$. Atmospheric pressure is pushing down on the surface of the water and a total vacuum is at the top of the pipe, so the maximum pressure the pump at the top of the well can exert on the water at the bottom of the well is atmospheric pressure, or 1.01×10^5 pascals:

$$\Delta P = 1.01 \times 10^5 \text{ Pa}$$

How far up a pipe can a pressure of 1.01×10^5 pascals pull water? Well, you know that $\Delta P = \rho g h$, so when you've pulled water up as far as it's going to go, $\rho g h$ of the column of water in the pipe equals 1.01×10^5 pascals:

$$\rho g h = 1.01 \times 10^5 \text{ Pa}$$

Solving for h gives you the formula for how far the water can rise:

$$h = \frac{1.01 \times 10^5 \text{ Pa}}{\rho g}$$

Plugging in the numbers (using the density value you know for water, 1,000 kg/m^3 at 4°C) gives you the height:

$$h = \frac{1.01 \times 10^5 \text{ Pa}}{\left(1{,}000 \text{ kg/m}^3\right)\left(9.8 \text{ m/s}^2\right)} \approx 10.3 \text{ m}$$

Therefore, the maximum height you can pump water out of a well with the pump at the top of the well is 10.3 meters. But in this case, the well is 20 meters deep. You turn to the slide park owners and say, "I have some bad news."

The solution? Put the pump at the bottom of the well and push the water up the pipe instead of trying to use air pressure to pull the water up the pipe.

Hydraulic machines: Passing on pressure with Pascal's principle

Pascal's principle says that given a fluid in a totally enclosed system, a change in pressure at one point in the fluid is transmitted to all points in the fluid, as well as to the enclosing walls. In other words, if you have a fluid enclosed in a pipe (with no air bubbles) and change the pressure in the fluid at one end of the pipe, the pressure changes all throughout the pipe to match.

The fact that pressure inside an enclosed system is the same (neglecting gravitational differences) has an interesting consequence. Because $P = F/A$, you get the following equation for force:

$$F = PA$$

So if the *pressure* is the same everywhere in an enclosed system but the *areas* you consider are different, can you get different forces?

To make this question clearer, look at Figure 8-2, which shows a system of enclosed fluid with two hydraulic pistons, one with a piston head of area A_1 and one with a piston head of area A_2. You apply a force of F_1 on the smaller piston. What is the force on the other piston, F_2?

Pressure at each point is F/A. According to Pascal's principle, the pressure is the same everywhere inside the fluid, so $F_1/A_1 = F_2/A_2$:

$$\frac{F_1}{A_1} = \frac{F_2}{A_2}$$

Solving for F_2 gives you the force at Point 2:

$$F_2 = \frac{F_1 A_2}{A_1}$$

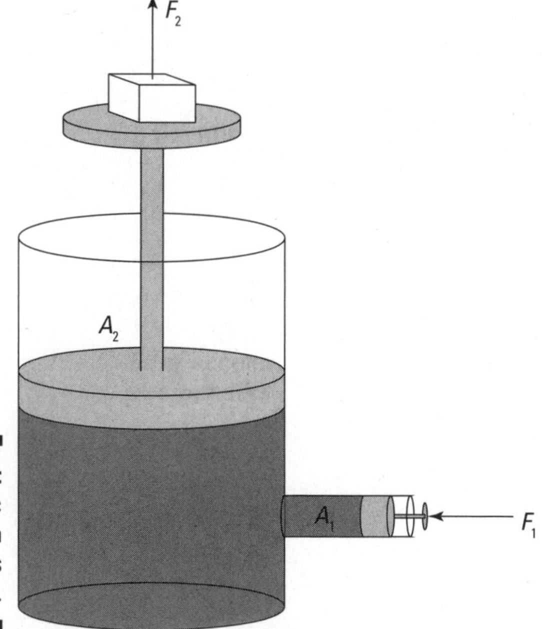

Figure 8-2: A hydraulic system magnifies force.

Cool. That means that you can develop a huge force from a small force if the ratio of the piston sizes is big. For example, say the area of Piston 2 is bigger than Piston 1 by a factor of 100. Does that mean that any force you apply to piston 1 will be multiplied by 100 times on piston 2?

Yes, indeed — that's how hydraulic equipment works. By using a small piston at one end and a large piston at the other, you can create huge forces. Backhoes and other hydraulic machines, such as garbage trucks and hydraulic lifts, use Pascal's principle to function.

What's the catch here? If you push on Piston 1 and get 100 times the force on Piston 2, you seem to be getting something for nothing. The catch is that you have to push the smaller piston 100 times as far as the second piston will move.

Buoyancy: Float Your Boat with Archimedes's Principle

Archimedes's principle says that any fluid exerts a buoyant force on an object wholly or partially submerged in it, and the magnitude of the buoyant force equals the weight of the fluid displaced by the object. An object that's less dense than water floats because the water it displaces weighs more than the object does. Therefore, as the object pushes down, the water pushes back up more strongly.

If you've ever tried to push a beach ball underwater, you've felt this principle in action. As you push the ball down, it pushes back up. In fact, a big beach ball can be tough to hold underwater. As a physicist in a bathing suit, you may wonder, "What's happening here?"

What is the buoyant force, F_b, the water exerts on the beach ball? To make this problem easier, you decide to consider the beach ball as a cube of height h and horizontal face with area A. So the buoyant force on the cubic beach ball is equal to the force at the bottom of the beach ball minus the force at the top:

$$F_{buoyancy} = F_{bottom} - F_{top}$$

And because $F = PA$, you can work pressure into the equation with a simple substitution:

$$F_{buoyancy} = (P_{bottom} - P_{top})A$$

You can also write the change in pressure, $P_{bottom} - P_{top}$, as ΔP:

$$F_{buoyancy} = \Delta PA$$

The change in pressure equals ρgh, so replace ΔP:

$$F_{buoyancy} = \rho ghA$$

Note that hA is the volume of the cube. Multiplying volume, V, by density, ρ, gives you the mass of the water displaced by the cube, m, so you can replace ρhA with m:

$$F_{buoyancy} = mg$$

You should recognize mg (mass times acceleration due to gravity) as the expression for weight, so the force of buoyancy is equal to the weight of the water displaced by the object you're submerging:

$$F_{buoyancy} = W_{water\ displaced}$$

That equation turns out to be Archimedes's principle.

Here's an example of how to use Archimedes's principle. Suppose the designers at Acme Raft Company have hired you to tell them how much of their new raft will be underwater when it's launched. You can see the new Acme raft in Figure 8-3. The density of the wood used in their rafts is 550 kilograms/meter^3, and the raft is 20 centimeters high.

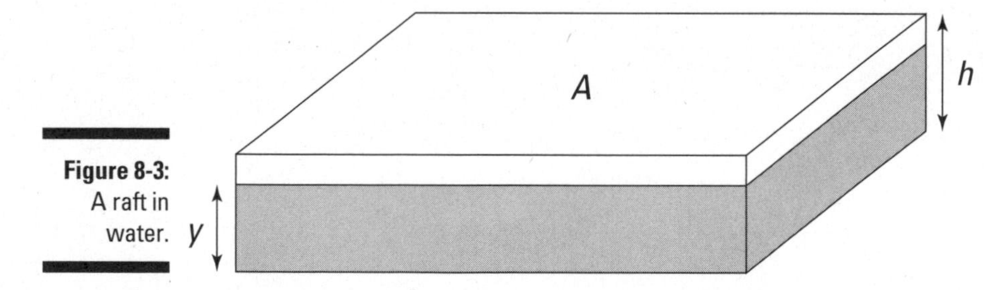

Figure 8-3: A raft in water.

You get out your clipboard and reason that to make the raft float, the weight of the raft must equal the buoyant force the water exerts on the raft.

Say the raft is of height h and horizontal surface area A; that would make its weight equal to the following:

$$W_{raft} = \rho_{raft}Ahg$$

Now what's the buoyant force that the water exerts on the raft? The buoyant force is equal to the weight of the water that the submerged part of the raft displaces. Say that when the raft floats, the bottom of the raft is a distance y underwater. Then the submerged volume of the raft is Ay. That makes the mass of the water displaced by the raft equal to the following:

$$m_{water\ displaced} = \rho_{water} Ay$$

The weight of the displaced water is just its mass multiplied by g, the acceleration due to gravity, so multiplying both sides of the equation by g gives you the weight, $W_{water\ displaced}$, on the left side of the equation. The displaced weight of water equals the following:

$$W_{water\ displaced} = \rho_{water} Ayg$$

For the raft to float, the weight of the displaced water must equal the weight of the raft, so set the values for raft weight and water weight equal to each other:

$$\rho_{raft} Ahg = \rho_{water} Ayg$$

A and g appear on both sides of the equation, so they cancel out. The equation simplifies to

$$\rho_{raft} h = \rho_{water} y$$

Solving for y gives you the equation for how much of the raft's height is underwater:

$$y = \frac{\rho_{raft} h}{\rho_{water}}$$

Plugging in the densities tells you how far the raft is submerged in terms of the raft's height:

$$y = \frac{550 \text{ kg/m}^3}{1{,}000 \text{ kg/m}^3} h = 0.550h$$

That means that 55 percent of the raft will be underwater. So if the raft is 20 centimeters (or 0.20 meters) high, how much is underwater when it's floating? You can plug in the value for the raft's height to find the answer:

$$y = 0.550(0.20 \text{ m}) = 0.11 \text{ m}$$

So 11 centimeters of the raft's height will be underwater.

Fluid Dynamics: Going with Fluids in Motion

Fluids move according to simple laws that are consistent with Newton's laws of motion. But even though these laws are simple, the range of possible fluid flows is enormous! As you can see all around you, fluids can do all kinds of motions: They can swirl like in a hurricane, they can be in steady uniform flows like when a tap is running, and they can tumble in the most complicated patterns like the steam rising from a boiling kettle. All these different kinds of flow can be characterized with certain properties, which is the subject of this section.

Characterizing the type of flow

Fluid flow has all kinds of aspects — it can be steady or unsteady, compressible or incompressible, and more. Some of these characteristics reflect properties of the liquid or gas itself, and others focus on how the fluid is moving. This section looks at the possibilities.

Note that fluid flow can actually get very complex when it becomes turbulent. Physicists haven't developed any elegant equations to describe turbulence because how turbulence works depends on the individual system — whether you have water cascading through a pipe or air streaming out of a jet engine. Usually, you have to resort to computers to handle problems that involve fluid turbulence.

Evenness: Steady or unsteady flow

Fluid flow can be steady or unsteady, depending on the fluid's velocity:

- **Steady:** In steady fluid flow, the velocity of the fluid is constant at any point.
- **Unsteady:** When the flow is unsteady, the fluid's velocity can differ between any two points.

For example, suppose you're sitting by the side of a stream and note that the water flow is not steady: You see eddies and backwash and all kinds of swirling. Imagine velocity vectors for a hundred points in the water, and you get a good picture of unsteady flow — the velocity vectors can be pointing all over the map, although the velocity vectors generally follow the stream's overall average flow. (Sometimes, in a complex flow, physicists divide the flow into a sum of a smooth average flow and complicated fluctuations, but you don't need to do that here.)

Squeezability: Compressible or incompressible flow

Fluid flow can be *compressible* or *incompressible,* depending on whether you can easily compress the fluid. Liquids are usually nearly impossible to compress, whereas gases (also considered a fluid) are very compressible.

A hydraulic system works only because liquids are incompressible — that is, when you increase the pressure in one location in the hydraulic system, the pressure increases to match everywhere in the whole system (for details, see the earlier section "Hydraulic machines: Passing on pressure with Pascal's principle"). Gases, on the other hand, are very compressible — even when your bike tire is stretched to its limit, you can still pump more air into it by pushing down on the plunger and squeezing it in. The laws of how gases behave when compressing and expanding in different situations can be found in Chapter 16.

Thickness: Viscous or nonviscous flow

Liquid flow can be *viscous* or *nonviscous*. *Viscosity* is a measure of the thickness of a fluid, and very gloppy fluids such as motor oil or shampoo are called *viscous fluids.*

Viscosity is actually a measure of friction in the fluid. When a fluid flows, the layers of fluid rub against one another, and in very viscous fluids, the friction is so great that the layers of flow pull against one other and hamper that flow.

Viscosity usually varies with temperature, because when the molecules of a fluid are moving faster (when the fluid is warmer), the molecules can more easily slide over each other. So when you pour pancake syrup, for example, you may notice that it's very thick in the bottle, but the syrup becomes quite runny when it spreads over the warm pancakes and heats up.

Spinning: Rotational or irrotational flow

Fluid flow can be rotational or irrotational. If, as you travel in a closed loop, you add up all the components of the fluid velocity vectors along your path and the end result is not zero, then the flow is *rotational.*

To test whether a flow has a rotational component, you can put a small object in the flow and let the flow carry it. If the small object spins, the flow is rotational; if the object doesn't spin, the flow is irrotational.

For example, look at the water flowing in a brook. It eddies around stones, curling around obstacles. At such locations, the water flow has a rotational component.

Some flows that you may think are rotational are actually irrotational. For example, away from the center, a vortex is actually an irrotational flow! You can see this if you look at the water draining from your bathtub. If you place a small floating object in the flow, it goes around the plug hole, but it does not spin about itself; therefore, the flow is irrotational.

On the other hand, flows that have no apparent rotation can actually be rotational. Take a shear flow, for example. In a *shear flow,* all the fluid is moving in the same direction, but the fluid is moving faster on one side. Suppose the fluid is moving faster on the left than on the right. The fluid isn't moving in a circle at all, but if you place a small floating object in this flow, the flow on the left side of the object is slightly faster, so the object begins to spin. The flow is rotational.

Picturing flow with streamlines

A handy way of visualizing the flow of a fluid is through streamlines. You draw a fluid's *streamline* so that a tangent to the streamline at any point is parallel to the fluid's velocity at that point. In other words, a streamline follows the fluid flow.

You can see an example in Figure 8-4, where the streamline is the darker line in the middle of the fluid flow.

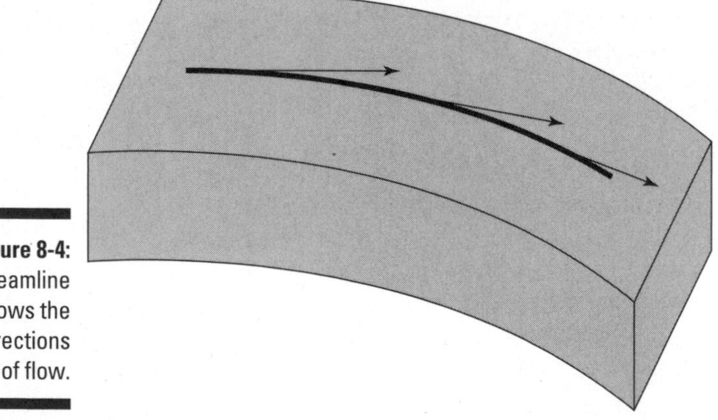

Figure 8-4: A streamline shows the directions of flow.

If you plot streamlines for any flow, you get an immediate picture of how that fluid is flowing. You can have as many or as few streamlines as you need to get an accurate picture of fluid flow.

When fluid flow is turbulent, streamlines can become all mixed up. That's why dealing with turbulent flows in a precise, mathematical way is very hard.

You can have a number of streamlines that form a *tube of flow*. That is, the streamlines form the walls of a tube. The interesting thing about tubes of flow is that fluid does not pass through the walls of such a tube — it's always conducted inside such a tube.

Getting Up to Speed on Flow and Pressure

However complicated a fluid flow may seem, fluids do obey some simple laws that can be expressed in equations. This section introduces the equation that describes the continuity of fluid flow (the result of the fact that matter is neither created nor destroyed) and the relation between speed and pressure. You also take a look at some of the consequences of these relations.

The equation of continuity: Relating pipe size and flow rates

If a fluid is flowing at a certain speed at a certain point in a system of pipes, you can predict what its speed will be at another point using the *equation of continuity*. Because the mass of the fluid is neither created nor destroyed, if mass moves away from one place at a certain rate, it must therefore move to the neighboring place at the same rate. With this idea expressed as an equation, you can find out how the speed changes in a narrowing pipe, for example.

Conserving mass with the equation of continuity

The equation of continuity comes from the idea that no mass disappears when fluid is flowing. In other words, the fluid you get out equals what you put in. You can find the equation of continuity by mixing a little geometry with the physics formulas for mass (which remains constant), density, and speed.

Imagine a cube of fluid flowing in a pipe with the rest of the fluid, as Figure 8-5 shows. The cube has an area A perpendicular to the fluid flow and has a length h along the fluid flow.

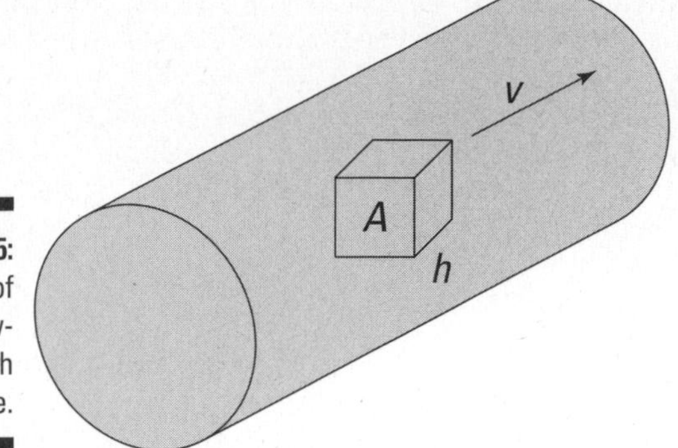

Figure 8-5: A cube of fluid flowing through a pipe.

Now say that the pipe narrows so much that the cube no longer fits. The boundary of the cube is going to change shape. What can you say stays constant between the original cube and the deformed cube? The mass of the fluid inside the box shape will stay constant because no fluid flows through the boundary. Therefore, you can say that

$$m_1 = m_2$$

where m_1 is the mass of fluid in the first cube and m_2 is the mass of the fluid in the deformed cube later on, where the pipe narrows.

If you instead look at the situation in terms of density, ρ, the mass of the fluid in the cube is the density multiplied by the volume of the cube, which is *Ah*. So you can restate the equation as

$$\rho_1 A_1 h_1 = \rho_2 A_2 h_2$$

where A_1 is the area of the front face of the cube originally, h_1 is the original length of the cube, and so on.

Now say you're measuring the amount of mass going by in time *t* to get the flow rate, so you divide the equation you just got by a time interval, *t:*

$$\rho_1 A_1 \frac{h_1}{t} = \rho_2 A_2 \frac{h_2}{t}$$

The length of the cube passing you by in time, *t,* gives you the speed of the fluid at that location, so *h*/*t* becomes the speed of the fluid at that location. Substituting *v,* the speed, for *h*/*t,* you get the following equation:

$$\rho_1 A_1 v_1 = \rho_2 A_2 v_2$$

And this quantity, ρAv, is called the *mass flow rate* — it's the mass of fluid that passes by a certain point per second. The MKS units of the mass flow rate are kilograms per second, or kg/s.

The mass flow rate has the same value at every point in a fluid conduit that has a single entry and single exit point. The mass flow rate at any two points along the conduit can be related like this, with the *equation of continuity:*

$$\rho_1 A_1 v_1 = \rho_2 A_2 v_2$$

Incompressible liquids: Changing the pipe size to change the flow rate

Because liquids are virtually incompressible, the density doesn't change along the flow. Therefore, if you come to a location where the same amount of water must squeeze through a smaller space than before, the water's velocity has to change — it goes faster. Just think of what you do to get water to squirt faster from a garden hose: You put your thumb over most of the end of the hose, forcing the water to squirt through a smaller area.

You can show this idea mathematically using the equation of continuity. For incompressible liquids, density must be the same at Point 1 and Point 2. Because $\rho_1 = \rho_2$, you have this equation, where ρ is the shared density:

$$\rho A_1 v_1 = \rho A_2 v_2$$

Dividing by ρ gives you

$$A_1 v_1 = A_2 v_2$$

where Av is called the *volume flow rate,* whose symbol is Q. So at any two points along the flow of an incompressible liquid, the following is true:

$$Q_1 = Q_2$$

Here's an example with some numbers. Say that you're playing with a fire hose, and you note that water is spraying from the hose at 7.7 meters/second. The cross-sectional area of the nozzle is 4.0×10^{-4} square meters. What is the speed of the water leaving the hydrant when it first enters the fire hose, which has a cross-sectional area of 1.0×10^{-2} square meters?

This is a good chance to use the fact that the volume rate of flow of an incompressible liquid is the same at any point along its flow. That means that the following is true:

$$A_1 v_1 = A_2 v_2$$

where A_1 is the cross-sectional area of the hose, v_1 is the speed of the water as it enters the hose, A_2 is the cross-sectional area of the nozzle, and v_2 is the speed at which water leaves the hose.

Solving for v_1, the speed of the water as it enters the hose, gives you the following equation:

$$v_1 = \frac{A_2 v_2}{A_1}$$

Plugging in the numbers gives you the answer:

$$v_1 = \frac{A_2 v_2}{A_1} = \frac{\left(4.0\times10^{-4}\text{m}^2\right)\left(7.7 \text{ m/s}\right)}{\left(1.0\times10^{-2}\text{m}^2\right)} \approx 0.31 \text{ m/s}$$

So water enters the hose at a comparatively leisurely speed of 0.31 meters/second and leaves the nozzle at a speedier 7.7 meters/second.

Bernoulli's equation: Relating speed and pressure

Now you come to the powerhouse of fluid flow — Daniel Bernoulli's equation, which lets you relate pressure, fluid speed, and height. Using Bernoulli's equation, you can find the difference in fluid pressure between two points if you know the fluid's speed and height at those two points.

Bernoulli's equation relates a moving fluid's pressure, density, speed, and height from Point 1 to Point 2 in this way:

$$P_1 + \frac{1}{2}\rho v_1^2 + \rho g y_1 = P_2 + \frac{1}{2}\rho v_2^2 + \rho g y_2$$

Here's what the variables stand for in this equation (where the subscripts indicate whether you're talking about Point 1 or Point 2):

- P is the pressure of the fluid.
- ρ is the fluid's density.
- g is the acceleration due to gravity.
- v is the fluid's speed.
- y is the height of the fluid.

The equation assumes that you're working with the steady flow of an incompressible, irrotational, nonviscous fluid (see the earlier section "Characterizing the type of flow" for details).

Getting a lift

You can easily demonstrate Bernoulli's principle at home. All you need are two pieces of regular copy paper. Hold the two pieces of paper by the very top so that they dangle freely; then hold them face to face at a separation of a few inches. The air pressure between the two pieces of paper is the same as the air pressure on the other side, so they hang there without moving. Now here comes the cool part. If you blow between the two pieces of paper, what do you think will happen? Most of your friends would probably say that the two pieces of paper will drift apart. But if you try it, you find that the two pieces of paper actually move together! You can work out why this happens because you know Bernoulli's principle.

When you blow between the two pieces of paper, the air increases its speed and reduces its pressure simultaneously. Because the pressure between the pieces of paper is now less than the pressure outside, the pieces of paper move together.

And the fun doesn't end there: If you amaze your friends with the paper trick, you can further boggle their minds by using the principle to explain how airplanes fly. The cross-section of an airplane wing has a kind of swept dome shape to it. Because of this particular shape, air moving toward the wing diverges at the leading edge. Some air goes over the wing, and some air goes under the wing before rejoining at the trailing edge of the wing. But because of the shape of the cross-section, the air that goes over has to travel a longer distance than the air that goes under; therefore, it has to travel faster. And as you have just demonstrated with two pieces of paper (according to Bernoulli's principle), faster air has a lower pressure. So the pressure of the air below the wing is greater than the pressure above the wing. This pressure difference provides the lift required for the airplane to fly.

One thing you can take immediately from this equation is what's called *Bernoulli's principle,* which says that increasing a fluid's speed can lead to a decrease in pressure.

Pipes and pressure: Putting it all together

Together, the equation of continuity and Bernoulli's equation allow you to relate the pressure in pipes to their changes in diameter. You often use the equation of continuity, which tells you that a particular volume of a liquid flows at a constant rate, to find the speeds you use in Bernoulli's equation, which relates speed to pressure.

Here's an example. The operating room is hushed as you're ushered into it. On the operating table lies a very important person who has an aneurism in the aorta, the principal artery leading from the heart. An aneurysm is an enlargement in a blood vessel where the walls have weakened.

The doctors tell you, "The cross-sectional area of the aneurysm is 2.0*A*, where *A* is the cross-sectional area of the normal aorta. We want to operate, but first we need to know how much higher the pressure is in the aneurysm before we cut into it."

Hmm, you think. You happen to know that the normal speed of blood through a person's aorta is 0.40 meters/second. And quickly checking Table 8-1, you see that the density of blood is 1,060 kg/m^3. But will that be enough information?

You'd like to use Bernoulli's equation here because it relates pressure and velocity:

$$P_1 + \frac{1}{2}\rho v_1^2 + \rho g y_1 = P_2 + \frac{1}{2}\rho v_2^2 + \rho g y_2$$

You can simplify Bernoulli's equation because the patient is lying on the operating table, which means that $y_1 = y_2$, so Bernoulli's equation becomes the following:

$$P_1 + \frac{1}{2}\rho v_1^2 = P_2 + \frac{1}{2}\rho v_2^2$$

You want to know how much more pressure is in the aneurysm than in the normal aorta, so you're looking for $P_2 - P_1$. Rearrange the equation:

$$P_2 - P_1 = \frac{1}{2}\rho v_1^2 - \frac{1}{2}\rho v_2^2$$

$$\Delta P = \frac{1}{2}\rho v_1^2 - \frac{1}{2}\rho v_2^2$$

That's looking better; you already know ρ (the density of the blood) and v_1 (the speed of blood in a normal person's aorta). But what's v_2, the speed of blood inside the aneurysm, equal to? You think hard — and you have an inspiration: The equation of continuity can come to the rescue because it relates speeds to cross-sectional areas:

$$\rho_1 A_1 v_1 = \rho_2 A_2 v_2$$

Because the density of blood is the same at Point 1 and Point 2, in the normal aorta and inside the aneurysm, you can divide out the density to get:

$$A_1 v_1 = A_2 v_2$$

Solving for v_2 gives you the following:

$$v_2 = \frac{A_1 v_1}{A_2}$$

Now plug in the numbers. Because the doctors told you $A_2 = 2.0A_1$ and you know that $v_1 = 0.4$ m/s, you get

$$v_2 = \frac{A_1(0.4 \text{ m/s})}{2.0A_1} = 0.2 \text{ m/s}$$

So now you're ready to work with the equation you derived:

$$\Delta P = \frac{1}{2}\rho v_1^2 - \frac{1}{2}\rho v_2^2$$

You can factor out ρ, the density, on the right side of the equation:

$$\Delta P = \frac{1}{2}\rho\left(v_1^2 - v_2^2\right)$$

Plugging in the numbers gives you the following:

$$\Delta P = \frac{1}{2}\left(1{,}060 \text{ kg/m}^3\right)\left[(0.4 \text{ m/s})^2 - (0.2 \text{ m/s})^2\right] \approx 64 \text{ Pa}$$

You tell the doctors that the pressure is 64 pascals higher in the aneurysm than in the normal aorta.

"How's that?" ask the doctors. "Give that to us in units we can understand."

"The pressure is about 0.01 pounds per square inch higher in the aneurysm."

"How's that? That's nothing," say the doctors. "We'll operate immediately — you just saved a very important person's life!"

All in a day's work for a physicist.

Part III
Manifesting the Energy to Work

In this part . . .

If you drive a car up a hill and park it, it still has energy — potential energy. If the brake slips and the car rolls down the hill, it has a different kind of energy at the bottom — kinetic energy. This part alerts you to what energy is and how the work you do when moving and stretching objects becomes energy. Thinking in terms of work and energy allows you to solve problems that Newton's laws of motion don't even let you attempt or that would be more difficult using Newton's laws. You also find out about simple harmonic motion, which is useful for things like springs and pendulums.

Chapter 9

Getting Some Work Out of Physics

In This Chapter

- Taking stock of the work force
- Evaluating kinetic and potential energy
- Walking the path of conservative and nonconservative forces
- Accounting for mechanical energy and power in work

You know all about work; it's what you do when you have to do physics problems. You sit down with your calculator, you sweat a little, and you get through it. You've done your work. Unfortunately, that doesn't count as work in physics terms.

In physics, *work* is done when a force moves an object through a displacement. That may not be your boss's idea of work, but it gets the job done in physics. Along with the basics of work, I use this chapter to introduce kinetic and potential energy, look at conservative and nonconservative forces, and examine mechanical energy and power.

Looking for Work

Holding heavy objects — like, say, a set of exercise weights — up in the air seems to take a lot of work. In physics terms, however, that isn't true. Even though holding up weights may take a lot of biological work, no mechanical work takes place if the weights aren't moving.

In physics, mechanical *work* is performed on an object when a force moves the object through a displacement. When the force is constant and the angle between the force and the displacement is θ, then the work done is given by $W = Fs \cos \theta$. In layman's terms, if you push a 1,000-pound hockey puck for some distance, physics says that the work you do is the component of the force you apply in the direction of travel multiplied by the distance you go.

To get a picture of the full work spectrum, you need to look across different systems of measurement. After you have the measurement units down, you can look at practical working examples, such as pushing and dragging. You can also figure out what negative work means.

Working on measurement systems

Work is a scalar, not a vector; therefore, it has only a magnitude, not a direction (more on scalars and vectors in Chapter 4). Because work is force times distance, $Fs \cos \theta$, it has the unit Newton-meter (N·m) in the MKS system (see Chapter 2 for info on systems of measurement).

Mechanical work done by a net force is equivalent to a transfer of energy (this is called the *work-energy theorem*), which has units called *joules.* Because of this, work and energy have the same units. For conversion purposes, 1 newton of force applied through a distance of 1 meter (where the force is applied along the line of the displacement) is equivalent to 1 joule, or 1 J, of work. (In the foot-pound-second system, work has the unit *foot-pound.* You can also discuss energy and work in terms of kilowatt-hours, which you may be familiar with from electric bills; 1 kilowatt-hour (kWh) = 3.6×10^6 joules.)

Pushing your weight: Applying force in the direction of movement

Motion is a requirement of work. For work to be done, a net force has to move an object through a displacement. Work is a product of force and displacement.

Here's an example: Say that you're pushing a huge gold ingot home, as Figure 9-1 shows. How much work do you have to do to get it home? To find work, you need to know both force and displacement. First, find out how much force pushing the ingot requires.

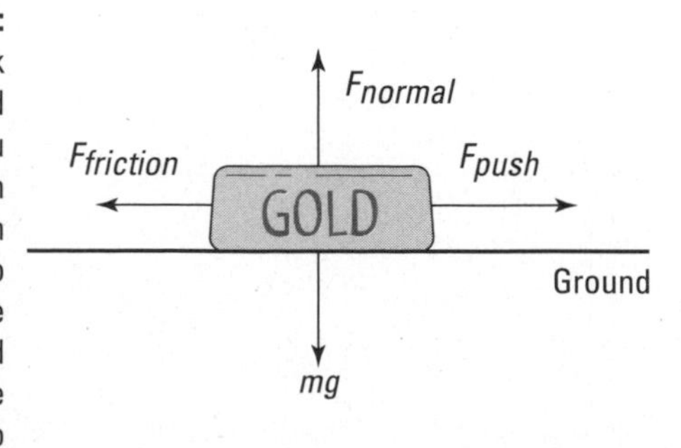

Figure 9-1: To do work on this gold ingot, you have to push with enough force to overcome friction and cause the ingot to move.

Suppose that the kinetic coefficient of friction (see Chapter 6), μ_k, between the ingot and the ground is 0.250 and that the ingot has a mass of 1,000 kilograms. What's the force you have to exert to keep the ingot moving without accelerating it? Start with this equation for the force of friction:

$$F_F = \mu_k F_N$$

Assuming that the road is flat, the magnitude of the normal force, F_N, is just mg (mass times the acceleration due to gravity). That means that

$$F_F = \mu_k mg$$

where m is the mass of the ingot and g is the acceleration due to gravity on the surface of the Earth. Plugging in the numbers gives you the following:

$$\begin{aligned} F_F &= \mu_k mg \\ &= (0.250)(1{,}000 \text{ kg})(9.8 \text{ m/s}^2) \\ &= 2{,}450 \text{ N} \end{aligned}$$

You have to apply a force of 2,450 newtons to keep the ingot moving without accelerating.

You know the force, so to find work, you need to know the displacement. Say that your house is 3 kilometers, or 3,000 meters, away. To get the ingot home, you have to do this much work:

$$W = Fs \cos\theta$$

Because you're pushing the ingot with a force that's parallel to the ground, the angle between F and s is 0°, and cos 0° = 1, so plugging in the numbers gives you the following:

$$\begin{aligned} W &= Fs \cos\theta \\ &= (2{,}450 \text{ N})(3{,}000 \text{ m})(1) \\ &= 7.35 \times 10^6 \text{ J} \end{aligned}$$

You need to do 7.35×10^6 joules of work to move your ingot home. Want some perspective? Well, to lift 1 kilogram 1 meter straight up, you have to supply a force of 9.8 newtons (about 2.2 pounds) over that distance, which takes 9.8 joules of work. To get your ingot home, you need 750,000 times that. Put another way, 1 kilocalorie equals 4,186 joules. A kilocalorie is commonly called a Calorie (capital C) in nutrition; therefore, to move the ingot home, you need to expend about 1,755 Calories. Time to get out the energy bars!

Using a tow rope: Applying force at an angle

You may prefer to drag objects rather than push them — dragging heavy objects may be easier, especially if you can use a tow rope, as Figure 9-2 shows.

When you're pulling at an angle θ, you're not applying a force in the exact same direction as the direction of motion. To find the work in this case, all you have to do is find the component of the force along the direction of travel. Work properly defined is the force along the direction of travel multiplied by the distance traveled:

$$W = F_{pull} s \cos \theta$$

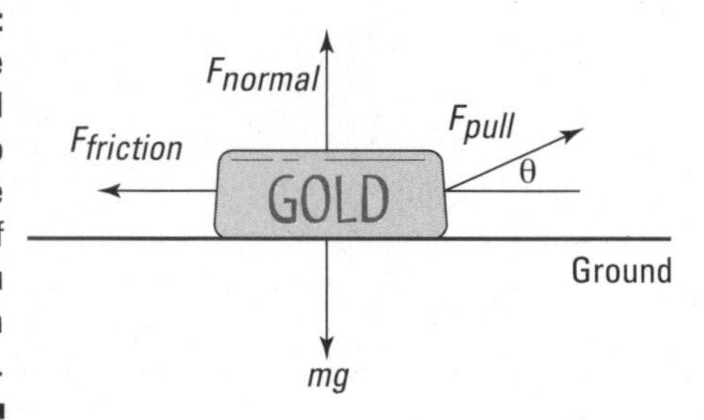

Figure 9-2: More force is required to do the same amount of work if you pull at a larger angle.

Pulling harder to do the same amount of work

If you apply force at an angle instead of parallel to the direction of motion, you have to supply more force to perform the same amount of work.

Say that instead of pushing your ingot, you choose to drag it along with rope that's at an angle of 10° from the ground instead of parallel. This time, $\theta = 10°$ instead of zero. If you want to do the same amount of work as when you pushed the ingot (7.35×10^6 joules), then you need the component of your force that is in the direction of the displacement to be the same as before — that is, 2,450 newtons. This means that

$$F_{pull} \cos \theta = 2{,}450 \text{ N}$$

If you solve for the magnitude of your force, you have

$$F_{pull} = \frac{2{,}450 \text{ N}}{\cos 10^\circ} \approx 2{,}490 \text{ N}$$

If you pull at a 10° angle, you have to supply about 40 extra newtons of force to do the same amount of work. But before you brace yourself to pull really hard, think about the situation a bit more — you don't have to do all that work.

Cutting down on your work by reducing friction

If you pull at an angle, the component of the force you apply that's directed along the floor — in the direction of the displacement — does the work. The component of the force you apply that's directed at right angles to this — straight up — does no work, but it does go some way toward lifting the ingot (or whatever you're towing). The force isn't big enough to lift the ingot clean off the ground, but it does reduce its normal force with the ground, and you know what that means: less friction.

Work out how much frictional force you have if you drag your ingot with a rope that's at a 10° angle. The coefficient of friction is the same as before, but now the normal force with the ground is given by the weight of the ingot minus the upward component of the force you supply. Therefore, the force of friction is given by

$$F_{friction} = \mu(mg - F_{pull} \sin \theta)$$

Here, the vertical component of the force you apply to the ingot is given by $F_{pull} \sin \theta$. The force of friction must be smaller than before because the normal force is smaller — you can already see that you need to do less work to move the ingot.

Because you want to do the least amount of work, you want to drag the ingot across the ground with the smallest force needed to overcome friction. So set the horizontal component of your force equal to the force of friction:

$$F_{pull} \cos \theta = F_{friction}$$

Now plug in the frictional force, which gives you the following:

$$F_{pull} \cos \theta = \mu(mg - F_{pull} \sin \theta)$$

If you rearrange this equation to solve for F_{pull}, you can find the magnitude of the force you need to apply:

$$F_{pull} = \frac{\mu mg}{\cos\theta + \mu \sin\theta}$$

$$= \frac{(0.25)(1{,}000 \text{ kg})(9.8 \text{ m/s}^2)}{\cos 10^\circ + (0.25)\sin 10^\circ} \approx 2{,}380 \text{ N}$$

This is slightly smaller than the force you'd have to apply if you pushed the ingot straight on. If the rope is at a 10° angle, the work you'd do in pulling the ingot over the horizontal distance of 3,000 meters would be

$$\begin{aligned} W &= F_{pull} s \cos\theta \\ &= (2{,}380 \text{ N})(3{,}000 \text{ m})(\cos 10^\circ) \\ &\approx 7.0 \times 10^6 \text{ J} \end{aligned}$$

So you see, you have to do less work if you pull at an angle because there's less frictional force to overcome.

Negative work: Applying force opposite the direction of motion

If the force moving the object has a component in the same direction as the motion, the work that force does on the object is positive. If the force moving the object has a component in the opposite direction of the motion, the work done by that force on the object is negative.

Consider this example: You've just gone out and bought the biggest television your house can handle. You finally get the TV home, and you have to lift it up the porch stairs. It's a heavy one — about 100 kilograms, or 220 pounds — and as you lift it up the first few stairs, a distance of about 0.50 meters, you think you should've gotten some help because of how much work you're doing (***Note:*** F equals mass times acceleration, or 100 kilograms times g, the acceleration due to gravity; θ is 0° because the force and the displacement are in the same direction, the direction in which the TV is moving.):

$$\begin{aligned} W_1 &= Fs \cos\theta \\ &= mgs \cos 0^\circ \\ &= (100 \text{ kg})(9.8 \text{ m/s}^2)(0.50 \text{ m})(1.0) = 490 \text{ J} \end{aligned}$$

However, as you get the TV to the top of the steps, your back decides that you're carrying too much weight and advises you to drop it. Slowly, you lower the TV back to its original position (with no acceleration so that the force you apply is equal and opposite to the weight of the TV) and take a breather. How much work did you do on the way down? Believe it or not, you did *negative* work on the TV, because the force you applied (still upward) was in the opposite direction of travel (downward). In this case, $\theta = 180^\circ$, and $\cos 180^\circ = -1$. Here's what you get when you solve for the work:

$$\begin{aligned} W_2 &= Fs \cos\theta \\ &= mgs \cos 180^\circ \\ &= (100 \text{ kg})(9.8 \text{ m/s}^2)(0.50 \text{ m})(-1.0) = -490 \text{ J} \end{aligned}$$

The net work you've done is $W = W_1 + W_2 = 0$ joules, or zero work. That makes sense, because the TV is right back where it started.

Because the force of friction always acts to oppose the motion, the work done by frictional forces is always negative.

Making a Move: Kinetic Energy

When you start pushing or pulling a stationary object with a constant force, it starts to move if the force you exert is greater than the net forces resisting the movement, such as friction and gravity. If the object starts to move at some speed, it will acquire kinetic energy. *Kinetic energy* is the energy an object has because of its motion. *Energy* is the ability to do work.

You know the ins and outs of kinetic energy. So how do you calculate it?

The work-energy theorem: Turning work into kinetic energy

A force acting on an object that undergoes a displacement does work on the object. If this force is a *net* force that accelerates the object (according to Newton's second law — see Chapter 5), then the velocity changes due to the acceleration (see Chapter 3). The change in velocity means that there is a change in the kinetic energy of the object.

The change in kinetic energy of the object is equal to the work done by the net force acting on it. This is a very important principle called the *work-energy theorem.*

After you know how work relates to kinetic energy, you're ready to take a look at how kinetic energy relates to the speed and mass of the object.

The equation to find kinetic energy, *KE,* is the following, where m is mass and v is velocity:

$$KE = \frac{1}{2}mv^2$$

Using a little math, you can show that work is also equal to $(1/2)mv^2$. Say, for example, that you apply a force to a model airplane in order to get it flying and that the plane is accelerating. Here's the equation for net force:

$$F = ma$$

The work done on the plane, which becomes its kinetic energy, equals the following:

$$W = Fs \cos \theta$$

Net force, *F*, equals mass times acceleration. Assume that you're pushing in the same direction that the plane is going; in this case, cos 0° = 1, so

$$W = Fs = mas$$

You can tie this equation to the final and original velocity of the object. Use the equation $v_f^2 - v_i^2 = 2as$ (see Chapter 3), where v_f equals final velocity and v_i equals initial velocity. Solving for *a* gives you

$$a = \frac{v_f^2 - v_i^2}{2s}$$

If you plug this value of *a* into the equation for work, $W = mas$, you get the following:

$$W = \frac{1}{2}m\left(v_f^2 - v_i^2\right)$$

If the initial velocity is zero, you get

$$W = \frac{1}{2}mv_f^2$$

This is the work that you put into accelerating the model plane — that is, into the plane's motion — and that work becomes the plane's kinetic energy, *KE:*

$$KE = \frac{1}{2}mv_f^2$$

This is just the work-energy theorem stated as an equation.

Using the kinetic energy equation

You normally use the kinetic energy equation to find the kinetic energy of an object when you know its mass and velocity. Say, for example, that you're at a firing range and you fire a 10-gram bullet with a velocity of 600 meters/second at a target. What's the bullet's kinetic energy? The equation to find kinetic energy is

$$KE = \frac{1}{2}mv^2$$

All you have to do is plug in the numbers, remembering to convert from grams to kilograms first to keep the system of units consistent throughout the equation:

$$KE = \frac{1}{2}mv^2$$
$$= \frac{1}{2}(0.010\text{ kg})(600\text{ m/s})^2 = 1{,}800\text{ J}$$

The bullet has 1,800 joules of energy, which is a lot of energy to pack into a 10-gram bullet.

You can also use the kinetic energy equation if you know how much work goes into accelerating an object and you want to find, say, its final speed. For example, say you're on a space station, and you have a big contract from NASA to place satellites in orbit. You open the station's bay doors and grab your first satellite, which has a mass of 1,000 kilograms. With a tremendous effort, you hurl it into its orbit, using a net force of 2,000 newtons, applied in the direction of motion, over 1 meter. What speed does the satellite attain relative to the space station? The work you do is equal to

$$W = Fs\cos\theta$$

Because $\theta = 0°$ here (you're pushing the satellite straight on), $W = Fs$:

$$W = Fs = (2{,}000\text{ N})(1.0\text{ m}) = 2{,}000\text{ J}$$

Your work goes into the kinetic energy of the satellite, so

$$W = \frac{1}{2}mv^2$$

Now solve for v and plug in some numbers. You know that m equals 1,000 kilograms and W equals 2,000 joules, so

$$v = \sqrt{\frac{2W}{m}} = \sqrt{\frac{2(2{,}000\text{ J})}{1{,}000\text{ kg}}} = 2\text{ m/s}$$

The satellite ends up with a speed of 2 meters/second relative to you — enough to get it away from the space station and into its own orbit.

Bear in mind that forces can also do negative work. If you want to catch a satellite and slow it to 1 meter/second with respect to you, the force you apply to the satellite is in the opposite direction of its motion. That means it loses kinetic energy, so you do negative work on it.

Calculating changes in kinetic energy by using net force

In everyday life, multiple forces act on an object, and you have to take them into account. If you want to find the change in an object's kinetic energy, you have to consider only the work done by the net force. In other words, you convert only the work done by the net force into kinetic energy.

For example, when you play tug-of-war against your equally strong friends, you pull against each other but nothing moves. Because there's no movement, no work is done and you have no net increase in kinetic energy from the two forces.

Take a look at Figure 9-3. You may want to determine the speed of the 100-kilogram refrigerator at the bottom of the ramp, using the fact that the net work done on the refrigerator goes into its kinetic energy. How do you do that? You start by determining the net force on the refrigerator and then find out how much work that force does. Converting that net-force work into kinetic energy lets you calculate what the refrigerator's speed will be at the bottom of the ramp.

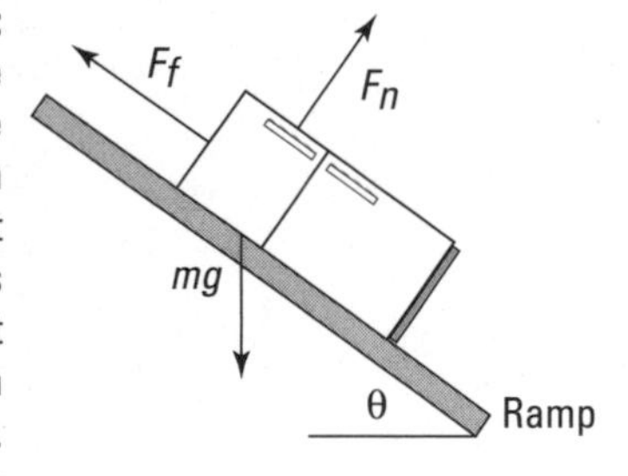

Figure 9-3: You find the net force acting on an object to find its speed at the bottom of a ramp.

What's the net force acting on the refrigerator? In Chapter 6, you find that the component of the refrigerator's weight acting along the ramp is

$$F_{g,\ ramp} = mg \sin \theta$$

where m is the mass of the refrigerator and g is the acceleration due to gravity. The normal force is

$$F_N = mg \cos \theta$$

which means that the kinetic force of friction is

$$F_F = \mu_k F_N = \mu_k mg \cos \theta$$

where μ_k is the kinetic coefficient of friction. The net force accelerating the refrigerator down the ramp, F_{net}, therefore, is

$$F_{net} = F_{g,\,ramp} - F_F = mg \sin\theta - \mu_k mg \cos\theta$$

You're most of the way there! If the ramp is at a 30° angle to the ground and there's a kinetic coefficient of friction of 0.57, plugging the numbers into this equation results in the following:

$$F_{net} = (100 \text{ kg})(9.8 \text{ m/s}^2)(\sin 30°) - (0.57)(100 \text{ kg})(9.8 \text{ m/s}^2)(\cos 30°) \approx 6.2 \text{ N}$$

The net force acting on the refrigerator is about 6.2 newtons. This net force acts over the entire 3.0-meter ramp, so the work done by this force is

$$\begin{aligned} W &= F_{net}s \\ &= (6.2 \text{ N})(3.0 \text{ m}) \approx 19 \text{ J} \end{aligned}$$

You find that 19 joules of work goes into the refrigerator's kinetic energy. That means you can find the refrigerator's kinetic energy like this:

$$W = \frac{1}{2}mv^2$$

You want the speed here, so solving for v and plugging in the numbers gives you

$$v = \sqrt{\frac{2W}{m}} = \sqrt{\frac{2(19 \text{ J})}{100 \text{ kg}}} \approx 0.61 \text{ m/s}$$

The refrigerator will be going 0.61 meters/second at the bottom of the ramp.

Energy in the Bank: Potential Energy

There's more to motion than kinetic energy — an object can also have *potential energy,* which is stored energy or the energy an object has because of its position. The energy is called *potential* because it can be converted to kinetic energy or other forms of energy, such as heat.

Objects can have potential energy from different sources. To give an object potential energy, all you need to do is perform work on an object against a force, such as when you pull back on an object connected to a spring. Gravity is a very common source of potential energy in physics problems.

Suppose you have the job of taking your little cousin Jackie to the park, and you put the little tyke on the slide. Jackie starts at rest and then accelerates, ending up with quite a bit of speed at the bottom of the slide. You sense physics at work here. Taking out your clipboard, you put Jackie higher up the slide and let go, watching carefully. Sure enough, Jackie ends up going even faster at the bottom of the slide. You decide to move Jackie even higher up. (Suddenly, Jackie's mother shows up and grabs him from you. That's enough physics for one day.)

What was happening on the slide? Where did Jackie's kinetic energy come from? It ultimately came from the work you did lifting Jackie against the force of gravity. Jackie sits at rest at the bottom of the slide, so he has no kinetic energy. If you lift him to the top of the slide and hold him, he waits for the next trip down the slide, so he has no motion and no kinetic energy. However, you did work lifting him up against the force of gravity, so he has potential energy. As Jackie slides down the (frictionless) slide, gravity turns your work — and Jackie's potential energy — into kinetic energy.

To new heights: Gaining potential energy by working against gravity

How much work do you do when you lift an object against the force of gravity? Suppose that you want to store a cannonball on an upper shelf at height h above where the cannonball is now. The work you do is

$$W = Fs \cos \theta$$

In this case, F equals the force required to overcome gravity, s equals distance, and θ is the angle between them. The gravitational force on an object is mg (mass times the acceleration due to gravity, 9.8 meters/second^2), and when you lift the cannonball straight up, $\theta = 0°$, so

$$W = Fs \cos \theta = mgh$$

The variable h here is the distance you lift the cannonball. To lift the ball, you have to do a certain amount of work, or m times g times h. The cannonball is stationary when you put it on the shelf, so it has no kinetic energy. However, it does have potential energy, which is the work you put into the ball to lift it to its present position.

If the cannonball rolls to the edge of the shelf and falls off, how much kinetic energy would it have just before it strikes the ground (which is where it started when you first lifted it)? It would have mgh joules of kinetic energy at that point. The ball's potential energy, which came from the work you put in lifting it, changes to kinetic energy thanks to the fall.

In general, you can say that if you have an object of mass m near the surface of the Earth (where the acceleration due to gravity is g), at a height h, then the potential energy of that mass compared to what it'd be at height 0 (where $h = 0$ at some reference height) is

$$PE = mgh$$

And if you move an object vertically against the force of gravity from height h_i to height h_f, its change in potential energy is

$$\Delta PE = mg(h_f - h_i)$$

The work you perform on the object changes its potential energy.

Achieving your potential: Converting potential energy into kinetic energy

Gravitational potential energy for a mass m at height h near the surface of the Earth is mgh more than the potential energy would be at height 0. (It's up to you where you choose height 0.)

For example, say that you lift a 40-kilogram cannonball onto a shelf 3.0 meters from the floor, and the ball rolls and slips off, headed toward your toes. If you know the potential energy involved, you can figure out how fast the ball will be going when it reaches the tips of your shoes. Resting on the shelf, the cannonball has this much potential energy with respect to the floor:

$$\begin{aligned} PE &= mgh \\ &= (40\ \text{kg})(9.8\ \text{m/s}^2)(3.0\ \text{m}) \\ &\approx 1{,}200\ \text{J} \end{aligned}$$

The cannonball has 1,200 joules of potential energy stored by virtue of its position in a gravitational field. What happens when it drops, just before it touches your toes? That potential energy is converted into kinetic energy. So how fast will the cannonball be going at toe impact? Because its potential energy is converted into kinetic energy, you can write the problem as the following (see the section "Making a Move: Kinetic Energy" earlier in this chapter for an explanation of the kinetic energy equation):

$$\begin{aligned} PE &= KE \\ PE &= \frac{1}{2}mv^2 \\ 1{,}200\ \text{J} &= \frac{1}{2}(40\ \text{kg})v^2 \end{aligned}$$

Plugging in the numbers and putting velocity on one side, you get the speed:

$$v = \sqrt{\frac{2(1{,}200\text{ J})}{40\text{ kg}}} \approx 7.7\text{ m/s}$$

The velocity of 7.7 meters/second converts to about 25 feet/second. You have a 40-kilogram cannonball — or about 88 pounds — dropping onto your toes at 25 feet/second. You play around with the numbers and decide you don't like the results. Prudently, you turn off your calculator and move your feet out of the way.

Choose Your Path: Conservative versus Nonconservative Forces

The work a *conservative force* does on an object is path-independent; the actual path taken by the object makes no difference. Fifty meters up in the air has the same gravitational potential energy whether you get there by taking the steps or by hopping on a Ferris wheel. That's different from the force of friction, which dissipates kinetic energy as heat. When friction is involved, the path you take does matter — a longer path will dissipate more kinetic energy than a short one. For that reason, friction is a *nonconservative force.*

For example, suppose you and some buddies arrive at Mt. Newton, a majestic peak that rises h meters into the air. You can take two ways up — the quick way or the scenic route. Your friends drive up the quick route, and you drive up the scenic way, taking time out to have a picnic and to solve a few physics problems. They greet you at the top by saying, "Guess what — our potential energy compared to before is mgh greater."

"Mine, too," you say, looking out over the view. You pull out this equation (originally presented in the section "To new heights: Gaining potential energy by working against gravity," earlier in this chapter):

$$\Delta PE = mg(h_f - h_i)$$

This equation basically states that the actual path you take when going vertically from h_i to h_f doesn't matter. All that matters is your beginning height compared to your ending height. Because the path taken by the object against gravity doesn't matter, gravity is a conservative force.

Here's another way of looking at conservative and nonconservative forces. Say that you're vacationing in the Alps and that your hotel is at the top of Mt. Newton. You spend the whole day driving around — down to a lake one minute, to the top of a higher peak the next. At the end of the day, you end up back at the same location: your hotel on top of Mt. Newton.

What's the change in your gravitational potential energy? In other words, how much net work did gravity perform on you during the day? Gravity is a conservative force, so the change in your gravitational potential energy is 0. Because you've experienced no net change in your gravitational potential energy, gravity did no net work on you during the day.

The road exerted a normal force on your car as you drove around (see Chapter 6), but that force was always perpendicular to the road, so it didn't do any work, either.

Conservative forces are easier to work with in physics because they don't "leak" energy as you move around a path — if you end up in the same place, you have the same amount of energy. If you have to deal with nonconservative forces such as friction, including air friction, the situation is different. If you're dragging something over a field carpeted with sandpaper, for example, the force of friction does different amounts of work on you depending on your path. A path that's twice as long will involve twice as much work to overcome friction.

What's really not being conserved around a track with friction is the total potential and kinetic energy, which taken together is *mechanical energy.* When friction is involved, the loss in mechanical energy goes into heat energy. You can say that the total amount of energy doesn't change if you include that heat energy. However, the heat energy dissipates into the environment quickly, so it isn't recoverable or convertible. For that and other reasons, physicists often work in terms of mechanical energy.

Keeping the Energy Up: The Conservation of Mechanical Energy

Mechanical energy is the sum of potential and kinetic energy, or the energy acquired by an object upon which work is done. The *conservation of mechanical energy,* which occurs in the absence of nonconservative forces (see the preceding section), makes your life much easier when solving physics problems, because the sum of kinetic energy and potential energy stays the same.

In this section, you examine the different forms of mechanical energy: kinetic and potential. You also find out how to relate the kinetic energy to the object's motion, how potential energy arises from the forces acting on the object, and how you can calculate the potential energy for the particular case of gravitational forces. And last, I explain how you can use mechanical energy to make calculations easier.

Shifting between kinetic and potential energy

Imagine a roller coaster car traveling along a straight stretch of track. The car has mechanical energy because of its motion: *kinetic energy*. Imagine that the track has a hill and that the car has just enough energy to get to the top before it descends the other side, back down to a straight and level track (see Figure 9-4). What happens? Well, at the top of the hill, the car is pretty much stationary, so where has all its kinetic energy gone? The answer is that it has been converted to *potential energy*. As the car begins its descent on the other side of the hill, the potential energy begins to be converted back to kinetic energy, and the car gathers speed until it reaches the bottom of the hill. Back at the bottom, all the potential energy the car had at the top of the hill has been converted back into kinetic energy.

An object's potential energy derives from work done by forces, and a label for a particular potential energy comes from the forces that are its source. For example, the roller coaster has potential energy because of the gravitational forces acting on it, so this is often called *gravitational potential energy*. For more on potential energy, see the section "Energy in the Bank: Potential Energy" earlier in this chapter.

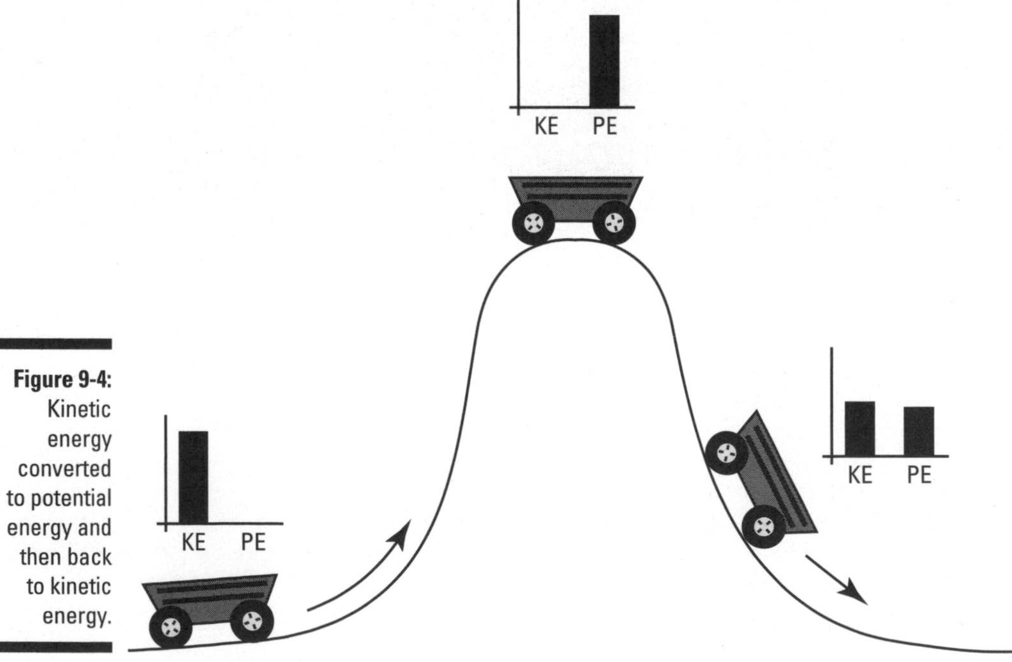

Figure 9-4: Kinetic energy converted to potential energy and then back to kinetic energy.

The roller coaster car's total mechanical energy, which is the sum of its kinetic and potential energies, remains constant at all points of the track. The combination of the kinetic and potential energies does vary, however. When no work is done on an object, its mechanical energy remains constant, whatever motions it may undergo.

Say, for example, that you see a roller coaster at two different points on a track — Point 1 and Point 2 — so that the coaster is at two different heights and two different speeds at those points. Because mechanical energy is the sum of the potential energy (mass × gravity × height) and kinetic energy (½mass × velocity²), the total mechanical energy at Point 1 is

$$ME_1 = mgh_1 + \frac{1}{2}mv_1^2$$

At Point 2, the total mechanical energy is

$$ME_2 = mgh_2 + \frac{1}{2}mv_2^2$$

What's the difference between ME_2 and ME_1? If there's no friction (or another nonconservative force), then $ME_1 = ME_2$, or

$$mgh_1 + \frac{1}{2}mv_1^2 = mgh_2 + \frac{1}{2}mv_2^2$$

These equations represent the *principle of conservation of mechanical energy.* The principle says that if the net work done by nonconservative forces is zero, the total mechanical energy of an object is conserved; that is, it doesn't change. (If, on the other hand, friction or another nonconservative force is present, the difference between ME_2 and ME_1 is equal to the net work the nonconservative forces do: $ME_2 - ME_1 = W_{nc}$.

Another way of rattling off the principle of conservation of mechanical energy is that at Point 1 and Point 2,

$$PE_1 + KE_1 = PE_2 + KE_2$$

You can simplify that mouthful to the following:

$$ME_1 = ME_2$$

where *ME* is the total mechanical energy at any one point. In other words, an object always has the same amount of energy as long as the net work done by nonconservative forces is zero.

You can cancel out the mass, *m,* in the previous equation, which means that if you know three of the values (heights and velocities), you can solve for the fourth:

$$gh_1 + \frac{1}{2}v_1^2 = gh_2 + \frac{1}{2}v_2^2$$

The mechanical-energy balance: Finding velocity and height

Breaking apart the equation for mechanical energy into potential and kinetic energy at two different points — $gh_1 + (1/2)v_1^2 = gh_2 + (1/2)v_2^2$ — allows you to solve for individual variables, such as velocity and height. Check out the following examples.

Determining final velocity with mechanical energy

You can use the principle of conservation of mechanical energy to find an object's final speed.

"Serving as a roller coaster test pilot is a tough gig," you say as you strap yourself into the Physics Park's new Bullet Blaster III coaster. "But someone has to do it." The crew closes the hatch and you're off down the totally frictionless track. Halfway down the 400-meter drop, however, the speedometer breaks. How can you record your top speed when you get to the bottom?

No problem; all you need is the principle of conservation of mechanical energy, which says that if the net work done by nonconservative forces is zero, the total mechanical energy of an object is conserved. You know that

$$mgh_1 + \frac{1}{2}mv_1^2 = mgh_2 + \frac{1}{2}mv_2^2$$

You can make this equation a little easier. Your initial velocity is 0 and your final height is 0, so two of the terms will drop out when you plug in the numbers. You can then divide both sides by *m,* so you get

$$mgh_1 = \frac{1}{2}mv_2^2$$
$$gh_1 = \frac{1}{2}v_2^2$$

Much nicer. Solve for v_2 by rearranging the terms and taking the square root of both sides:

$$v_2 = \sqrt{2gh_1}$$

Then plugging in the numbers gives you the speed:

$$v_2 = \sqrt{2\left(9.8 \text{ m/s}^2\right)\left(400 \text{ m}\right)} \approx 89 \text{ m/s}$$

The coaster travels at 89 meters/second, or about 198 miles/hour, at the bottom of the track — should be fast enough for most kids.

Determining final height with mechanical energy

Besides determining variables such as final speed with the principle of conservation of mechanical energy, you can determine final height. At this very moment, for example, suppose Tarzan is swinging on a vine over a crocodile-infested river at a speed of 13.0 meters/second. He needs to reach the opposite river bank 9.0 meters above his present position in order to be safe. Can he swing it? The principle of conservation of mechanical energy gives you the answer:

$$mgh_1 + \frac{1}{2}mv_1^2 = mgh_2 + \frac{1}{2}mv_2^2$$

At Tarzan's maximum height at the end of the swing, his speed, v_2, will be 0 meters/second, and assuming $h_1 = 0$ meters, you can relate h_2 to v_1 like this:

$$\frac{1}{2}v_1^2 = gh_2$$

Solving for h_2, this means that

$$h_2 = \frac{v_1^2}{2g}$$
$$= \frac{\left(13.0 \text{ m/s}\right)^2}{\left(2\right)\left(9.8 \text{ m/s}^2\right)} \approx 8.6 \text{ m}$$

Tarzan will come up 0.4 meters short of the 9.0 meters he needs to be safe, so he needs some help.

Powering Up: The Rate of Doing Work

Sometimes, it isn't just the amount of work you do but the rate at which you do work that's important. The concept of power gives you an idea of how much work you can expect in a certain amount of time.

Power in physics is the amount of work done divided by the time it takes, or the *rate* of work. Here's what that looks like in equation form:

$$P = \frac{W}{t}$$

Assume you have two speedboats of equal mass, and you want to know which one will get you up to a speed of 120 miles per hour faster. Ignoring silly details like friction, you'll need the same amount of work to get up to that speed, but how long it will take? If one boat takes three weeks to get you up to 120 miles per hour, that may not be the one you take to the races. In other words, the amount of work you do in a certain amount of time can make a big difference.

If the work done at any one instant varies, you may want to work out the average work done over the time *t*. An average quantity in physics is often written with a bar over it, as in the following equation for average power:

$$\bar{P} = \frac{\overline{W}}{t}$$

This section covers what units you're dealing with and the various ways to find power.

Using common units of power

Power is work or energy divided by time, so power has the units of joules/second, which is called the *watt* — a familiar term for just about anybody who uses anything electrical. You abbreviate a watt as simply W, so a 100-watt light bulb converts 100 joules of electrical energy into light and heat every second.

You occasionally run across symbol conflicts in physics, such as the W for watts and the *W* for work. This conflict isn't serious, however, because one symbol is for units (watts) and one is for a concept (work). Capitalization is standard, so be sure to pay attention to units versus concepts.

Note that because work and time are scalar quantities (they have no direction), power is a scalar as well.

Other units of power include foot-pounds per second (ft·lbs/s) and horsepower (hp). One hp = 550 ft·lbs/s = 745.7 W.

Say, for example, that you're in a horse-drawn sleigh on the way to your grandmother's house. At one point, the horse accelerates the sleigh with you on it, with a combined mass of 500 kilograms, from 1.0 meter/second to 2.0 meters/second in 2.0 seconds. How much power does the move take? Assuming no friction on the snow, the total work done on the sleigh, from the work-energy theorem, is

$$W = \frac{1}{2}mv_2^2 - \frac{1}{2}mv_1^2$$

Plugging in the numbers gives you

$$\begin{aligned} W &= \frac{1}{2}mv_2^2 - \frac{1}{2}mv_1^2 \\ &= \frac{1}{2}(500\text{ kg})(2.0\text{ m/s})^2 - \frac{1}{2}(500\text{ kg})(1.0\text{ m/s})^2 \\ &= 750\text{ J} \end{aligned}$$

Because the horse does this work in 2.0 seconds, the power needed is

$$P = \frac{750\text{ J}}{2.0\text{ s}} \approx 380\text{ W}$$

One horsepower is 745.7 watts, so the horse is giving you about one-half horsepower — not too bad for a one-horse open sleigh.

Doing alternate calculations of power

Because work equals force times distance, you can write the equation for power the following way, assuming that the force acts along the direction of travel:

$$P = \frac{W}{t} = \frac{Fs}{t}$$

where *s* is the distance traveled. However, the object's speed, *v,* is just *s* divided by *t,* so the equation breaks down to

$$P = \frac{W}{t} = \frac{Fs}{t} = Fv$$

That's an interesting result — power equals force times speed? Yep, that's what it says. However, because you often have to account for acceleration when you apply a force, you usually write the equation in terms of average power and average speed:

$$\overline{P} = F\overline{v}$$

Here's an example. Suppose your brother got himself a snappy new car. You think it's kind of small, but he claims it has over 100 horsepower. "Okay," you say, getting out your clipboard. "Let's put it to the test."

Your brother's car has a mass of 1.10×10^3 kilograms. On the big Physics Test Track on the edge of town, you measure its acceleration as 4.60 meters/second^2 over 5.00 seconds when the car started from rest. How much horsepower is that?

You know that $\bar{P} = F\bar{v}$, so all you need to calculate is the average speed and the net applied force. Take the net force first. You know that $F = ma$, so you can plug in the values to get

$$F = (1.10 \times 10^3 \text{ kg})(4.60 \text{ m/s}^2) = 5{,}060 \text{ N}$$

Okay, so the force applied to accelerate the car steadily is 5,060 newtons. Now all you need is the average speed. Say the starting speed was v_i and the ending speed v_f. You know that $v_i = 0$ m/s, so what is v_f? Well, you also know that because the acceleration was constant, the following equation is true:

$$v_f = v_i + at$$

As it happens, you know that acceleration and the time the car was accelerated over:

$$v_f = 0 \text{ m/s} + (4.60 \text{ m/s}^2)(5.00 \text{ s}) = 23.0 \text{ m/s}$$

Because the acceleration was constant, the average speed is

$$\bar{v} = \frac{v_i + v_f}{2}$$

Because $v_i = 0$ m/s, this breaks down to

$$\bar{v} = \frac{v_f}{2}$$

Plugging in the numbers gives you the average velocity:

$$\bar{v} = \frac{23.0 \text{ m/s}}{2} = 11.5 \text{ m/s}$$

Great — now you know the force applied and the average speed. You can use the equation $\bar{P} = F\bar{v}$ to find the average power. In particular

$$\bar{P} = (5{,}060 \text{ N})(11.5 \text{ m/s}) \approx 5.82 \times 10^4 \text{ W}$$

You still need to convert to horsepower. One horsepower = 745.7 watts, so

$$\bar{P} = \frac{5.82 \times 10^4 \cancel{\text{W}}}{1} \times \frac{1 \text{ hp}}{745.7 \cancel{\text{W}}} \approx 78.0 \text{ hp}$$

Therefore, the car developed an average of 78.0 horsepower, not 100 horsepower. "Rats," says your brother. "I demand a recount."

Okay, so you agree to calculate power another way. You know you can also calculate average power as work divided by time:

$$\bar{P} = \frac{W}{t}$$

And the work done by the car is the difference in the beginning and ending kinetic energies:

$$W = KE_f - KE_i$$

The car started at rest, so $KE_i = 0$ J. That leaves only the final kinetic energy to calculate:

$$KE_f = \frac{1}{2} m v_f^2$$

Plugging in the numbers gives you:

$$\begin{aligned} KE_f &= \frac{1}{2}\left(1.10 \times 10^3 \text{kg}\right)\left(23.0 \text{ m/s}\right)^2 \\ &\approx 2.91 \times 10^5 \text{J} \end{aligned}$$

So because $\bar{P} = W/t$ and the work done was 2.91×10^5 joules in 5.00 seconds, you get the following:

$$\bar{P} = \frac{2.91 \times 10^5 \text{J}}{5.00 \text{ s}} = 5.82 \times 10^4 \text{W}$$

And, as before

$$\bar{P} = \frac{5.82 \times 10^4 \cancel{\text{W}}}{1} \times \frac{1 \text{ hp}}{745.7 \cancel{\text{W}}} = 78.0 \text{ hp}$$

"Double rats," your brother says.

Chapter 10

Putting Objects in Motion: Momentum and Impulse

In This Chapter

- Checking your impulse
- Gaining knowledge about momentum
- Intertwining impulse and momentum
- Utilizing the conservation of momentum
- Examining different types of collisions

Both momentum and impulse are very important to *kinematics,* or the study of objects in motion. After you have these topics under your belt, you can start talking about what happens when objects collide (hopefully not your car or bike). Sometimes they bounce off each other (like when you hit a tennis ball with a racket), and sometimes they stick together (like when a dart hits a dart board). With the knowledge of impulse and momentum you pick up in this chapter, you can handle either case.

Looking at the Impact of Impulse

In physics terms, impulse tells you how much the momentum of an object will change when a force is applied for a certain amount of time. Say, for example, that you're shooting pool. Instinctively, you know how hard to tap each ball to get the results you want. The nine ball in the corner pocket? No problem — tap it and there it goes. The three ball bouncing off the side cushion into the other corner pocket? Another tap, this time a little stronger.

The taps you apply are called *impulses.* Take a look at what happens on a microscopic scale, millisecond by millisecond, as you tap a pool ball. The force you apply with your cue appears in Figure 10-1. The tip of each cue has a cushion, so the impact of the cue is spread out over a few milliseconds as the cushion squashes slightly. The impact lasts from the time when the cue touches the ball, t_i, to the time when the ball loses contact with the cue, t_f. As you can see from Figure 10-1, the force exerted on the ball changes during that time; in fact, it changes drastically, and figuring out what the force was doing at any 1 millisecond would be hard without some fancy equipment.

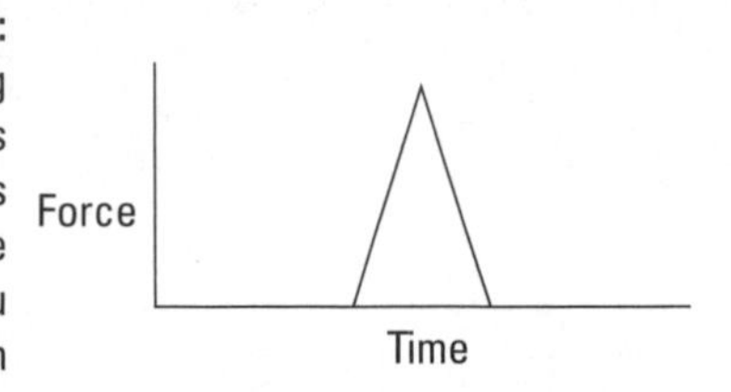

Figure 10-1: Examining force versus time gives you the impulse you apply on objects.

Because the pool ball doesn't come with any fancy equipment, you have to do what physicists normally do, which is to talk in terms of the average force over time. You can see what that average force looks like in Figure 10-2. Speaking as a physicist, you say that the impulse — or the tap — that the pool cue provides is the average force multiplied by the time that you apply the force.

Here's the equation for impulse:

$$\text{Impulse} = \mathbf{J} = \bar{\mathbf{F}}\Delta t$$

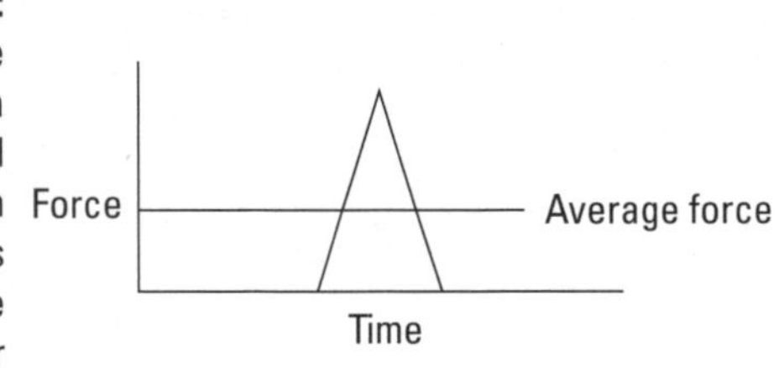

Figure 10-2: The average force over a time interval depends on the values the force has over that time.

Note that this equation is a vector equation, meaning it deals with both direction and magnitude (see Chapter 4). Impulse, **J,** is a vector, and it's in the same direction as the average force (which itself may be a net vector sum of other forces).

You get impulse by multiplying a quantity with units of newtons by a quantity with units of seconds, so the units of impulse are *newton-seconds* in the MKS system and *pound-seconds* in the FPS system. (See Chapter 2 for details on systems of measurement.)

Gathering Momentum

In physics terms, momentum is proportional to both mass and velocity, and to make your job easy, physics defines *momentum* as the product of mass times velocity. Momentum is a big concept both in introductory physics and in some advanced topics such as high-energy particle physics, where the components of atoms zoom around at high speeds. When the particles collide, you can often predict what will happen based on your knowledge of momentum.

Even if you're unfamiliar with the physics of momentum, you're already familiar with the general idea. Catching a runaway car going down a steep hill is a problem because of its momentum. If a car without any brakes is speeding toward you at 40 miles per hour, trying to stop it by standing in its way may not be a great idea. The car has a lot of momentum, and bringing it to a stop requires plenty of effort. Now think of an oil tanker. Its engines aren't strong enough to make it turn or stop on a dime. Therefore, an oil tanker can take 20 miles or more to come to a stop, all because of the ship's momentum.

The more mass that's moving, the more momentum the mass has. The greater the magnitude of its velocity (think of an even faster oil tanker), the more momentum it has. The symbol for momentum is **p,** so you can say that

$$\mathbf{p} = m\mathbf{v}$$

Momentum is a vector quantity, meaning that it has a magnitude and a direction (see Chapter 4). The magnitude is in the same direction as the velocity — all you have to do to get the momentum of an object is to multiply its mass by its velocity. Because you multiply mass by velocity, the units for momentum are kilogram-meters per second (kg·m/s) in the MKS system.

The Impulse-Momentum Theorem: Relating Impulse and Momentum

You can connect the impulse you give to an object — like striking a pool ball with a cue — with the object's change in momentum; all you need is a little algebra and the process you explore in this section, called the *impulse-momentum theorem.* What makes the connection easy is that you can play with the equations for impulse and momentum to simplify them so you can relate the two topics. What equations does physics have in its arsenal to connect these two? Relating force and velocity is a start. For example, force equals mass times acceleration (see Chapter 5), and the definition of average acceleration is

$$\bar{\mathbf{a}} = \frac{\Delta \mathbf{v}}{\Delta t}$$

where **v** stands for velocity and *t* stands for time. Now you may realize that if you multiply that equation by the mass, you get force, which brings you closer to working with impulse:

$$\mathbf{F} = m\mathbf{a} = m\left(\frac{\Delta\mathbf{v}}{\Delta t}\right)$$

Now you have force in the equation. To get impulse, multiply the force equation by Δt, the time over which you apply the force:

$$\mathbf{F}\Delta t = m\left(\frac{\Delta\mathbf{v}}{\Delta t}\right)\Delta t = m\Delta\mathbf{v}$$

Take a look at the final expression, $m\Delta\mathbf{v}$. Because momentum equals $m\mathbf{v}$ (see the section "Gathering Momentum" earlier in this chapter), this is just the difference in the object's initial and final momentum: $\mathbf{p}_f - \mathbf{p}_i = \Delta\mathbf{p}$. Therefore, you can add that to the equation:

$$\mathbf{F}\Delta t = \Delta\mathbf{p}$$

Now take a look at the term on the left, $\mathbf{F}\Delta t$. That's the impulse, **J** (see the section "Looking at the Impact of Impulse" earlier in this chapter), or the force applied to the object multiplied by the time that force was applied. Therefore, you can write this equation as

$$\mathbf{J} = \mathbf{F}\Delta t = \Delta\mathbf{p}$$

Getting rid of everything in the middle finally gives you the impulse-momentum theorem, which says that impulse equals change in momentum:

$$\mathbf{J} = \Delta\mathbf{p}$$

The rest of this section provides some examples so you can practice this equation. But before you set off, think about what the formula means for the relation among impulse, force, and momentum. The impulse-momentum theorem defines a very simple relation between the impulse and momentum, namely that impulse is equal to the change in momentum. You can also see how a constant force applied over a time is equal to an impulse that is given by the force multiplied by the time:

$$\mathbf{J} = \mathbf{F}\Delta t$$

Last, you can tie the force and momentum together through the impulse, which gives you

$$\mathbf{F}\Delta t = \Delta\mathbf{p}$$

The meaning of this relation may become clearer if you divide both side by Δt:

$$\mathbf{F} = \frac{\Delta \mathbf{p}}{\Delta t}$$

So you see that the force is given by the rate of change of momentum. This is a whole new way of thinking about force! Wherever you see momentum changing with time, you know a force is acting, and if calculating the momentum is easier, it can lead to an easier way of calculating the force. Check out the following examples.

Shooting pool: Finding force from impulse and momentum

With the equation $\mathbf{J} = \Delta \mathbf{p}$, you can relate the impulse with which you hit an object to its consequent change in momentum. How about using the equation the next time you hit a pool ball? You line up the shot that the game depends on. You figure that the end of your cue will be in contact with the ball for 5 milliseconds (a millisecond is a thousandth of a second).

You measure the ball at 200 grams (or 0.200 kilograms). After testing the side cushion with calipers, spectroscope, and tweezers, you figure that you need to give the ball a speed of 2.0 meters per second. What average force do you have to apply? To find the average force, first find the impulse you have to supply. You can relate that impulse to the change in the ball's momentum this way:

$$\mathbf{J} = \Delta \mathbf{p} = \mathbf{p}_f - \mathbf{p}_i$$

Because the pool ball doesn't change direction, you can use this equation for the component of the pool ball's momentum in the direction in which you strike it. Because you're using a component of the vector, you remove the bold from *p*.

So what's the change in the ball's momentum? The speed you need, 2.0 meters per second, is the magnitude of the pool ball's final velocity. Assuming the pool ball starts at rest, the change in the ball's momentum will be the following:

$$\Delta p = p_f - p_i = m(v_f - v_i)$$

Where v is the component of the ball's velocity in the direction in which you strike it. Plugging in the numbers gives you the change in momentum:

$$\Delta p = m(v_f - v_i) = (0.200 \text{ kg})(2.0 \text{ m/s} - 0.0 \text{ m/s}) = 0.40 \text{ kg}\cdot\text{m/s}$$

You need a change in momentum of 0.40 kilogram-meters per second, which is also the impulse you need. Because $J = F\Delta t$, this equation becomes the following for the component of the force in the direction of motion:

$$F\Delta t = 0.40 \text{ kg·m/s}$$

Therefore, the force you need to apply works out to be

$$F = \frac{0.40 \text{ kg} \cdot \text{m/s}}{\Delta t}$$

In this equation, the time your cue ball is in contact with the ball is 5 milliseconds, or 5.0×10^{-3} seconds. Plug in the time to find the force:

$$F = \frac{0.40 \text{ kg} \cdot \text{m/s}}{5.0 \times 10^{-3} \text{ s}} = 80 \text{ N}$$

You have to apply about 80 newtons (or about 18 pounds) of force, which sounds like a huge amount. However, you apply it over such a short time, 5.0×10^{-3} seconds, that it seems like much less.

Singing in the rain: An impulsive activity

After a triumphant evening at the pool hall, you decide to leave and discover that it's raining. You grab your umbrella from your car, and the handy rain gauge on the umbrella's top tells you that 100 grams of water are hitting the umbrella each second at an average speed of 10 meters per second. If the umbrella has a total mass of 1.0 kilograms, what force do you need to hold it upright in the rain?

Figuring the force you usually need to hold the weight of the umbrella is no problem — you just figure mass times the acceleration due to gravity ($F = ma$), or $(1.0 \text{ kg})(9.8 \text{ m/s}^2) = 9.8 \text{ N}$.

But what about the rain falling on your umbrella? Even if you assume that the water falls off the umbrella immediately, you can't just add the weight of the water, because the rain is falling with a speed of 10 meters per second; in other words, the rain has *momentum*. What can you do? You know that you're facing 100 grams (0.10 kilograms) of water falling onto the umbrella each second at a velocity of 10 meters per second downward. When that rain hits your umbrella, the water comes to rest, so the change in momentum of rain every second is

$$\Delta p = m\Delta v$$

You're considering only the vertical components of the vectors, so the variables aren't in bold. Plugging in numbers gives you the change in momentum:

$$\Delta p = m\Delta v = (0.10 \text{ kg})(10 \text{ m/s}) = 1.0 \text{ kg·m/s}$$

The rain's change in momentum, every second, as it hits your umbrella is 1.0 kilogram-meter per second. You can relate that change to force with the impulse-momentum theorem, which tells you that

$$F\Delta t = \Delta p$$

Dividing both sides by Δt allows you to solve for the force, F:

$$F = \frac{\Delta p}{\Delta t}$$

You know that Δp = 1.0 kg·m/s in 1.0 second, so plugging in Δp and setting Δt to 1.0 second gives you the force of the rain:

$$F = \frac{\Delta p}{\Delta t} = \frac{1.0 \text{ kg} \cdot \text{m/s}}{1.0 \text{ s}} = 1.0 \text{ kg} \cdot \text{m/s}^2 = 1.0 \text{ N}$$

In addition to the 9.8 newtons of the umbrella's weight, you need 1.0 newton to stand up to the falling rain as it drums on the umbrella, for a total of 10.8 newtons, or about 2.4 pounds of force.

When Objects Go Bonk: Conserving Momentum

The *principle of conservation of momentum* states that when you have an isolated system with no external forces, the initial total momentum of objects before a collision equals the final total momentum of the objects after the collision. In other words, $\Sigma \mathbf{p}_i = \Sigma \mathbf{p}_f$.

You may have a hard time dealing with the physics of impulses because of the short times and the irregular forces. But with the principle of conservation, items that are hard to measure — for example, the force and time involved in an impulse — are out of the equation altogether. Thus, this simple principle may be the most useful idea I provide in this chapter.

Deriving the conservation formula

You can derive the principle of conservation of momentum from Newton's laws, what you know about impulse, and a little algebra.

Say that two careless space pilots are zooming toward the scene of an interplanetary crime. In their eagerness to get to the scene first, they collide. During the collision, the average force the second ship exerts on the first ship is $\mathbf{F}_{12}$. By the impulse-momentum theorem (see the section "The Impulse-Momentum

Theorem: Relating Impulse and Momentum"), you know the following for the first ship:

$$\mathbf{F}_{12}\Delta t = \Delta\mathbf{p}_1 = m_1\Delta\mathbf{v}_1 = m_1(\mathbf{v}_{f1} - \mathbf{v}_{i1})$$

And if the average force exerted on the second ship by the first ship is $\mathbf{F}_{21}$, you also know that

$$\mathbf{F}_{21}\Delta t = \Delta\mathbf{p}_2 = m_2\Delta\mathbf{v}_2 = m_2(\mathbf{v}_{f2} - \mathbf{v}_{i2})$$

Now you add these two equations together, which gives you the resulting equation:

$$\mathbf{F}_{12}\Delta t + \mathbf{F}_{21}\Delta t = m_1(\mathbf{v}_{f1} - \mathbf{v}_{i1}) + m_2(\mathbf{v}_{f2} - \mathbf{v}_{i2})$$

Distribute the mass terms and rearrange the terms on the right until you get the following:

$$\mathbf{F}_{12}\Delta t + \mathbf{F}_{21}\Delta t = m_1\mathbf{v}_{f1} - m_1\mathbf{v}_{i1} + m_2\mathbf{v}_{f2} - m_2\mathbf{v}_{i2}$$
$$\mathbf{F}_{12}\Delta t + \mathbf{F}_{21}\Delta t = (m_1\mathbf{v}_{f1} + m_2\mathbf{v}_{f2}) - (m_1\mathbf{v}_{i1} + m_2\mathbf{v}_{i2})$$

This is an interesting result, because $m_1\mathbf{v}_{i1} + m_2\mathbf{v}_{i2}$ is the *initial total momentum* of the two rocket ships $(\mathbf{p}_{1i} + \mathbf{p}_{i2})$ and $m_1\mathbf{v}_{f1} + m_2\mathbf{v}_{f2}$ is the *final total momentum* $(\mathbf{p}_{1f} + \mathbf{p}_{2f})$ of the two rocket ships. Therefore, you can write this equation as follows:

$$\mathbf{F}_{12}\Delta t + \mathbf{F}_{21}\Delta t = (\mathbf{p}_{1f} + \mathbf{p}_{2f}) - (\mathbf{p}_{1i} + \mathbf{p}_{i2})$$

If you write the initial total momentum as $\mathbf{p_f}$ and the final total momentum as $\mathbf{p}_i$, the equation becomes

$$\mathbf{F}_{12}\Delta t + \mathbf{F}_{21}\Delta t = \mathbf{p_f} - \mathbf{p_i}$$

Where do you go from here? Both terms on the left include Δt, so you can rewrite $\mathbf{F}_{12}\Delta t + \mathbf{F}_{21}\Delta t$ as the sum of the forces involved, $\Sigma\mathbf{F}$, multiplied by the change in time:

$$\Sigma\mathbf{F}\Delta t = \mathbf{p_f} - \mathbf{p_i}$$

If you're working with what's called an *isolated* or *closed system*, you have no external forces to deal with. Such is the case in space. If two rocket ships collide in space, there are no external forces that matter, so by Newton's third law (see Chapter 5), $\mathbf{F}_{12}\Delta t = -\mathbf{F}_{21}\Delta t$. In other words, when you have a closed system, you get the following:

$$\Sigma\mathbf{F}\Delta t = \mathbf{p_f} - \mathbf{p_i}$$
$$0 = \mathbf{p_f} - \mathbf{p_i}$$

This converts to

$$\mathbf{p_f} = \mathbf{p_i}$$

The equation $\mathbf{p_f} = \mathbf{p_i}$ says that when you have an isolated system with no external forces, the initial total momentum before a collision equals the final total momentum after a collision, giving you the principle of conservation of momentum.

Finding velocity with the conservation of momentum

You can use the principle of conservation of momentum to measure other characteristics of motion, such as velocity. Say, for example, that you're out on a physics expedition and you happen to pass by a frozen lake where a hockey game is taking place. You measure the speed of one player as 11.0 meters per second just as he collides, rather brutally for a pick-up game, with another player initially at rest. You watch with interest, wondering how fast the resulting mass of hockey players will slide across the ice. After asking a few friends in attendance, you find out that the first player has a mass of 100 kilograms and the bulldozed player (who turns out to be his twin) also has a mass of 100 kilograms. So what's the final speed of the player tangle?

You're dealing with a closed system, because you neglect the force of friction here, and although the players are exerting a force downward on the ice, the normal force (see Chapter 5) is exerting an equal and opposite force on them; therefore, the vertical force sums to zero.

But what about the resulting horizontal speed along the ice? Due to the principle of conservation of momentum, you know that

$$\mathbf{p_f} = \mathbf{p_i}$$

Imagine that the collision is head on, so all the motion occurs in one dimension — along a line. So you only need to examine the components of the vector quantities in this single dimension. The component of a vector in one dimension is just a number, so I don't write them in bold.

The victim isn't moving before the hit, so he starts without any momentum. Therefore, the initial momentum, p_i, is simply the initial momentum of the enforcer, Player 1. To put this equation into more helpful terms, substitute Player 1's mass and initial velocity $(m_1 v_{i1})$ for the initial momentum (p_i):

$$p_i = m_1 v_{i1}$$

After the hit, the players tangle up and move with the same final velocity. Therefore, the final momentum, p_f, must equal the combined mass of the two players multiplied by their final velocity, $(m_1 + m_2)v_f$, which gives you the following equation:

$$(m_1 + m_2)v_f = m_1 v_{i1}$$

Solving for v_f gives you the equation for their final velocity:

$$v_f = \frac{m_1 v_{i1}}{m_1 + m_2}$$

Plugging in the numbers gives you the answer:

$$\begin{aligned} v_f &= \frac{m_1 v_{i1}}{m_1 + m_2} \\ &= \frac{(100 \text{ kg})(11 \text{ m/s})}{100 \text{ kg} + 100 \text{ kg}} \\ &= \frac{1{,}100 \text{ kg} \cdot \text{m/s}}{200 \text{ kg}} \\ &= 5.5 \text{ m/s} \end{aligned}$$

The speed of the two players together will be half the speed of the original player. That may be what you expected, because you end up with twice the moving mass as before; because momentum is conserved, you end up with half the speed. Beautiful. You note the results down on your clipboard.

Finding firing velocity with the conservation of momentum

The principle of conservation of momentum comes in handy when you can't measure velocity with a simple stopwatch. Say, for example, that you accept a consulting job from an ammunition manufacturer that wants to measure the muzzle velocity of its new bullets. No employee has been able to measure the velocity yet, because no stopwatch is fast enough. What do you do? You decide to arrange the setup shown in Figure 10-3, where you fire a bullet of mass m_1 into a hanging wooden block of mass m_2.

The directors of the ammunition company are perplexed — how can your setup help? Each time you fire a bullet into a hanging wooden block, the bullet kicks the block into the air. So what? You decide they need a lesson on the principle of conservation of momentum. The original momentum, you explain, is the momentum of the bullet:

$$p_i = mv_i$$

Because the bullet sticks in the wooden block, the final momentum is the product of the total mass, $m_1 + m_2$, and the final velocity of the bullet/wooden block combination:

$$p_f = (m_1 + m_2)v_f$$

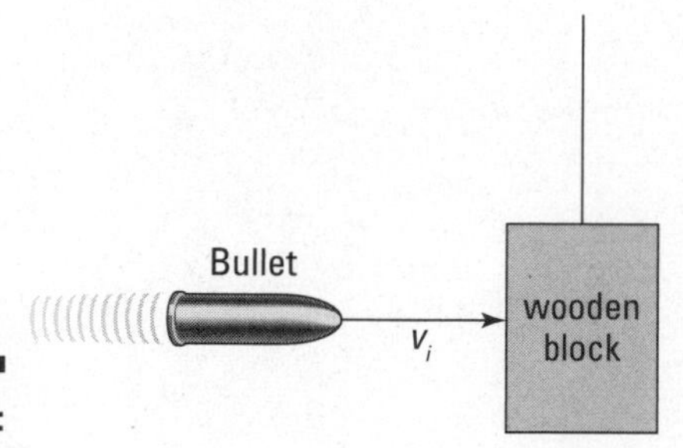

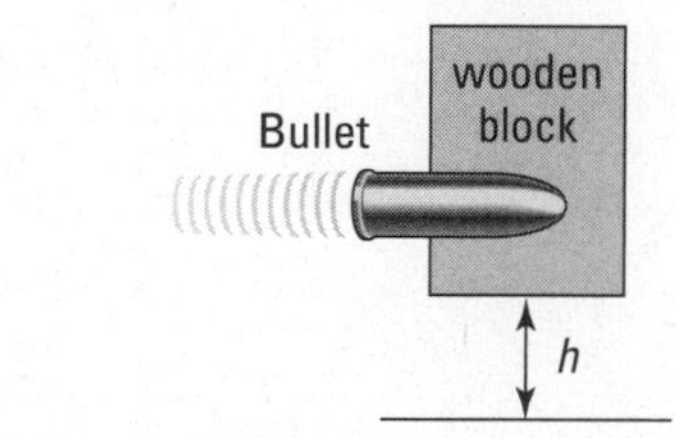

Figure 10-3: Shooting a wooden block on a string allows you to experiment with velocity, but don't try this at home!

Because of the principle of conservation of momentum, you can say that

$$p_f = p_i$$

Therefore, you can plug in the earlier expressions for final and initial momentum:

$$(m_1 + m_2)v_f = m_1 v$$

$$v_f = \frac{m_1 v_i}{m_1 + m_2}$$

The directors start to get dizzy, so you explain how the kinetic energy of the block when it's struck goes into its final potential energy when it rises to height h (see Chapter 9). Here's how you can represent the bullet's kinetic energy and the bullet-and-block's change in potential energy:

$$\Delta KE = \Delta PE$$

$$\frac{1}{2}mv^2 = mgh$$

$$\frac{1}{2}(m_1 + m_2)v_f^2 = (m_1 + m_2)gh$$

You can plug in the value of v_f, which gives you

$$\frac{1}{2}(m_1 + m_2)\frac{m_1^2 v_i^2}{(m_1 + m_2)^2} = (m_1 + m_2)gh$$

$$\frac{m_1^2 v_i^2}{2(m_1 + m_2)} = (m_1 + m_2)gh$$

$$v_i^2 = \frac{2(m_1 + m_2)(m_1 + m_2)gh}{m_1^2}$$

$$v_i = \sqrt{\frac{2(m_1 + m_2)^2 gh}{m_1^2}}$$

With a flourish, you explain that solving for v_i gives you the bullet's initial velocity:

$$v_i = \sqrt{\frac{2(m_1 + m_2)^2 gh}{m_1^2}}$$

You measure that the bullet has a mass of 50 grams, that the wooden block has a mass of 10.0 kilograms, and that upon impact, the block rises 50.0 centimeters into the air. Plugging in those values gives you your result:

$$v_i = \sqrt{\frac{2(0.050 \text{ kg} + 10.0 \text{ kg})^2 (9.81 \text{ m/s}^2)(0.500 \text{ m})}{(0.050 \text{ kg})^2}} \approx 630 \text{ m/s}$$

The initial velocity is 630 meters per second, which converts to about 2,070 feet per second. "Brilliant!" the directors cry as they hand you a big check.

When Worlds (Or Cars) Collide: Elastic and Inelastic Collisions

Examining collisions in physics can be pretty entertaining, especially because the principle of conservation of momentum makes your job so easy (see the earlier section "When Objects Go Bonk: Conserving Momentum"). But when you're dealing with collisions, there's often more to the story than impulse and momentum. Sometimes, kinetic energy is also conserved, which gives you the extra edge you need to figure out what happens in all kinds of collisions, even across two dimensions.

Collisions are important in many physics problems. Two cars collide, for example, and you need to find the final velocity of the two when they stick together. You may even run into a case where two railway cars going at different velocities collide and couple together, and you need to determine the final velocity of the two cars.

But what if you have a more general case where the two objects don't stick together? Say, for example, you have two pool balls that hit each other at different speeds and at different angles and bounce off with different speeds and different angles. How the heck do you handle that situation? You have a way to handle these collisions, but you need more than just the principle of conservation of momentum gives you. In this section, I explain the difference between elastic and inelastic collisions and then do a few elastic-collision problems.

Determining whether a collision is elastic

When bodies collide in the real world, they sometimes squash and deform to some degree. The energy to perform the deformation comes from the objects' original kinetic energy. In other cases, friction turns some of the kinetic energy into heat. Physicists classify collisions in *closed systems* (where the net forces add up to zero) based on whether colliding objects lose kinetic energy to some other form of energy:

- **Elastic collision:** In an elastic collision, the total kinetic energy in the system is the same before and after the collision. If losses to heat and deformation are much smaller than the other energies involved, such as when two pool balls collide and go their separate ways, you can generally ignore the losses and say that kinetic energy was conserved.

- **Inelastic collision:** In an inelastic collision, the collision changes the total kinetic energy in a closed system. In this case, friction, deformation, or some other process transforms the kinetic energy. If you can observe appreciable energy losses due to nonconservative forces (such as friction), kinetic energy isn't conserved.

  You see inelastic collisions when objects stick together after colliding, such as when two cars crash and weld themselves into one. However, objects don't need to stick together in an inelastic collision; all that has to happen is the loss of some kinetic energy. For example, if you smash your car into a car and deform it, the collision is inelastic, even if you can drive away after the accident.

Regardless of whether a collision is elastic or inelastic, momentum is *always* the same before and after the collision, as long as you have a closed system.

Colliding elastically along a line

When a collision is elastic, kinetic energy is conserved. The most basic way to look at elastic collisions is to examine how the collisions work along a straight line. If you run your bumper car into a friend's bumper car along a straight line, you bounce off and kinetic energy is conserved along the line. But the behavior of the cars depends on the mass of the objects involved in the elastic collision.

Bumping into a heavier mass

You take your family to the Physics Amusement Park for a day of fun and calculation, and you decide to ride the bumper cars. You wave to your family as you speed your 300-kilogram car-and-driver up to 10.0 meters per second. Suddenly, Bonk! What happened? The 400-kilogram car-and-driver in front of you had come to a complete stop, and you rear-ended the car elastically; now you're traveling backward and the other car is traveling forward. "Interesting," you think. "I wonder if I can solve for the final velocities of both bumper cars."

You know that the momentum was conserved, and you know that the car in front of you was stopped when you hit it, so if your car is Car 1 and the other is Car 2, you get the following:

$$m_1 v_{f1} + m_2 v_{f2} = m_1 v_{i1}$$

However, this doesn't tell you what v_{f1} and v_{f2} are, because you have two unknowns and only one equation here. You can't solve for v_{f1} or v_{f2} exactly in this case, even if you know the masses and v_{i1}. You need some other equations relating these quantities. How about using the conservation of kinetic energy? The collision was elastic, so kinetic energy was conserved. $KE = (1/2)mv^2$, so here's your equation for the two cars' final and initial kinetic energies:

$$\frac{1}{2}m_1 v_{f1}^2 + \frac{1}{2}m_2 v_{f2}^2 = \frac{1}{2}m_1 v_{i1}^2$$

Now you have two equations and two unknowns, v_{f1} and v_{f2}, which means you can solve for the unknowns in terms of the masses and v_{i1}. You have to dig through a lot of algebra here because the second equation has many squared velocities, but when the dust settles, you get the following two equations:

$$v_{f1} = \frac{(m_1 - m_2)v_{i1}}{m_1 + m_2}$$

$$v_{f2} = \frac{2m_1 v_{i1}}{m_1 + m_2}$$

Now you have v_{f1} and v_{f2} in terms of the masses and v_{i1}. Plugging in the numbers gives you the two bumper cars' final velocities. Here's the velocity of your car:

$$\begin{aligned} v_{f1} &= \frac{(m_1 - m_2)v_{i1}}{m_1 + m_2} \\ &= \frac{(300\text{ kg} - 400\text{ kg})(10.0\text{ m/s})}{300\text{ kg} + 400\text{ kg}} \\ &\approx -1.43\text{ m/s} \end{aligned}$$

And here's the final velocity of the other guy:

$$\begin{aligned} v_{f2} &= \frac{2m_1 v_{i1}}{m_1 + m_2} \\ &= \frac{2(300\text{ kg})(10.0\text{ m/s})}{300\text{ kg} + 400\text{ kg}} \\ &\approx 8.57\text{ m/s} \end{aligned}$$

The two speeds tell the whole story. You started off at 10.0 meters per second in a bumper car of 300 kilograms, and you hit a stationary bumper car of 400 kilograms in front of you. Assuming the collision took place directly and the second bumper car took off in the same direction you were going before the collision, you rebounded at –1.43 meters per second — backward, because this quantity is negative and the bumper car in front of you had more mass — and the bumper car in front of you took off at a speed of 8.57 meters per second.

Bumping into a lighter mass

After having a bad experience in a previous trip to the bumper car pit — where your light bumper car rear-ended a heavy bumper car (see the preceding section for the calculation) — you decide to go back and pick on some poor light cars in a monster bumper car. What happens if your bumper car (plus driver) has a mass of 400 kilograms and you rear-end a stationary 300-kilogram car? In this case, you use the equation for conservation of kinetic energy, the same formula you use in the preceding section. Here's what your final velocity comes out to:

$$\begin{aligned} v_{f1} &= \frac{(m_1 - m_2)v_{i1}}{m_1 + m_2} \\ &= \frac{(400\text{ kg} - 300\text{ kg})(10.0\text{ m/s})}{300\text{ kg} + 400\text{ kg}} \\ &\approx 1.43\text{ m/s} \end{aligned}$$

The little car's final velocity comes out to

$$\begin{aligned} v_{f2} &= \frac{2m_1 v_{i1}}{m_1 + m_2} \\ &= \frac{2(400\text{ kg})(10.0\text{ m/s})}{300\text{ kg} + 400\text{ kg}} \\ &\approx 11.4\text{ m/s} \end{aligned}$$

In this case, you don't bounce backward. The lighter, stationary car takes off after you hit it, but not all your forward momentum is transferred to the other car. Is momentum still conserved? Here are your formulas for the initial and final momentums:

- $p_i = m_1 v_{i1}$
- $p_f = m_1 v_{f1} + m_2 v_{f2}$

Putting in the numbers, here's the initial momentum:

$$p_i = m_1 v_{i1} = (400\text{ kg})(10.0\text{ m/s}) = 4{,}000\text{ kg·m/s}$$

And here's the final momentum:

$$\begin{aligned} p_f &= m_1 v_{f1} + m_2 v_{f2} \\ &= (400\text{ kg})(1.43\text{ m/s}) + (300\text{ kg})(11.4\text{ m/s}) \\ &\approx 4{,}000\text{ kg·m/s} \end{aligned}$$

The numbers match, so momentum is conserved in this collision, just as it is for your collision with a heavier car.

Colliding elastically in two dimensions

Collisions don't always occur along a straight line. For example, balls on a pool table can travel in two dimensions, both *x* and *y*, as they roll around. Collisions along two dimensions introduce variables such as angle and direction.

Say, for example, that your physics travels take you to the golf course, where two players are lining up for their final putts of the day. The players are tied, so these putts are the deciding shots. Unfortunately, the player closer to the hole breaks etiquette, and both golfers putt at the same time. Their 45-gram golf balls collide! You can see what happens in Figure 10-4.

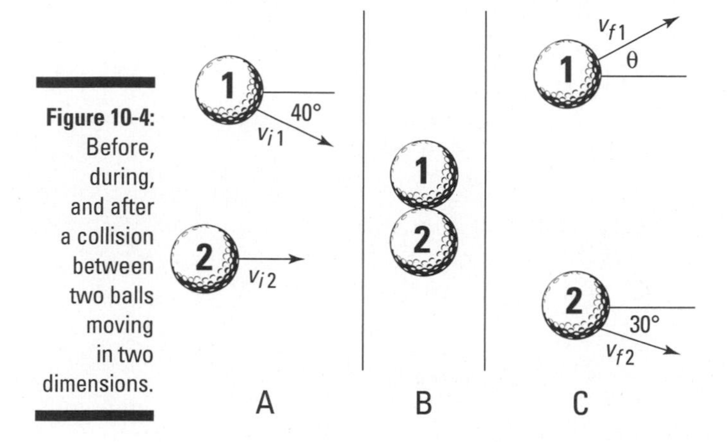

Figure 10-4: Before, during, and after a collision between two balls moving in two dimensions.

You quickly stoop down to measure all the angles and velocities involved in the collision. You measure the speeds: v_{i1} = 1.0 meter per second, v_{i2} = 2.0 meters per second, and v_{f2} = 1.2 meters per second. You also measure most of the angles, as Figure 10-4 shows. However, you can't get the final angle and speed of Ball 1.

Because the golf balls collide elastically, both momentum and kinetic energy are conserved. In particular, momentum is conserved in both the *x* and *y* directions, and total kinetic energy is conserved as well. You need both of these conservations to find the final speed and direction of Ball 1.

First work out the final speed of Ball 1. Because the masses of the balls are equal, you can call each mass *m*. The initial total kinetic energy (for both balls) is

$$\begin{aligned} KE_i &= \frac{1}{2}mv_{i1}^2 + \frac{1}{2}mv_{i2}^2 \\ &= \frac{1}{2}m(1.0\text{ m/s})^2 + \frac{1}{2}m(2.0\text{ m/s})^2 \\ &= \frac{5.0m}{2}\text{ m}^2/\text{s}^2 \end{aligned}$$

Then the final kinetic energy is given by

$$\begin{aligned} KE_f &= \frac{1}{2}mv_{f1}^2 + \frac{1}{2}mv_{f2}^2 \\ &= \frac{1}{2}mv_{f1}^2 + \frac{1}{2}m(1.2\text{ m/s})^2 \end{aligned}$$

Then because kinetic energy is conserved, the final kinetic energy must be equal to the initial kinetic energy, and so you can write

$$\frac{5.0m}{2}\ \text{m}^2/\text{s}^2 = \frac{1}{2}mv_{f1}^2 + \frac{1}{2}m(1.2\ \text{m/s})^2$$

Then you can rearrange the equation to isolate the term with the final velocity of Ball 1, v_{f1}:

$$\frac{1}{2}mv_{f1}^2 = \frac{5.0m}{2}\ \text{m}^2/\text{s}^2 - \frac{1}{2}m(1.2\ \text{m/s})^2$$

$$\frac{1}{2}mv_{f1}^2 = 2.0m\ \text{m}^2/\text{s}^2$$

Then if you solve for v_{f1} (divide both sides by *m,* multiply both sides by 2, and take the square root), you get

$$v_{f1} = 2.0\ \text{m/s}$$

So there you have it: The final speed of Ball 1 is 2.0 meters per second.

To work out the angle of Ball 1's velocity, you use the conservation of momentum. Momentum is conserved in both the *x* and *y* directions, so the following equations are true:

- $p_{fx} = p_{ix}$
- $p_{fy} = p_{iy}$

In other words, the final momentum in the *x* direction is the same as the initial momentum in the *x* direction, and the final momentum in the *y* direction is the same as the initial momentum in the *y* direction. Here's what the initial momentum in the *x* direction looks like:

$$p_{ix} = mv_{i1}\cos 40° + mv_{i2}$$

You can see that this is the sum of the *x* momenta of both balls.

The final momentum in the *x* direction is given by

$$p_{fx} = mv_{f1}\cos\theta + mv_{f2}\cos 30°$$

The *x* component of momentum is conserved, so you can equate the initial and final momenta in the *x* direction:

$$mv_{i1}\cos 40° + mv_{i2} = mv_{f1}\cos\theta + mv_{f2}\cos 30°$$

Divide both sides by *m* to get

$$v_{i1}\cos 40° + v_{i2} = v_{f1}\cos\theta + v_{f2}\cos 30°$$

If you rearrange this equation to put the term with the unknown angle, θ, on one side, you get

$$v_{f1}\cos\theta = v_{i1}\cos 40° + v_{i2} - v_{f2}\cos 30°$$

Dividing by v_{f1} gives you

$$\cos\theta = \frac{v_{i1}\cos 40° + v_{i2} - v_{f2}\cos 30°}{v_{f1}}$$

Plug in the measured values and the final speed of Ball 1 you calculated previously to get

$$\cos\theta = \frac{(1.0\text{ m/s})\cos 40° + (2.0\text{ m/s}) - (1.2\text{ m/s})\cos 30°}{2.0\text{ m/s}}$$

Finally, you can take the inverse cosine of each side to find the angle:

$$\theta \approx 30°$$

So there you have it: After the collision, Ball 1 moves with a velocity of 2.0 m/s at an angle of 30° from horizontal. You have combined the use of conservation of kinetic energy (in an elastic collision) and conservation of momentum (as in all collisions) to work out the final velocity of Ball 1.

Chapter 11

Winding Up with Angular Kinetics

In This Chapter

- Changing gears from linear motion to rotational motion
- Calculating tangential speed and acceleration
- Understanding angular acceleration and velocity
- Identifying the torque involved in rotational motion
- Maintaining rotational equilibrium

This chapter is the first of two on handling objects that rotate, from space stations to marbles. Rotation is what makes the world go round — literally — and if you know how to handle linear motion and Newton's laws (see the first two parts of the book if you don't), the rotational equivalents I present in this chapter and in Chapter 12 are pieces of cake. And if you don't have a grasp on linear motion, no worries. You can get a firm grip on the basics of rotation here. You see all kinds of rotational ideas in this chapter: angular acceleration, tangential speed and acceleration, torque, and more. Kinetics deals not only with the motions of objects but the forces behind those motions. Rotational kinetics deals with rotational motions and the forces behind them (torque). But enough spinning the wheels. Read on!

Going from Linear to Rotational Motion

You need to change equations when you go from linear motion to rotational motion. Here are the angular equivalents (or analogs) for the linear motion equations:

	Linear	***Angular***
Velocity	$v = \frac{\Delta s}{\Delta t}$	$\omega = \frac{\Delta \theta}{\Delta t}$
Acceleration	$a = \frac{\Delta v}{\Delta t}$	$\alpha = \frac{\Delta \omega}{\Delta t}$
Displacement	$s = v_i \Delta t + \frac{1}{2} a \Delta t^2$	$\theta = \omega_i \Delta t + \frac{1}{2} \alpha \Delta t^2$
Motion with time canceled out	$v_f^2 - v_i^2 = 2as$	$\omega_f^2 - \omega_i^2 = 2\alpha\theta$

In all these equations, t stands for time, Δ means "change in," $_f$ means final, and $_i$ means initial. In the linear equations, v is velocity, s is displacement, and a is acceleration. In the angular equations, ω is angular velocity (measured in radians/second), θ is angular displacement in radians, and α is angular acceleration (in radians/second^2).

You know that the quantities displacement, velocity, and acceleration are all vectors; well, their angular equivalents are vectors, too. First, consider angular displacement, $\Delta\theta$ — this is a measure of the angle through which an object has rotated. The magnitude tells you the size of the angle of the rotation, and the direction is parallel to the axis of the rotation. Similarly, angular velocity, ω, has a magnitude equal to the angular speed and a direction that defines the axis of rotation. The angular acceleration, α, has a magnitude equal to the rate at which the angular velocity is changing; it's also directed along the axis of rotation.

If you consider only motion in a plane, then you have only one possible direction for the axis of rotation: perpendicular to the plane. In this case, these vector quantities have only one component — this vector component is just a number, and the sign of the number indicates all you need to know about the direction. For example, a positive angular displacement may be a clockwise rotation, and a negative angular displacement may be a counterclockwise rotation.

Just as the magnitude of the velocity is the speed, the magnitude of the angular velocity is the angular speed. Just as the magnitude of a displacement is a distance, the magnitude of an angular displacement is an angle — that is, the magnitude of the vector quantity is a scalar quantity.

Note: In the next section, I begin by looking at the motion in a plane considering only the single component of the vectors — which are scalar numbers (I identify the vector with its single component). So for that section, the quantities, $\Delta\theta$, ω, and α don't appear in bold type because they represent the single component of a rotation in a plane. In the section "Applying Vectors to Rotation," I take a closer look at the vector nature of the angular displacement, velocity, and acceleration.

Understanding Tangential Motion

Tangential motion is motion that's perpendicular to *radial motion,* or motion along a radius. Given a central point, vectors in the surrounding space can be broken into two components: *radial direction,* which points directly away from the center of the circle, and *tangential direction,* which follows the circle and is directed perpendicular to the radial direction. Motion in the tangential direction is referred to as *tangential motion.*

You can tie angular quantities such as angular displacement (θ), angular velocity (ω), and angular acceleration (α) to their associated tangential quantities. All you have to do is multiply by the radius, using these equations:

- $s = r\theta$
- $v = r\omega$
- $a = r\alpha$

These equations rely on using radians as the measure of angles; they don't work if you try to use degrees.

Say you're riding a motorcycle, for example, and the wheels' angular speed is $\omega = 21.5\pi$ radians per second. What does this mean in terms of your motorcycle's speed? To determine your motorcycle's velocity, you need to relate angular velocity, ω, to linear velocity, v. The following sections explain how you can make such relations.

Finding tangential velocity

At any point on a circle, you can pick two special directions: The direction that points directly away from the center of the circle (along the radius) is called the *radial* direction, and the direction that's perpendicular to this is called the *tangential* direction.

When an object moves in a circle, you can think of its *instantaneous velocity* (the velocity at a given point in time) at any particular point on the circle as an arrow drawn from that point and directed in the tangential direction. For this reason, this velocity is called the *tangential velocity*. The magnitude of the tangential velocity is the *tangential speed,* which is simply the speed of an object moving in a circle.

Given an angular velocity of magnitude ω, the tangential velocity at any radius is of magnitude $r\omega$. The idea that the tangential velocity increases as the radius increases makes sense, because given a rotating wheel, you'd expect a point at radius r to be going faster than a point closer to the hub of the wheel.

Take a look at Figure 11-1, which shows a ball tied to a string. The ball is whipping around with angular velocity of magnitude ω.

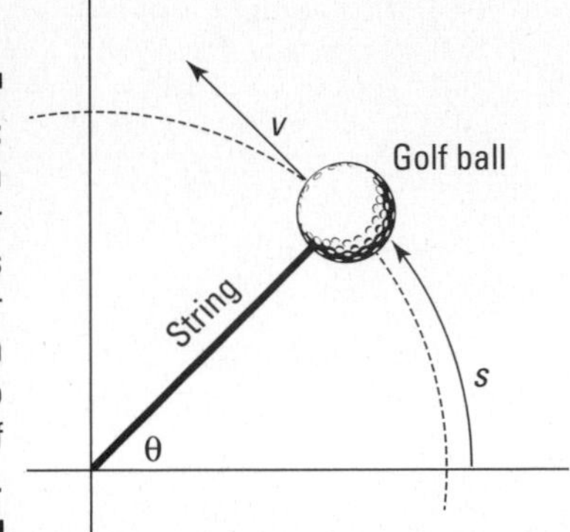

Figure 11-1: A ball in circular motion has angular speed with respect to the radius of the circle.

You can easily find the magnitude of the ball's velocity, v, if you measure the angles in radians. A circle has 2π radians; the complete distance around a circle — its circumference — is $2\pi r$, where r is the circle's radius. In general, therefore, you can connect an angle measured in radians with the distance you cover along the circle, s, like this:

$$s = r\theta$$

where r is the radius of the circle. Now, you can say that $v = s/t$, where v is magnitude of the velocity, s is the distance, and t is time. You can substitute for s to get

$$v = \frac{s}{t} = \frac{r\theta}{t}$$

Because $\omega = \theta/t$, you can say that

$$v = \frac{s}{t} = \frac{r\theta}{t} = r\omega$$

In other words,

$$v = r\omega$$

Now you can find the magnitude of the velocity. The wheels of a motorcycle are turning with an angular velocity of 21.5π radians/second. If you can find the tangential velocity of any point on the outside edges of the wheels, you can find the motorcycle's speed. Say, for example, that the radius of one of your motorcycle's wheels is 40 centimeters. You know that $v = r\omega$, so just plug in the numbers:

$$v = r\omega = (0.40 \text{ m/s})21.5\pi \approx 27 \text{ m/s}$$

Converting 27 meters/second to miles/hour gives you about 60 mph.

Finding tangential acceleration

Tangential acceleration is a measure of how the tangential velocity of a point at a certain radius changes with time. Tangential acceleration is just like linear acceleration (see Chapter 3), but it's particular to the tangential direction, which is relevant to circular motion. Here you look at the magnitude of the angular acceleration, α, which tells you how the speed of the object in the tangential direction is changing.

For example, when you start a lawn mower, a point on the tip of one of its blades starts at a tangential velocity of zero and ends up with a tangential velocity with a pretty large magnitude. So how do you determine the point's tangential acceleration? You can use the following equation from Chapter 3, which relates velocity to acceleration (where Δv is the change in velocity and Δt is the change in time) to relate tangential quantities like tangential velocity to angular quantities such as angular velocity:

$$a = \frac{\Delta v}{\Delta t}$$

Tangential velocity, v, equals $r\omega$ (as you see in the preceding section), so you can plug in this information:

$$a = \frac{\Delta(r\omega)}{\Delta t}$$

Because the radius is constant here, the equation becomes

$$a = \frac{r\Delta\omega}{\Delta t}$$

However, $\Delta\omega/\Delta t = \alpha$, the angular acceleration, so the equation becomes

$$a = r\alpha$$

Translated into layman's terms, this says tangential acceleration equals angular acceleration multiplied by the radius.

Finding centripetal acceleration

Newton's first law says that when there are no net forces, an object in motion will continue to move uniformly in a straight line (see Chapter 5). For an object to move in a circle, a force has to cause the change in direction — this force is called the *centripetal force*. Centripetal force is always directed toward the center of the circle.

The *centripetal acceleration* is proportional to the centripetal force (obeying Newton's second law; see Chapter 5). This is the component of the object's acceleration in the radial direction (directed toward the center of the circle), and it's the rate of change in the object's velocity that keeps the object moving in a circle; this force does not change the magnitude of the velocity, only the direction.

You can connect angular quantities, such as angular velocity, to centripetal acceleration. Centripetal acceleration is given by the following equation (for more on the equation, see Chapter 7):

$$a_c = \frac{v^2}{r}$$

where v is the velocity and r is the radius. Linear velocity is easy enough to tie to angular velocity because $v = r\omega$ (see the section "Finding tangential velocity"). Therefore, you can rewrite the acceleration formula as

$$a_c = \frac{(r\omega)^2}{r}$$

REMEMBER

The centripetal-acceleration equation simplifies to

$$a_c = r\omega^2$$

Nothing to it. The equation for centripetal acceleration means that you can find the centripetal acceleration needed to keep an object moving in a circle given the circle's radius and the object's angular velocity.

Say that you want to calculate the centripetal acceleration of the moon around the Earth. Start with the old equation

$$a_c = \frac{v^2}{r}$$

First you have to calculate the tangential velocity of the moon in its orbit. Alternatively, you can calculate the tangential velocity from the angular velocity. Using the new version of the equation, $a_c = r\omega^2$, is easier because the moon orbits the Earth in about 28 days, so you can easily calculate the moon's angular velocity.

Because the moon makes a complete orbit around the Earth in about 28 days, it travels 2π radians around the Earth in that period, so its angular velocity is

$$\omega = \frac{\Delta\theta}{\Delta t} = \frac{2\pi \text{ rad}}{28 \text{ days}}$$

Converting 28 days to seconds gives you the following:

$$\frac{28 \cancel{\text{days}}}{1} \times \frac{24 \cancel{\text{hours}}}{1 \cancel{\text{day}}} \times \frac{60 \cancel{\text{minutes}}}{1 \cancel{\text{hour}}} \times \frac{60 \text{ seconds}}{1 \cancel{\text{minute}}} \approx 2.42 \times 10^6 \text{ seconds}$$

Therefore, you get the following angular velocity:

$$\begin{aligned}\omega &= \frac{\Delta\theta}{\Delta t} \\ &= \frac{2\pi \text{ rad}}{2.42 \times 10^6 \text{ s}} \\ &= 2.60 \times 10^{-6} \text{ rad/s}\end{aligned}$$

You now have the moon's angular velocity, 2.60×10^{-6} radians per second. The average radius of the moon's orbit is 3.85×10^8 meters, so its centripetal acceleration is

$$\begin{aligned}a_c &= r\omega^2 \\ &= (3.85 \times 10^8 \text{ m})(2.60 \times 10^{-6} s^{-1})^2 \\ &\approx 2.60 \times 10^{-3} \text{ m/s}^2\end{aligned}$$

In the preceding equation, the units of angular velocity, radians per second, are written as s^{-1} because the radian is a *dimensionless* unit. A *radian* is the angle swept by an arc that has a length equal to the radius of the circle. Think of it as a particular portion of the whole circle; as such, it has no dimensions. So when you have "radians per second," you can omit "radians," which leaves you with "per second." Another way to write this is to use the exponent –1, so you can represent radians per second as s^{-1}.

Just for kicks, you can also find the force needed to keep the moon going around in its orbit. Force equals mass times acceleration (see Chapter 5), so you multiply acceleration by the mass of the moon, 7.35×10^{22} kilograms:

$$F_c = ma_c = (7.35 \times 10^{22} \text{ kg})(2.60 \times 10^{-3} \text{ m/s}^2) \approx 1.91 \times 10^{20} \text{ N}$$

The force in newtons, 1.91×10^{20} N, converts to about 4.3×10^{19} pounds of force needed to keep the moon going around in its orbit.

Applying Vectors to Rotation

Angular displacement, angular velocity, and angular acceleration are each vector quantities. When you consider circular motion in a plane, these vectors only have one component, which is a scalar number; in that case, you don't have to consider the direction very much. However, when you have circular motion in more than one plane (as with the motions of the planets, which orbit on very slightly different planes) or when the plane of rotation changes (like in a wobbling spinning top, for example), then the direction of these vectors becomes significant.

Angular velocity and angular acceleration are vectors that are directed along the axis of the rotation.

In this section, you hear more about the directions of the angular vectors. For the rest of this section, the quantities $\boldsymbol{\Delta\theta}$, $\boldsymbol{\omega}$, and $\boldsymbol{\alpha}$ appear in bold type because you're explicitly dealing with vectors.

Calculating angular velocity

When a wheel is spinning, it has not only an angular speed but also a direction. Here's what the angular velocity vector tells you:

- The size of the angular velocity vector tells you the angular speed.
- The direction of the vector tells you the axis of the rotation, as well as whether the rotation is clockwise or counterclockwise.

Say that a wheel has a constant angular speed, ω — which direction does its angular velocity, $\boldsymbol{\omega}$, point? It can't point along the rim of the wheel, as tangential velocity does, because its direction would then change every second. In fact, the only real choice for its direction is perpendicular to the wheel.

The direction of the angular velocity always takes people by surprise: Angular velocity, $\boldsymbol{\omega}$, points along the axle of a wheel (see Figure 11-2). Because the angular velocity vector points the way it does, it has no component along the wheel. The wheel is spinning, so the tangential (linear) velocity at any point on the wheel is constantly changing direction — except for at the very center point of the wheel, where the base of the angular velocity vector sits. If the wheel is lying flat on the ground, the vector's head points up or down, away from the wheel, depending on which direction the wheel is rotating.

You can use the right-hand rule to determine the direction of the angular velocity vector. Wrap your right hand around the wheel so that your fingers point in the direction of the tangential motion at any point — the fingers on your right hand should go in the same direction as the wheel's rotation. When you wrap your right hand around the wheel, your thumb points in the direction of the angular velocity vector, $\boldsymbol{\omega}$.

Figure 11-2 shows a wheel lying flat, turning counterclockwise when viewed from above. Wrap your fingers in the direction of rotation. Your thumb, which represents the angular velocity vector, points up; it runs along the wheel's axle. If the wheel were to turn clockwise instead, your thumb — and the vector — would have to point down, in the opposite direction.

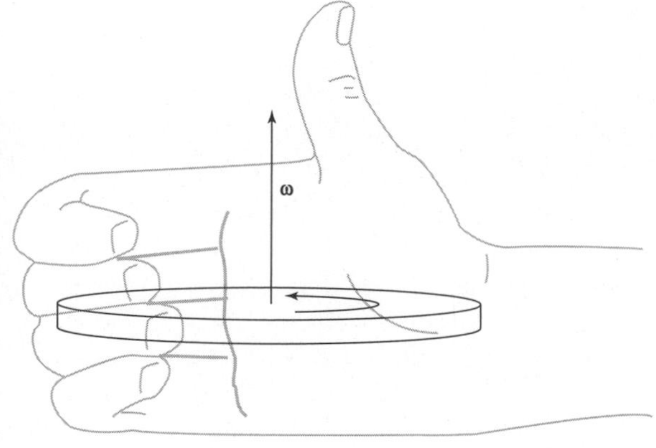

Figure 11-2: Angular velocity points in a direction perpendicular to the wheel.

Figuring angular acceleration

In this section, you find out how the angular acceleration and angular velocity relate to each other in terms of their magnitude and direction. You first see what happens in the simplest case, where the angular acceleration and velocity are in the same direction or in opposite directions. Then you look at a situation in which angular acceleration and angular velocity are at an angle to each other, leading to a tilting of the rotational axis.

Changing the speed and reversing direction

If the angular velocity vector points out of the plane of rotation (see the preceding section), what happens when the angular velocity changes — when the wheel speeds up or slows down? A change in velocity signifies the presence of angular acceleration. Like angular velocity, $\boldsymbol{\omega}$, angular acceleration, $\boldsymbol{\alpha}$, is a vector, meaning it has a magnitude and a direction. Angular acceleration is the rate of change of angular velocity:

$$\alpha = \frac{\Delta\omega}{\Delta t}$$

For example, look at Figure 11-3, which shows what happens when angular acceleration affects angular velocity. In this case, α points in the same direction as ω in 11-3A. When the angular acceleration vector, α, points along the angular velocity, ω, the magnitude of ω will increase as time goes on, as Figure 11-3B shows.

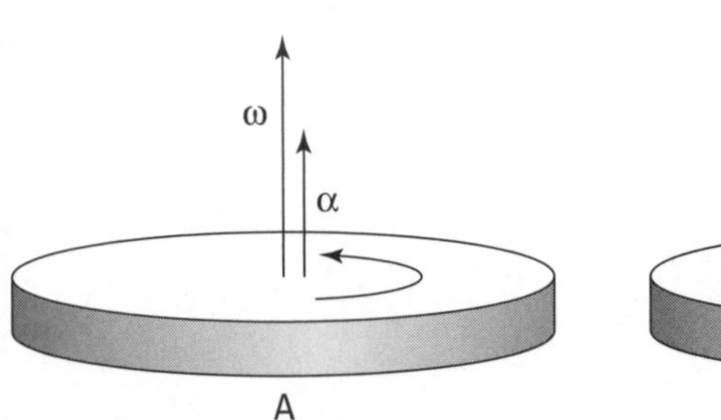

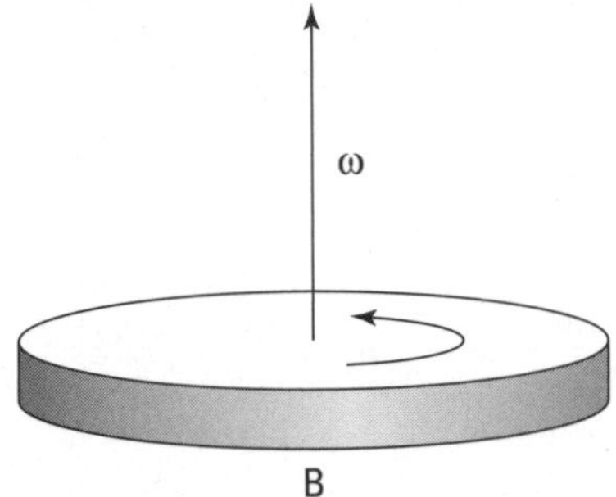

Figure 11-3: Angular acceleration in the same direction as the angular velocity.

Just as an object's linear velocity and linear acceleration may be in opposite directions, the angular acceleration also doesn't have to be in the same direction as the angular velocity vector (as Figure 11-4A shows). If the angular acceleration is directed in the opposite direction of the angular velocity, then the magnitude of the angular velocity decreases at a rate given by the magnitude of the angular acceleration.

Just as in the case of linear velocity and acceleration, the angular acceleration gives the rate of change of angular velocity: The magnitude of the angular acceleration gives the rate at which the angular velocity changes, and the direction gives the direction of the change. You can see a decreased angular velocity in Figure 11-4B.

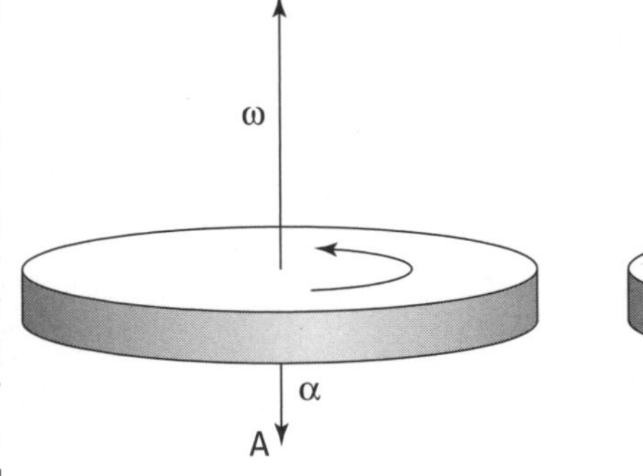

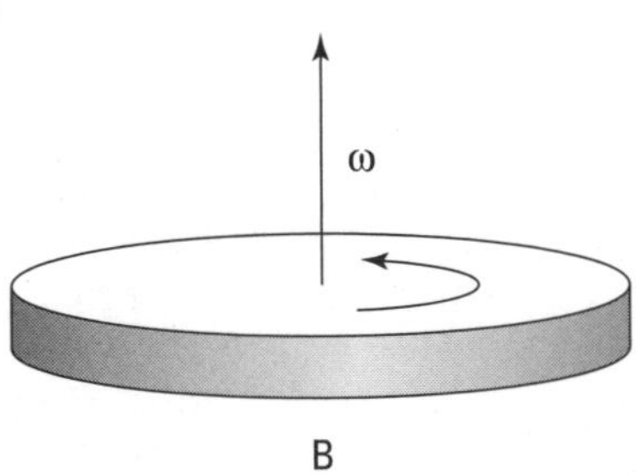

Figure 11-4: Angular acceleration in the direction opposite the angular velocity reduces the angular speed.

Tilting the axle

The *angular acceleration* is the rate of change of angular velocity — the change can be to the direction instead of the magnitude. For example, suppose you take hold of the axle of the spinning wheel in Figure 11-3 and tilt it. You'd change the angular velocity of the wheel but not by changing its magnitude (the angular speed of the wheel would remain constant); rather, you'd change the direction of the angular velocity by changing the axis of rotation — this is an angular acceleration that's directed perpendicular to the angular velocity, as in Figure 11-5.

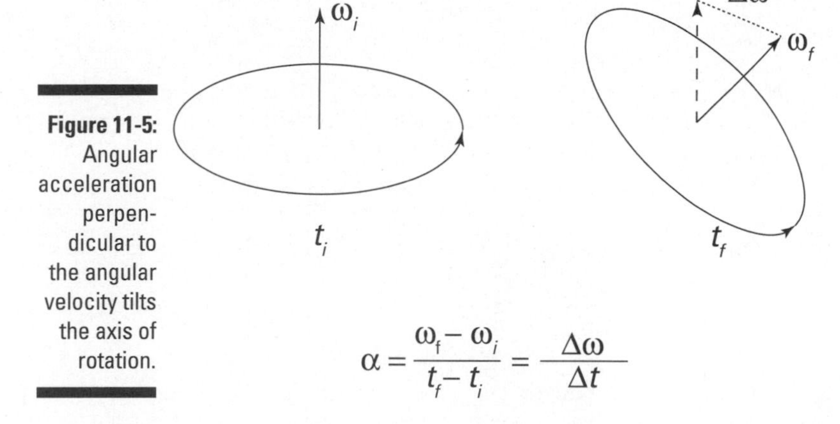

Figure 11-5: Angular acceleration perpendicular to the angular velocity tilts the axis of rotation.

Doing the Twist: Torque

For extended objects (rods, disks, or cubes, for example), which, unlike point objects, have their mass distributed through space, you have to take into account where the force is applied. Enter torque. *Torque* is a measure of the ability of a force to cause rotation. In physics terms, the torque exerted on an object depends on the force itself (its magnitude and direction) and where you exert the force. You go from the strictly linear idea of force as something that acts in a straight line (such as when you push a refrigerator up a ramp) to its angular counterpart, torque.

Just as force causes acceleration, torque causes angular acceleration, so you can think of torque as the angular equivalent of force (see Chapter 12 for more info on that aspect of torque).

Torque brings forces into the rotational world. Most objects aren't just points or rigid masses, so if you push them, they not only move but also turn. For example, if you apply a force tangentially to a merry-go-round, you don't move the merry-go-round away from its current location — you cause it to start spinning. Rotational motions and the forces behind them are the focus of this chapter and Chapter 12.

Look at Figure 11-6, which shows a seesaw with a mass m on it. If you want to balance the seesaw, you can't have a larger mass, M, placed on a similar spot on the other side of the seesaw. Where you put the larger mass M determines whether the seesaw balances. As you can see in Figure 11-6A, if you put the mass M on the pivot point — also called the *fulcrum* — of the seesaw, you don't have balance. The larger mass exerts a force on the seesaw, but the force doesn't balance it.

As you can see in Figure 11-6B, as you increase the distance you put the mass M away from the fulcrum, the balance improves. In fact, if $M = 2m$, you need to put the mass M exactly half as far from the fulcrum as the mass m is.

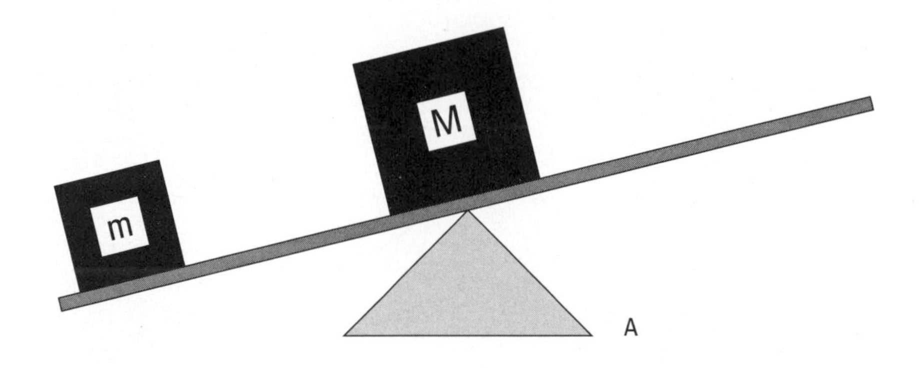

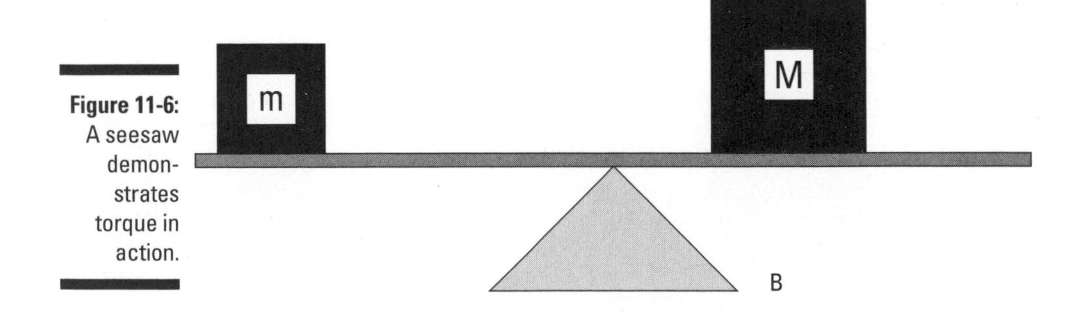

Figure 11-6: A seesaw demonstrates torque in action.

The torque is a vector. The magnitude of the torque tells you the ability of the torque to generate rotation; more specifically, the magnitude of the torque is proportional to the angular acceleration it generates. The direction of the torque is along the axis of this angular acceleration. This section starts by considering torques and forces that are in a plane, so you only need to think about the magnitude of the torque and not the full vector. Later, I explain a little more about the direction of the torque vector.

Mapping out the torque equation

How much torque you exert on an object depends on the following:

- The force you exert, F
- Where you apply the force; the *lever arm* — also called the *moment arm* — is the perpendicular distance from the pivot point to the point at which you exert your force and is related to the distance from the axis, r, by $l = r \sin \theta$, where θ is the angle between the force and a line from the axis to the point where the force is applied.

Assume that you're trying to open a door, as in the various scenarios in Figure 11-7. You know that if you push on the hinge, as in diagram A, the door won't open; if you push the middle of the door, as in diagram B, the door will open; but if you push the edge of the door, as in diagram C, the door will open more easily.

In Figure 11-7, the lever arm, l, is distance r from the hinge at which you exert your force. The torque is the product of the magnitude of the force multiplied by the lever arm. It has a special symbol, the Greek letter τ (tau):

$$\tau = Fl$$

The units of torque are force units multiplied by distance units, which is newton-meters in the MKS system and foot-pounds in the foot-pound-second system (see Chapter 2 for more on these measurement systems).

For example, the lever arm in Figure 11-7 is distance r (because this is the distance perpendicular to the force), so $\tau = Fr$. If you push with a force of 200 newtons and r is 0.5 meters, what's the torque you see in the figure? In diagram A, you push on the hinge, so your distance from the pivot point is zero, which means the lever arm is zero. Therefore, the magnitude of the torque is zero. In diagram B, you exert the 200 newtons of force at a distance of 0.5 meters perpendicular to the hinge, so

$$\tau = Fl = (200 \text{ N})(0.5 \text{ m}) = 100 \text{ N·m}$$

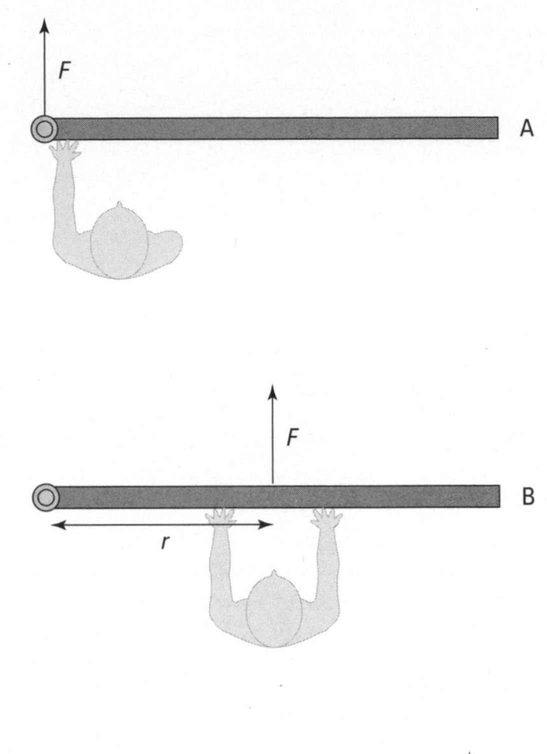

Figure 11-7: The torque you exert on a door depends on where you push it.

The magnitude of the torque here is 100 newton-meters. But now take a look at diagram C. You push with 200 newtons of force at a distance of $2r$ perpendicular to the hinge, which makes the lever arm $2r$ or 1.0 meter, so you get this torque:

$$\tau = Fl = (200\text{ N})(1.0\text{ m}) = 200\text{ N}\cdot\text{m}$$

Now you have 200 newton-meters of torque, because you push at a point twice as far away from the pivot point. In other words, you double the magnitude of your torque. But what would happen if, say, the door were partially open when you exerted your force? Well, you would calculate the torque easily, if you have lever-arm mastery.

Understanding lever arms

If you push a partially open door in the same direction as you push a closed door, you create a different torque because of the non-right angle between your force and the door.

Take a look at Figure 11-8A to see a person obstinately trying to open a door by pushing along the door toward the hinge. You know this method won't produce any turning motion, because the person's force has no lever arm to produce the needed turning force. In this case, the lever arm is zero, so it's clear that even if you apply a force at a given distance away from a pivot point, you don't always produce a torque. The direction you apply the force also counts, as you know from your door-opening expertise.

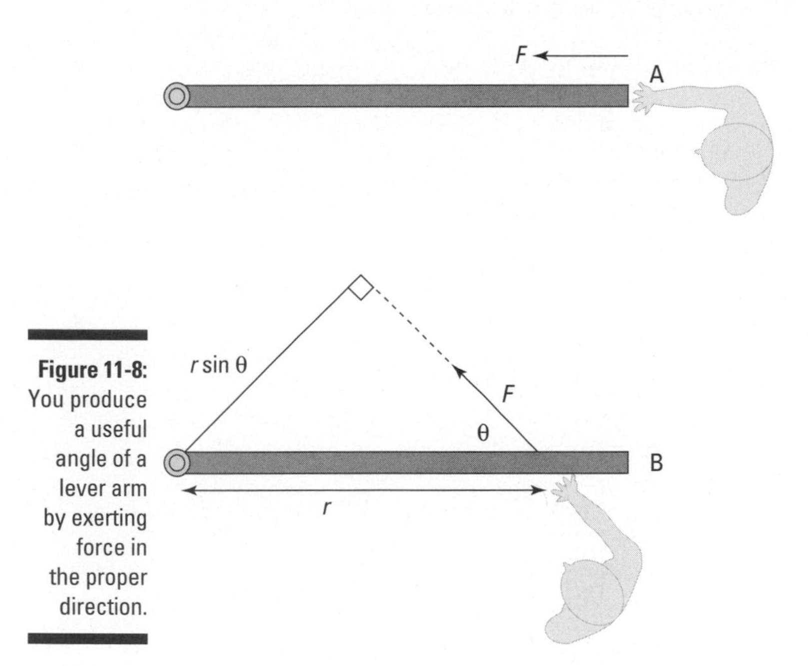

Figure 11-8: You produce a useful angle of a lever arm by exerting force in the proper direction.

Figuring out the torque generated

Generating torque is how you open doors, whether you have to quickly pop a car door or slowly pry open a bank-vault door. But how do you find out how much torque you generate? First, you calculate the lever arm, and then you multiply that lever arm by the force to get the torque.

Take a look at Figure 11-8B. You apply a force to the door at some angle, θ. The force may open the door, but it isn't a sure thing, because as you can tell from the figure, you apply less of a turning force here. What you need to do is find the lever arm first. As you can see in Figure 11-8B, you apply the force at a distance *r* from the hinge. If you apply that force perpendicularly to the door, the lever arm's length would be *r,* and you'd get

$$\tau = Fr$$

However, that's not the case here, because the force isn't perpendicular to the door.

The *lever arm* is the effective distance from the pivot point at which the force acts perpendicularly. Picture moving the point where the force is applied, carrying the force vector along without changing its direction. When you move to a spot where the force is perpendicular to the direction of the axis of rotation, the distance to the axis is the lever arm.

To see how this works, take a look at diagram B in Figure 11-8, where you can draw a lever arm from the pivot point so that the force is perpendicular to the lever arm. To do this, extend the force vector until you can draw a line from the pivot point that's perpendicular to the force vector. You create a new triangle. The lever arm and the force are at right angles with respect to each other, so you create a right triangle. The angle between the force and the door is θ, and the distance from the hinge at which you apply the force is *r* (the hypotenuse of the right triangle), so the lever arm becomes

$$l = r \sin \theta$$

When θ goes to zero, so does the lever arm, so there's no torque (see diagram A in Figure 11-8). You know that $\tau = Fl$, so you can now find $\tau = Fr \sin \theta$, where θ is the angle between the force and the door.

This is a general equation; if you apply a force with a magnitude of *F* at a distance *r* from a pivot point, where the angle between that displacement vector **r** and the force vector **F** is θ, the torque you produce will have a magnitude of $\tau = Fr \sin \theta$. So, for example, if $\theta = 45°$, $F = 200$ newtons, and $r = 1.0$ meter, you get

$$\tau = Fr \sin \theta = (200 \text{ N})(1.0 \text{ m})(0.707) \approx 140 \text{ N·m}$$

This number is less than you'd expect if you just push perpendicularly to the door (which would be 200 newton-meters).

Recognizing that torque is a vector

Torque is a vector, so it has not only magnitude but also direction. The direction of the torque is the same as the angular acceleration that it causes. It is perpendicular to the force and the lever arm in a right-hand fashion.

To get a little more technical, torque is given by the *cross-product* of the vector that points from the axis of rotation to the point at which the force is applied, **r,** and the force vector, **F.** The cross-product is written as an ×, so mathematically, the torque vector is the following:

$$\boldsymbol{\tau} = \mathbf{r} \times \mathbf{F}$$

This equation is really a fancy mathematical way of saying that the torque vector has a magnitude of $rF \sin \theta$ and that the direction of the torque vector is as Figure 11-9 shows.

The right-hand rule is a useful way of remembering the direction of torque. If you point the thumb of your right hand in the direction of the radius vector **r** and your fingers in the direction of the force vector **F,** then your palm faces the direction of the torque vector $\boldsymbol{\tau}$.

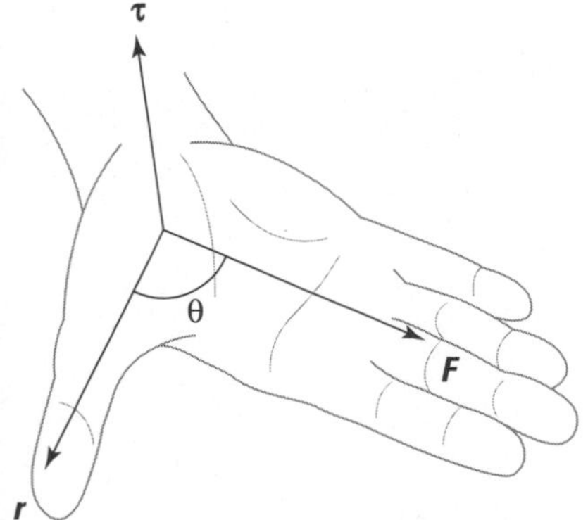

Figure 11-9: A turning motion toward larger positive angles indicates a positive vector.

Spinning at Constant Velocity: Rotational Equilibrium

You may know equilibrium as a state of balance, but what's equilibrium in physics terms? When you say an object has *equilibrium,* you mean that the motion of the object isn't changing; in other words, the object has no acceleration (it can have motion, however, as in constant velocity and/or constant angular velocity). As far as linear motion goes, the vector sum of all forces acting on the object must be zero for the object to be in equilibrium. The net force acting on the object is zero: $\Sigma\mathbf{F} = 0$.

Equilibrium occurs in rotational motion in the form of rotational equilibrium. When an object is in *rotational equilibrium,* it has no angular acceleration — the object may be rotating, but it isn't speeding up or slowing down or changing directions (its tilt angle), which means its angular velocity is constant. When an object has rotational equilibrium, you see no net turning force on the object, which means that the net torque on the object must be zero:

$$\Sigma\tau = 0$$

This equation represents the rotational equivalent of linear equilibrium. Rotational equilibrium is a useful idea because given a set of torques operating on an extended object, you can determine what torque is necessary to stop the object from rotating. In this section, you try out three problems that involve objects in rotational equilibrium.

Determining how much weight Hercules can lift

Say that Hercules wants to lift a massive dumbbell using the deltoid (shoulder) muscle in his right arm and hold the weight at arm's length. His arm, which has a weight of magnitude F_a = 28.0 newtons, can exert a force F of 1,840 newtons. His deltoid muscle is attached to the arm at 13.0°, as Figure 11-10 shows. The figure also shows the distances between the pivot point and the points of application of the forces: The distance to the muscle is 0.150 m, to the effective point of application of the weight of the arm is 0.310 m (half the length of the arm), and to the dumbbell is 0.620 m. The weight of the dumbbell has magnitude F_d.

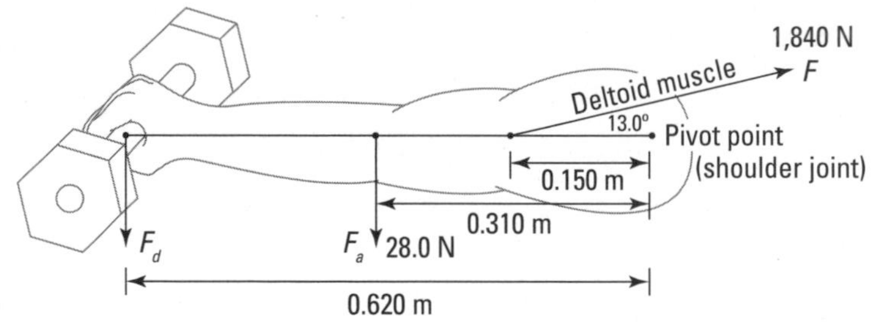

Figure 11-10: A schematic of the forces acting on Hercules's arm.

What is the maximum weight of the dumbbell Hercules can hold at arm's length, and what are the two components of the force F_b, the force against his body? Because Hercules is holding the dumbbell without accelerating, the net force acting must be zero, so F_b must cancel out the sum of the forces in Figure 11-10.

Hercules's arm is not moving, so $\Sigma\mathbf{F} = 0$ and $\Sigma\tau = 0$. Look at $\Sigma\mathbf{F} = 0$ first. In the x direction, that gives you the following force against Hercules's body:

$$\Sigma\mathbf{F}_x = F_{bx} + F\cos 13.0° = 0$$

$$F_{bx} = -F\cos 13.0°$$

Plugging in the value of force F gives you

$$F_{bx} = (-1{,}840\text{ N})\cos 13.0° \approx -1{,}790\text{ N}$$

That was pretty easy. Already you have F_{bx}, which is –1,790 newtons. Now find the force against Hercules's body in the y direction:

$$\Sigma\mathbf{F}_y = F_{by} + F\sin 13.0° - F_a - F_d = 0$$

$$F_{by} + (1{,}840\text{ N})\sin 13.0° - 28.0\text{ N} - F_d = 0$$

$$F_{by} = -(1{,}840\text{ N})\sin 13.0° + 28.0\text{ N} + F_d$$

Well, that gives you one equation in two variables, F_{by} and F_d, so you need more information to solve for those variables.

Torque to the rescue. You can get that additional information with the equation $\Sigma\tau = 0$. If you look at Figure 11-10, you see that three forces are acting on the arm to cause torques around the arm joint: the y component of F (the pull of Hercules's deltoid muscle), F_a (the weight of his arm), and F_d (the weight of the dumbbell).

The component of F in the y direction is $F_y = (1{,}841\text{ N})\sin 13.0°$. The magnitude of the weight of Hercules's arm is $F_a = 28.0$ newtons, and you don't yet know the magnitude of the weight of the dumbbell, F_d.

So what are the torques due to these three torque-causing forces? The direction of the torque is in the direction perpendicular to the plane of Figure 11-10. Consider the component of the torque in this direction, such that positive values correspond to counterclockwise-acting torques and negative values correspond to clockwise-acting torques. Because this component of the torque vectors is a number (scalar), I don't write it in bold type. The torque from the y component of the muscle-pull F is the following:

$$\tau_M = F_y\,(0.150\text{ m})$$

$$= (1{,}840\text{ N})\sin 13.0°\,(0.150\text{ m})$$

This torque is positive because it leads to a turning force in the counterclockwise direction, as Figure 11-10 shows (or you can reason that the torque is positive because the angle between the force and the lever is $\theta = 90°$, so $l = r \sin \theta = (0.150 \text{ m}) \sin 90° = 0.150 \text{ m}$). The torque from the weight of Hercules's arm is

$$\tau_a = (28.0 \text{ N})(-0.310 \text{ m})$$

This torque is negative because the lever arm is negative, so the force causes a clockwise torque, as Figure 11-10 shows (or you can find that the torque is negative because the angle between the force and the lever is $\theta = 90°$, so $l = r \sin \theta = (0.310 \text{ m}) \sin -90° = -0.310 \text{ m}$). The torque due to the weight of the dumbbell is

$$\tau_d = -F_d(0.620 \text{ m})$$

This is obviously negative for the same reason that τ_a is negative.

Because $\Sigma\tau = 0$, that means that

$$\tau_M + \tau_a + \tau_d = \Sigma\tau$$
$$(1{,}840 \text{ N})(0.150 \text{ m}) \sin 13.0° + (-31.0 \text{ N})(0.280 \text{ m}) + (-F_d)(0.620 \text{ m}) = 0$$

Calculating the products and solving for F_d gives you the following:

$$(62.0) + (-8.7) + (-F_d)(0.620 \text{ m}) = 0$$
$$(53.3 \text{ N} \cdot \text{m}) - F_d(0.620 \text{ m}) = 0$$
$$F_d(0.620 \text{ m}) = 53.3 \text{ N} \cdot \text{m}$$
$$F_d = \frac{53.3 \text{ N} \cdot \text{m}}{0.620 \text{ m}}$$
$$F_d \approx 86.0 \text{ N}$$

Great — you have the force on the arm socket in the x direction, F_{bx}, and now you know the weight of the maximum dumbbell that Hercules could hold at arm's length indefinitely.

That leaves only F_{by}, the force on the arm socket in the y direction, to calculate. Earlier, you found that

$$F_{by} = -(1{,}840 \text{ N}) \sin 13.0° + 28.0 \text{ N} + F_d$$

You now know that $F_d = 86.0$ N, so plug in that value. You get the following:

$$F_{by} = -(1{,}840 \text{ N}) \sin 13.0° + 28.0 \text{ N} + 86.0 \text{ N}$$
$$F_{by} = -413.9 \text{ N} + 28.0 \text{ N} + 86.0 \text{ N}$$
$$\approx -300 \text{ N}$$

Here, the negative sign indicates that the net vertical force is in the downward direction.

Therefore, due to the shallowness of the angle between arm and muscle, Hercules can hold a dumbbell of an 86.0-newton weight at arm's length — if he doesn't mind a horizontal force on his arm socket of 1,790 newtons and a vertical force of 300 newtons.

Hanging a flag: A rotational equilibrium problem

The manager at the hardware store you work at asks you to help hang a flag over the top of the store. The store is extra-proud of the flag because it's an extra-big one (to check it out, see Figure 11-11). The problem is that the bolt holding the flagpole in place seems to break all the time, and both the flag and pole go hurtling over the edge of the building, which doesn't help the store's image.

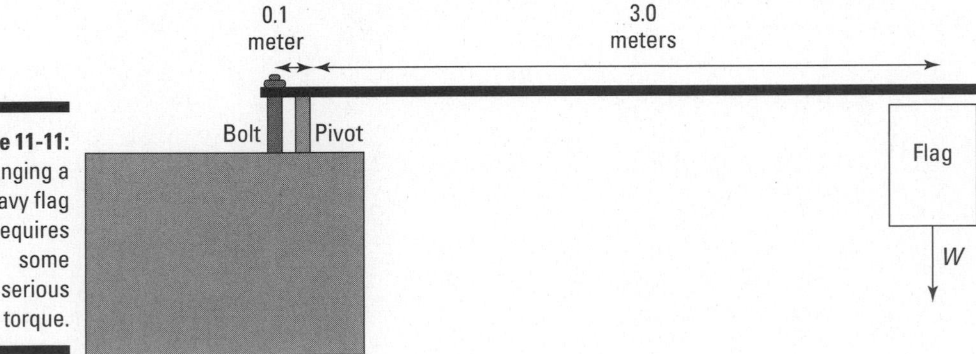

Figure 11-11: Hanging a heavy flag requires some serious torque.

To find out how much force the bolt needs to provide, you start taking measurements and note that the flag has a mass of 50 kilograms — much more than the mass of the pole, so you can neglect that. The manager had previously hung the flag 3.0 meters from the pivot point, and the bolt is 10 centimeters from the pivot point. To get rotational equilibrium, you need to have zero net torque:

$$\Sigma\tau = 0$$

In other words, if the torque due to the flag is τ_1 and the torque due to the bolt is τ_2, then the following is true:

$$0 = \tau_1 + \tau_2$$

What are the torques involved here? The direction of all the torque vectors is perpendicular to the plane of Figure 11-11, so consider only the component of these vectors in that direction (a positive component would correspond to a counterclockwise rotational force in Figure 11-11, and a negative component would correspond to a clockwise rotational force). Because you're dealing with the components of the vector, which are numbers (not directions), I don't write them in bold type. You know that the flag's weight provides a torque τ_1 around the pivot point, where

$$\tau_1 = mgl_1$$

where m is the mass of the pole, g is the acceleration due to gravity, and l_1 is the lever arm for the flag. Plugging in the numbers gives you the following:

$$\tau_1 = mgl_1 = (50 \text{ kg})(9.8 \text{ m/s}^2)(-3.0 \text{ N}) = -1{,}470 \text{ N·m}$$

Note that this is a negative torque because the lever arm is negative — the force causes a clockwise turning force, as Figure 11-11 shows. (You can check this mathematically: The angle between the force and the lever is $\theta = -90°$, so $l = r \sin\theta = (3.0 \text{ m}) \sin -90° = -3.0 \text{ m}$.) What about the torque τ_2 due to the bolt? As with any torque, you can write τ_2 as

$$\tau_2 = F_2 l_2$$

where F_2 is the magnitude of the force at the bolt.

Plugging in as many numbers as you know gives you

$$\tau_2 = F_2(0.10 \text{ m})$$

The lever arm is positive because the bolt provides a counterclockwise turning force (or mathematically, the angle between the force and the lever is $\theta = 90°$, so $l = r \sin\theta = (0.10 \text{ m}) \sin 90° = 0.10 \text{ m}$. Because you want rotational equilibrium, the following condition must hold:

$$\Sigma\tau = \tau_1 + \tau_2 = 0$$

In other words, the torques must balance out, so

$$\tau_2 = -\tau_1 = 1{,}470 \text{ N·m}$$

Now you can finally find F_2, because you know both τ_2 and l. Plug the known values into the equation $\tau_2 = F_2 l_2$ and solve for F_2:

$$\tau_2 = F_2 l_2 = F_2(0.10) = 1{,}470 \text{ N·m}$$

Putting F_2 on one side and solving the equation gives you

$$\tau_2 = F_2 l_2$$
$$1{,}470\ \text{N·m} = F_2(0.10\ \text{m})$$
$$F_2 = 14{,}700\ \text{N}$$

The bolt needs to provide at least 14,700 newtons of force, or about 330 pounds.

Ladder safety: Introducing friction into rotational equilibrium

A hardware store owner has come to you for help with another problem. A clerk has climbed near the top of a ladder to hang a sign for the company's upcoming sale. The owner doesn't want the ladder to slip — lawsuits, he explains — so he asks you whether the ladder is going to fall.

The situation appears in Figure 11-12. Here's the question: Will the force of friction keep the ladder from moving if θ is 45° and the static coefficient of friction (see Chapter 6) with the floor is 0.7?

You have to work with net forces to determine the overall torque. Write down what you know (you can assume that the weight of the ladder is concentrated at its middle and that you can neglect the force of friction of the ladder against the wall because the wall is very smooth):

- $\mathbf{F}_W$ = Force exerted by the wall on the ladder
- $\mathbf{F}_C$ = Weight of the clerk = 450 N
- $\mathbf{F}_L$ = Weight of the ladder = 200 N
- $\mathbf{F}_F$ = Force of friction holding the ladder in place
- $\mathbf{F}_N$ = Normal force (see Chapter 5)

You need to determine the needed force of friction here, and you want the ladder to be in both linear and rotational equilibrium. Linear equilibrium tells you that the force exerted by the wall on the ladder, $\mathbf{F}_W$, must be the same as the force of friction in magnitude but opposite in direction, because those are the only two horizontal forces. Therefore, if you can find $\mathbf{F}_W$, you know what the force of friction, $\mathbf{F}_F$, needs to be.

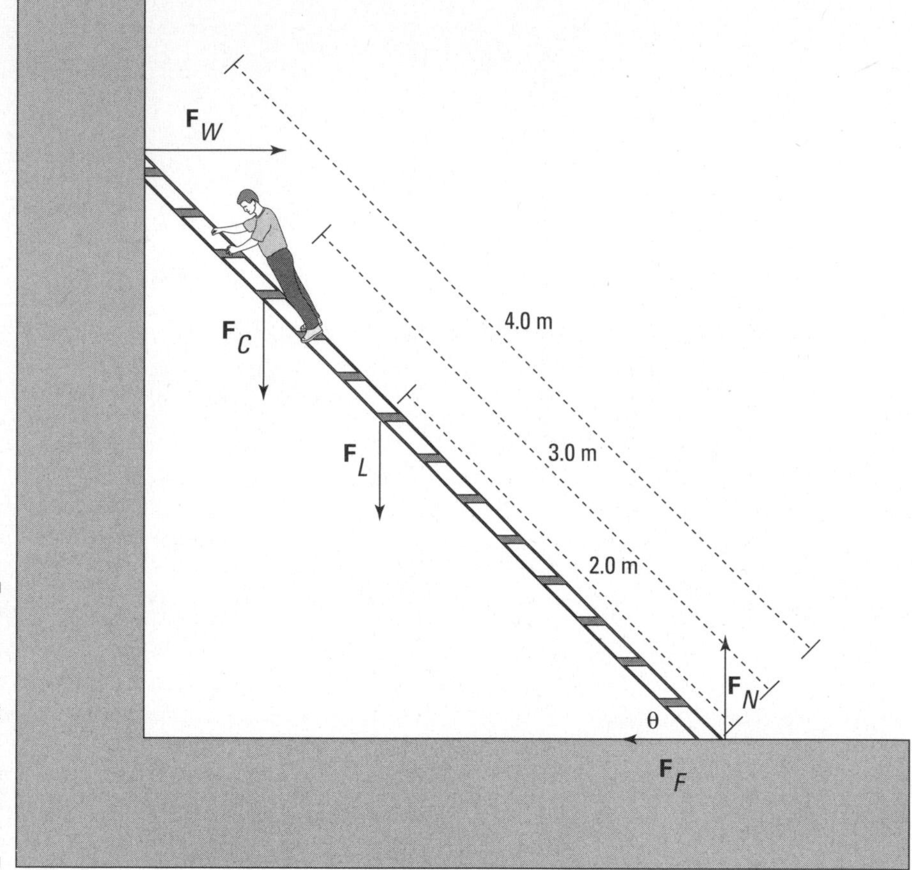

Figure 11-12: Keeping a ladder upright requires friction and rotational equilibrium.

You know that the ladder is in rotational equilibrium, which means that

$$\Sigma\tau = 0$$

To find $\mathbf{F}_W$, take a look at the torques around the bottom of the ladder, using that point as the pivot point. All the torques around the pivot point have to add up to zero. The direction of all the torque vectors is in the plane perpendicular to the one in Figure 11-12, so consider only the component of these vectors in that direction (a positive component would correspond to a counterclockwise rotational force in Figure 11-12, and a negative component would correspond to a clockwise rotational force). Because you're dealing with the components of the vector, which are numbers, I don't write them in bold type.

Here's how to find the three torques around the bottom of the ladder:

- **Torque due to the force from the wall against the ladder:** Here, r is the full length of the ladder:

  $$F_W(4.0 \text{ m}) \sin -45° = (-2.83 \text{ m})F_W$$

  Note that the torque due to the force from the wall is negative because it tends to produce a clockwise motion.

- **Torque due to the clerk's weight:** In this case, r is 3.0 meters, the distance from the bottom of the ladder to the clerk's location:

  $$F_C(3.0 \text{ m}) \sin 45° = (450 \text{ N})(3.0 \text{ m}) \sin 45° \approx 954 \text{ N·m}$$

- **Torque due to the ladder's weight:** You can assume that the ladder's weight is concentrated in the middle of the ladder, so $r = 2.0$ meters, half the total length of the ladder. Therefore, the torque due to the ladder's weight is

  $$F_L(2.0 \text{ m}) \sin 45° = (200 \text{ N})(2.0 \text{ m}) \sin 45° \approx 283 \text{ N·m}$$

  These last two torques are positive because the lever arms are positive, and therefore the forces generate a counterclockwise turning force, as Figure 11-12 shows.

Now, because $\Sigma\tau = 0$, you get the following result when you add all the torques together:

$$\Sigma\tau = 954 \text{ N·m} + 283 \text{ N·m} - (2.83 \text{ m})F_W$$

$$0 = 1{,}237 \text{ N·m} - (2.83 \text{ m})F_W$$

$$(2.83 \text{ m})F_W = 1{,}237 \text{ N·m}$$

$$F_W \approx 437 \text{ N}$$

The force the wall exerts on the ladder is 437 newtons, which is also equal to the frictional force of the bottom of the ladder on the floor, because F_W and the frictional force are the only two horizontal forces in the whole system. Therefore,

$$F_F = 437 \text{ N}$$

You know the force of friction that you need. But how much friction do you actually have? The basic equation for friction (as outlined in Chapter 6) tells you that

$$F_{F\,actual} = \mu_s F_N$$

where μ_s is the coefficient of static friction and F_N is the normal force of the floor pushing up on the ladder, which must balance all the downward-pointing forces in this problem because of linear equilibrium. This means that

$$F_N = W_C + W_L = 450 \text{ N} + 200 \text{ N} = 650 \text{ N}$$

Plugging this into the equation for $F_{F\,actual}$ and using the value of μ_s, 0.700, gets you the following:

$$F_{F\,actual} = \mu_s F_N = (0.700)(650) = 455 \text{ N}$$

You need 437 newtons of force, and you actually have 455 newtons. Good news — the ladder isn't going to slip.

Chapter 12

Round and Round with Rotational Dynamics

In This Chapter

- Converting Newton's linear thinking into rotational thinking
- Utilizing the moment of inertia
- Finding the angular equivalent of work
- Looking at the rotational kinetic energy caused by work
- Conserving angular momentum

This chapter is all about applying forces and seeing what happens in the rotational world. You find out what Newton's second law (force equals mass times acceleration) becomes for rotational motion, you see how inertia comes into play in rotational motion, and you get the story on rotational kinetic energy, rotational work, and angular momentum.

Rolling Up Newton's Second Law into Angular Motion

Newton's second law, force equals mass times acceleration ($\mathbf{F} = m\mathbf{a}$; see Chapter 5), is a physics favorite in the linear world because it ties together the vector's force and acceleration. But if you have to talk in terms of angular kinetics rather than linear motion, what happens? Can you get Newton spinning?

Angular kinetics has equivalents (or *analogs*) for linear equations (see Chapter 11). So what's the angular analog for $\mathbf{F} = m\mathbf{a}$? You may guess that **F**, the linear force, becomes $\boldsymbol{\tau}$. And you may also guess that **a**, linear acceleration, becomes $\boldsymbol{\alpha}$, angular acceleration. But what the heck is the angular analog of *m*, mass? The answer is *rotational inertia, I,* and you come to this answer by converting tangential acceleration to angular acceleration. As I show you in this section, your final formula is $\Sigma\boldsymbol{\tau} = I\boldsymbol{\alpha}$, the angular form of Newton's second law.

Switching force to torque

You can start the linear-to-angular conversion process with a simple example. Say that you're whirling a ball in a circle on the end of a string, as in Figure 12-1. You apply a tangential force (along the circle) to the ball, making it speed up (keep in mind that this force is not directed toward the center of the circle, as when you have a centripetal force; see Chapter 11). You want to write Newton's second law in terms of torque rather than force.

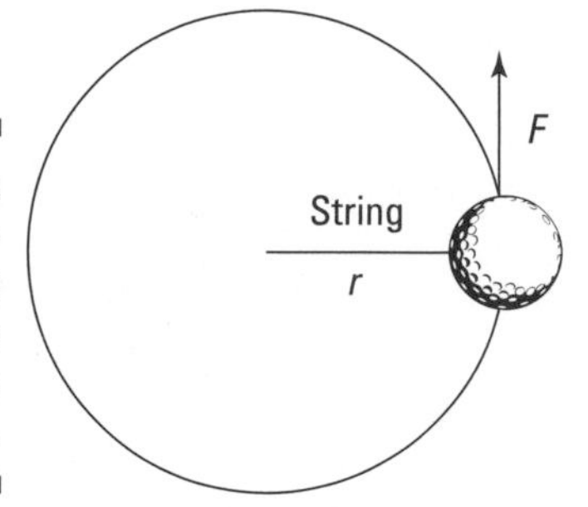

Figure 12-1: A tangential force applied to a ball on a string.

Start by working with only the magnitudes of the vector quantities, saying that

$$F = ma$$

To put this equation terms of angular quantities, such as torque, multiply by the radius of the circle, r (see Chapter 11 for details on the relationship between angular and linear quantities):

$$Fr = mra$$

Because you're applying tangential force to the ball, the force and the circle's radius are at right angles (see Figure 12-1), so you can replace Fr with torque:

$$\tau = mra$$

You're now partly done making the transition to rotational motion. Instead of working with linear force, you're working with torque, which is linear force's rotational analog.

Converting tangential acceleration to angular acceleration

To move from linear motion to angular motion, you have to convert a, tangential acceleration, to α, angular acceleration. Great, but how do you make the conversion? You can multiply angular acceleration by the radius to get the linear equivalent, which is the magnitude of the tangential acceleration (see Chapter 11): $a = r\alpha$. Substitute $r\alpha$ for a in the equation for the angular equivalent of Newton's second law, $\tau = mra$:

$$\tau = mr(r\alpha) = mr^2\alpha$$

Now you've related the magnitude of the torque to magnitude of the angular acceleration. The direction of the angular acceleration and the torque turns out to be the same, so this equation is true for vectors, too:

$$\boldsymbol{\tau} = mr^2\boldsymbol{\alpha}$$

Factoring in the moment of inertia

To go from linear force, $\mathbf{F} = m\mathbf{a}$, to torque (linear force's angular equivalent), you have to find the angular equivalent of acceleration and mass. In the preceding section, you find angular acceleration, giving you the equation $\boldsymbol{\tau} = mr^2\boldsymbol{\alpha}$.

In this equation, mr^2 is the rotational analog of mass, officially called the *moment of inertia* (sometimes referred to as the *rotational inertia*). The *moment of inertia* is a measure of how resistant an object is to changes in its rotational motion.

In physics, the symbol for inertia is I, so you can write the equation for torque as follows:

$$\Sigma\boldsymbol{\tau} = I\boldsymbol{\alpha}$$

The symbol Σ means *sum of*, so $\Sigma\boldsymbol{\tau}$ means *net torque*. The units of moment of inertia are kilogram-square meters ($kg \cdot m^2$). Note how close the torque equation is to the equation for net force, which follows:

$$\Sigma\mathbf{F} = m\mathbf{a}$$

$\Sigma\boldsymbol{\tau} = I\boldsymbol{\alpha}$ is the angular form of Newton's second law for rotating bodies: net torque equals moment of inertia multiplied by angular acceleration.

Now you can put the equation to work. Say, for example, that you're whirling the 45-gram ball from Figure 12-1 in a 1.0-meter circle, and you want to speed it up at a rate of 2π radians per second^2. What magnitude of torque do you need? You know that

$$\tau = I\alpha$$

You can drop the symbol Σ from the angular version of the equation for Newton's second law when you're dealing with only one torque. The "sum of" the torques is the value of the only torque you're dealing with.

The moment of inertia equals mr^2, so

$$\tau = I\alpha = mr^2\alpha$$

Plugging in the numbers (after converting grams to kilograms) gives you

$$\tau = mr^2\alpha = (0.045 \text{ kg})(1.0 \text{ m})^2(2\pi \text{ s}^{-1}) = 9.0\pi \times 10^{-2} \text{ N}\cdot\text{m}$$

Your answer, $9.0\pi \times 10^{-2}$ N·m, is about 0.28 newton-meters of torque. Solving for the torque required in angular motion is much like being given a mass and a required acceleration and solving for the needed force in linear motion.

Moments of Inertia: Looking into Mass Distribution

The moment of inertia depends not only on the mass of the object but also on how the mass is distributed. For example, if two disks have the same mass but one has all the mass around the rim and one is solid, then the disks would have different moments of inertia.

Calculating moments of inertia is fairly simple if you only have to examine the orbital motion of small point-like objects, where all the mass is concentrated at one particular point at a given radius *r*. For instance, for a golf ball you're whirling around on a string, the moment of inertia depends on the radius of the circle the ball is spinning in:

$$I = mr^2$$

Here, *r* is the radius of the circle, from the center of rotation to the point at which all the mass of the golf ball is concentrated.

Crunching the numbers can get a little sticky when you enter the non–golf ball world, however, because you may not be sure of which radius to use. What if you're spinning a rod around? All the mass of the rod isn't concentrated at a single radius. When you have an extended object, such as a rod, each bit of mass is at a different radius. You don't have an easy way to deal with this, so you have to sum up the contribution of each particle of mass at each different radius like this:

$$I = \Sigma mr^2$$

You can use this concept of adding up the moments of inertia of all the elements to get the total in order to work out the moment of inertia of any distribution of mass. Here's an example using two point masses, which is a bit more complex than a single point mass. Say you have two golf balls, and you want to know what their combined moment of inertia is. If you have a golf ball at radius r_1 and another at r_2, the total moment of inertia is

$$I = \Sigma mr^2 = m(r_1^2 + r_2^2)$$

So how do you find the moment of inertia of, say, a disk rotating around an axis stuck through its center? You have to break the disk up into tiny balls and add them all up. Trusty physicists have already completed this task for many standard shapes; I provide a list of objects you're likely to encounter, and their moments of inertia, in Table 12-1. Figure 12-2 depicts the shapes that these moments of inertia correspond to.

Table 12-1 Moments of Inertia for Various Shapes and Solids

Shape	*Moment of Inertia*
(a) Solid cylinder or disk of radius *r*	$I = \frac{1}{2}mr^2$
(b) Hollow cylinder of radius *r*	$I = mr^2$
(c) Solid sphere of radius *r*	$I = \frac{2}{5}mr^2$
(d) Hollow sphere of radius *r*	$I = \frac{2}{3}mr^2$
(e) Rectangle rotating around an axis along one edge, where the other edge has length *r*	$I = \frac{1}{3}mr^2$
(f) Rectangle with sides r_1 and r_2 rotating around a perpendicular axis through the center	$I = \left(\frac{1}{12}\right)m\left(r_1^2 + r_2^2\right)$
(g) Thin rod of length *r* rotating about its middle	$I = \frac{1}{12}mr^2$
(h) Thin rod of length *r* rotating about one end	$I = \frac{1}{3}mr^2$

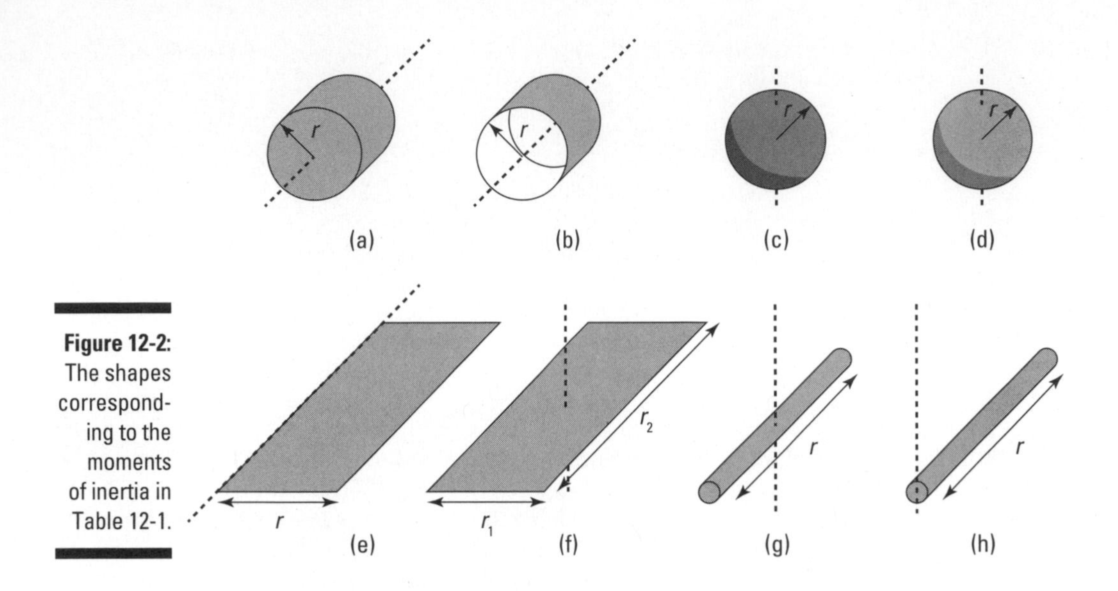

Figure 12-2: The shapes corresponding to the moments of inertia in Table 12-1.

Check out the following examples to see advanced moments of inertia in action.

DVD players and torque: A spinning-disk inertia example

Here's an interesting fact about DVD players: They actually change the angular speed of the DVD to keep the section of the DVD under the laser head moving at constant linear speed.

Say that a DVD has a mass of 30 grams and a diameter of 12 centimeters. It starts at 700 revolutions per second when you first hit *play* and winds down to about 200 revolutions per second at the end of the DVD 50 minutes later. What's the average torque needed to create this acceleration? You start with the torque equation:

$$\tau = I\alpha$$

A DVD is a disk shape rotating around its center, so from Table 12-1, you know that its moment of inertia is

$$I = \frac{1}{2}mr^2$$

The diameter of the DVD is 12 centimeters, so the radius is 6.0 centimeters. Putting in the numbers gives you the moment of inertia:

$$I = \frac{1}{2}mr^2 = \frac{1}{2}(0.030 \text{ kg})(0.060 \text{ m})^2 = 5.4 \times 10^{-5} \text{ kg} \cdot \text{m}^2$$

How about the angular acceleration, α? Here's the angular equivalent of the equation for linear acceleration (see Chapter 11 for details):

$$\alpha = \frac{\Delta \omega}{\Delta t}$$

But because the angular velocity always stays along the same axis, you can consider just the components of the angular velocity and angular acceleration along this axis. They are then related by

$$\alpha = \frac{\Delta \omega}{\Delta t}$$

The time, Δt, is 50 minutes, or 3,000 seconds. So what about $\Delta\omega$ (which equals $\omega_f - \omega_i$)? First, you need to express angular velocity in radians per second, not revolutions per second. You know that the initial angular velocity is 700 revolutions per second, so in terms of radians per second, you get

$$\omega_i = \frac{700 \text{ revolutions}}{1 \text{ s}} \times \frac{2\pi \text{ rad}}{1 \text{ revolution}} = 1{,}400\pi \text{ rad/s}$$

Similarly, you can get the final angular velocity this way:

$$\omega_f = \frac{200 \text{ revolutions}}{1 \text{ s}} \times \frac{2\pi \text{ rad}}{1 \text{ revolution}} = 400\pi \text{ rad/s}$$

Now you can plug the angular velocities and time into the angular acceleration formula:

$$\begin{aligned} \alpha &= \frac{\Delta \omega}{\Delta t} \\ &= \frac{(w_f - w_i)}{\Delta t} \\ &= \frac{(400\pi - 1{,}400\pi)\text{rad/s}}{3{,}000 \text{ s}} \\ &= \frac{-1{,}000\pi \text{ rad}}{3{,}000 \text{ s}^2} \\ &\approx -1.047 \text{ rad/s}^2 \end{aligned}$$

The angular acceleration is negative because the disk is slowing down. As previously defined, the component of the angular velocity along the axis of rotation is positive. The negative acceleration then leads to a reduction in this angular velocity.

You've found the moment of inertia and the angular acceleration, so now you can plug those values into the torque equation:

$$\tau = I\alpha = (5.4 \times 10^{-5}\ \text{kg·m}^2)(-1.047\ \text{s}^{-2}) \approx -5.65 \times 10^{-5}\ \text{N·m}$$

The average torque is -5.65×10^{-5} N·m. To get an impression of how easy or difficult this torque may be to achieve, you may ask how much force is this when applied to the outer edge — that is, at a 6-centimeter radius. Torque is force times the radius, so

$$F = \frac{\tau}{r} = \frac{-5.65 \times 10^{-5}\ \text{N} \cdot \text{m}}{0.06\ \text{m}} \approx 9 \times 10^{-4}\ \text{N}$$

This converts to about 2×10^{-4} pounds, or about 3×10^{-3} ounces of force. Slowing down the DVD doesn't take much force.

Angular acceleration and torque: A pulley inertia example

You may not always look at an object in motion and think "angular motion," as you do when looking at a spinning DVD. Take someone lifting an object with a rope on a pulley system, for example. The rope and the object are moving in a linear fashion, but the pulley has angular motion.

Say that you're using a pulley that has a mass of 1 kilogram and a radius of 10 centimeters to pull a 16-kilogram mass vertically (see Figure 12-3). You apply a force of 200 newtons. What's the angular acceleration of the pulley?

You use the equation for torque, including the sum symbol, Σ, because you're dealing with more than one torque in this problem (you always use the net torque in a problem, but many problems only have one torque, so the symbol drops off):

$$\Sigma\boldsymbol{\tau} = I\alpha$$

where $\Sigma\boldsymbol{\tau}$ means net torque.

In this case, there are two torques, $\boldsymbol{\tau}_1$ and $\boldsymbol{\tau}_2$. The direction of these torque vectors is perpendicular to the plane of Figure 12-3. Consider the component of the torque vectors in this direction, written τ_1 and τ_2, such that a positive value corresponds to a clockwise rotation.

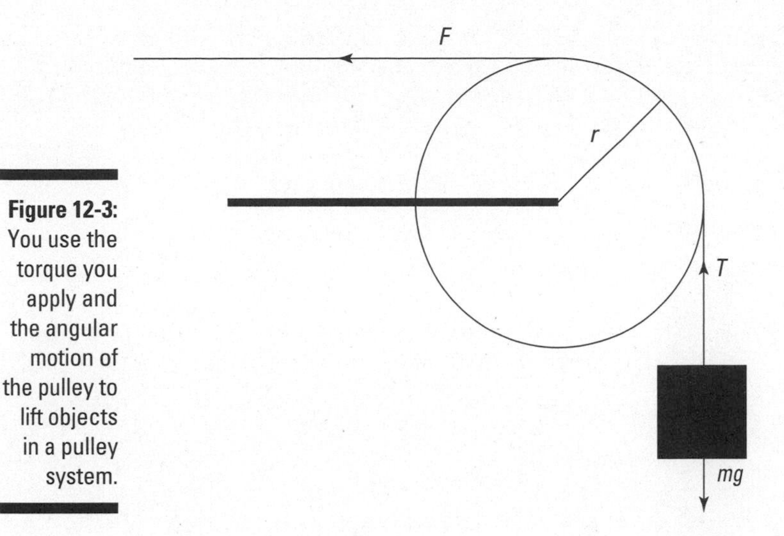

Figure 12-3: You use the torque you apply and the angular motion of the pulley to lift objects in a pulley system.

Solve the torque equation for angular acceleration, α, and write $\Sigma\tau$ as the sum of τ_1 and τ_2:

$$\alpha = \frac{\Sigma\tau}{I} = \frac{(\tau_1 + \tau_2)}{I}$$

where α is the component of the pulley's angular acceleration and τ is the torque on the pulley in the direction perpendicular to the plane of Figure 12-3.

First concentrate on the torques. The two forces act at radius of 10.0 centimeters, so the two torques are

- $\tau_1 = Fr$, with F as the force and r as the radius of the pulley
- $\tau_2 = -Tr$, where T is the tension in the rope between the mass m and the pulley

The pulley's support goes through the axis of rotation, so no torque comes from it.

You need to work out the tension in the rope, T, which is providing torque τ_2. The forces acting on the 16-kilogram mass, m, are its weight acting downward and the tension in the rope acting upward, so you can use Newton's second law to write the following:

$$-mg + T = ma$$

where a is the acceleration of the mass m. You want to find tension, so solve for T:

$$T = ma + mg$$

Because the string does not stretch, the acceleration of the mass m must be equal to the tangential acceleration of the edge of the pulley wheel. The tangential acceleration is related to the linear acceleration by $a = r\alpha$ (see Chapter 11 for details), so you can replace a to write the tension in the rope as

$$\begin{aligned} T &= ma + mg \\ &= m(r\alpha) + mg \\ &= m(r\alpha + g) \end{aligned}$$

Knowing the tension allows you to find τ_2, which equals $-Tr$. You know that τ_1 equals Fr, so you can work out the total torque acting on the pulley wheel:

$$\begin{aligned} \tau_1 + \tau_2 &= Fr - Tr \\ &= Fr - m(r\alpha + g)r \end{aligned}$$

If you consider the rotating part of the pulley to be a circular disk of radius r and mass M, then you can use Table 12-1 to look up the moment of inertia, which is $I = (1/2)\,Mr^2$. Because the total torque is equal to the moment of inertia times the angular acceleration, you can write the following:

$$Fr - m(r\alpha + g)r = \frac{1}{2}Mr^2\alpha$$

Then you can rearrange this equation to give the angular acceleration, like so:

$$\alpha = \frac{F - mg}{\frac{1}{2}(M + 2m)r}$$

Plugging in the numbers gives you the answer:

$$\begin{aligned} \alpha &= \frac{(200\text{ N}) - (16\text{ kg})(9.8\text{ m/s}^2)}{\frac{1}{2}(1.0\text{ kg} + 2(16\text{ kg}))(0.10\text{ m})} \\ &\approx 26\text{ rad/s}^2 \end{aligned}$$

So the angular acceleration is 26 radians per second^2, which is about 4 revolutions per second.

Wrapping Your Head around Rotational Work and Kinetic Energy

One major player in the linear-force game is *work* (see Chapter 9); the equation for work is work equals force times distance, or $W = Fs$. Work has a rotational analog. To relate a linear force acting for a certain distance with the idea of rotational work, you convert force to torque (its angular equivalent) and distance to angle. I show you how to derive the rotational-work equation in this section. I also show you what happens when you do work by turning an object, creating rotational motion — your work goes to increasing the kinetic energy.

Putting a new spin on work

When force moves an object through a distance, work is done on the object (refer to Chapter 9). Similarly, when a torque rotates an object through an angle, work is done. In this section, you work out how much work is done when you rotate a wheel by pulling a string attached to the wheel's outside edge (see Figure 12-4).

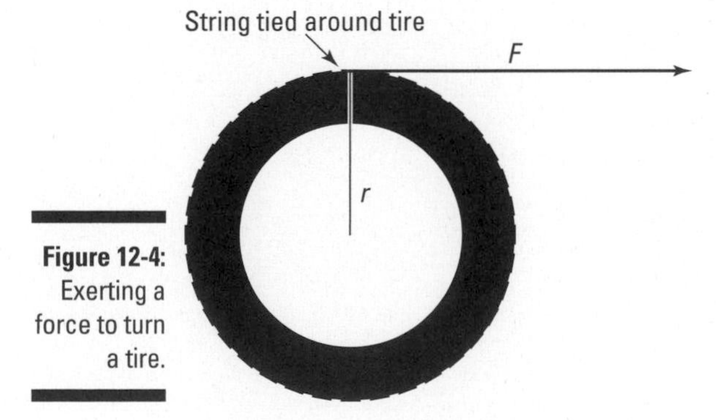

Figure 12-4: Exerting a force to turn a tire.

Work is the amount of force applied to an object multiplied by the distance it's applied. In this case, a force F is applied with the string. Bingo! The string lets you make the handy transition between linear and rotational work. So how much work is done? Use the following equation:

$$W = Fs$$

where s is the distance the person pulling the string applies the force over. In this case, the distance s equals the radius multiplied by the angle through which the wheel turns, $s = r\theta$, so you get

$$W = Fr\theta$$

However, the torque, τ, equals Fr in this case, because the string is acting at right angles to the radius (see Chapter 11). So you're left with

$$W = \tau\theta$$

When the string is pulled, applying a constant torque that turns the wheel, the work done equals $\tau\theta$. This makes sense, because linear work is Fs, and to convert to rotational work, you convert from force to torque and from distance to angle. The units here are the standard units for work — joules in the MKS system.

You have to give the angle in radians for the conversion between linear work and rotational work to come out right.

Say that you have a plane that uses propellers, and you want to determine how much work the plane's engine does on a propeller when applying a constant torque of 600 newton-meters over 100 revolutions. You start with the work equation in terms of torque:

$$W = \tau\theta$$

A full revolution is 2π radians, so θ equals 2π times 100, the number of revolutions. Plugging the numbers into the equation gives you the work:

$$W = \tau\theta = (600\ \text{N}\cdot\text{m})(100 \times 2\pi) \approx 3.77 \times 10^5\ \text{J}$$

The plane's engine does 3.77×10^5 joules of work.

Moving along with rotational kinetic energy

If you put a lot of work into turning an object, the object starts spinning. And when an object is spinning, all its pieces are moving, which means that it has kinetic energy. For spinning objects, you have to convert from the linear concept of kinetic energy to the rotational concept of kinetic energy.

You can calculate the kinetic energy of a body in linear motion with the following equation (see Chapter 9):

$$KE = \frac{1}{2}mv^2$$

where m is the mass of the object and v is the speed. This formula applies to every bit of the object that's rotating — each bit of mass has this kinetic energy.

To go from the linear version to the rotational version, you have to go from mass to moment of inertia, I, and from velocity to angular velocity, ω. You can tie an object's tangential speed to its angular speed like this (see Chapter 11):

$$v = r\omega$$

where r is the radius and ω is its angular speed. Plugging v's equivalent into the kinetic-energy equation gives you the following:

$$KE = \frac{1}{2}mv^2 = \frac{1}{2}m\left(r^2\omega^2\right)$$

The equation looks okay so far, but it holds true only for the one single bit of mass under discussion — each other bit of mass may have a different radius, so you're not finished. You have to sum up the kinetic energy of every bit of mass like this:

$$KE = \frac{1}{2}\Sigma\left(mr^2\omega^2\right)$$

You can simplify this equation. Start by noticing that even though each bit of mass may be different and be at a different radius, each bit has the same angular speed (they all turn through the same angle in the same time). Therefore, you can take the ω out of the summation:

$$KE = \frac{1}{2}\Sigma\left(mr^2\right)\omega^2$$

This makes the equation much simpler, because $\Sigma(mr^2)$ equals the moment of inertia, I (see the section "Rolling Up Newton's Second Law into Angular Motion," earlier in this chapter). Making this substitution takes all the dependencies on the individual radius of each bit of mass out of the equation, giving you

$$KE = \frac{1}{2}I\omega^2$$

Now you have a simplified equation for rotational kinetic energy. The equation proves useful because rotational kinetic energy is everywhere. A satellite spinning around in space has rotational kinetic energy. A barrel of beer rolling down a ramp from a truck has rotational kinetic energy. The latter example (not always with beer trucks, of course) is a common thread in physics problems.

Let's roll! Finding rotational kinetic energy on a ramp

Objects can have both linear and rotational kinetic energy. This fact is an important one, if you think about it, because when objects start rolling down ramps, any previous ramp expertise you have goes out the window. Why? Because when an object rolls down a ramp instead of sliding, some of its gravitational potential energy (see Chapter 9) goes into its linear kinetic energy, and some of it goes into its rotational kinetic energy.

Look at Figure 12-5, where you're pitting a solid cylinder against a hollow cylinder in a race down the ramp. Each object has the same mass. Which cylinder is going to win? In other words, which cylinder will have the higher speed at the bottom of the ramp? When looking only at linear motion, you can handle a problem like this by setting the potential energy equal to the final kinetic energy (assuming no friction!) like this:

$$PE = KE$$

$$mgh = \frac{1}{2}mv^2$$

where m is the mass of the object, g is the acceleration due to gravity, and h is the height at the top of the ramp. This equation would let you solve for the final speed.

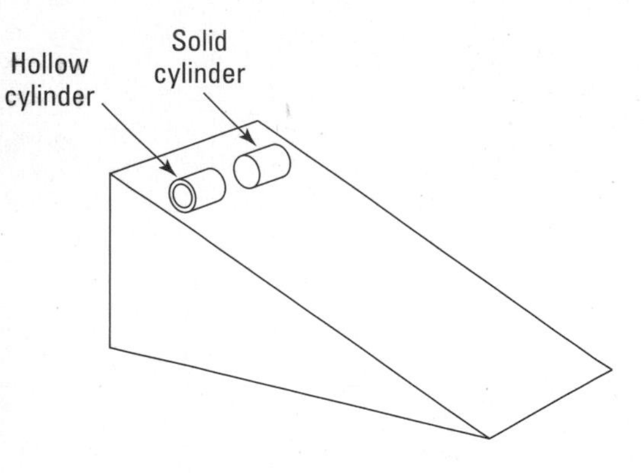

Figure 12-5: A solid cylinder and a hollow cylinder ready to race down a ramp.

But the cylinders are rolling in this case, which means that the initial gravitational potential energy becomes *both* linear kinetic energy and rotational kinetic energy. You can now write the equation as

$$mgh = \frac{1}{2}mv^2 + \frac{1}{2}I\omega^2$$

You can relate v and ω together with the equation $v = r\omega$, which means that $\omega = v/r$, so

$$mgh = \frac{1}{2}mv^2 + \frac{1}{2}I\left(\frac{v^2}{r^2}\right)$$

You want to solve for v, so try grouping things together. You can factor $(1/2)v^2$ out of the two terms on the right:

$$mgh = \frac{1}{2}mv^2 + \frac{1}{2}I\left(\frac{v^2}{r^2}\right)$$

$$mgh = \frac{1}{2}\left(m + \frac{I}{r^2}\right)v^2$$

Isolating v, you get the following:

$$v = \sqrt{\frac{2mgh}{m + I/r^2}}$$

For the hollow cylinder, the moment of inertia equals mr^2, as you can see in Table 12-1. For a solid cylinder, on the other hand, the moment of inertia equals $(1/2)mr^2$. Substituting for I for the hollow cylinder gives you the hollow cylinder's velocity:

$$v = \sqrt{gh}$$

Substituting for I for the solid cylinder gives you the solid cylinder's velocity:

$$v = \sqrt{\frac{4gh}{3}}$$

Now the answer becomes clear. The solid cylinder will be going $\sqrt{\frac{4}{3}}$ times as fast as the hollow cylinder, or about 1.15 times as fast, so the solid cylinder will win.

The hollow cylinder has as much mass concentrated at a large radius as the solid cylinder has distributed from the center all the way out to that radius, so this answer makes sense. With that large mass way out at the edge, the hollow cylinder doesn't need to go as fast to have as much rotational kinetic energy as the solid cylinder.

Can't Stop This: Angular Momentum

Picture a small child on a spinning playground ride, such as a merry-go-round, and she's yelling that she wants to get off. You have to stop the spinning ride, but it's going to take some effort. Why? Because it has *angular momentum.*

Linear momentum, **p,** is defined as the product of mass and velocity:

$$\mathbf{p} = m\mathbf{v}$$

This is a quantity that is conserved when there are no external forces acting. The more massive and faster moving an object, the greater the magnitude of momentum.

Physics also features angular momentum, **L.** The equation for angular momentum looks like this:

$$\mathbf{L} = I\boldsymbol{\omega}$$

where I is the moment of inertia and $\boldsymbol{\omega}$ is the angular velocity.

Note that angular momentum is a vector quantity, meaning it has a magnitude and a direction. The vector points in the same direction as the $\boldsymbol{\omega}$ vector (that is, in the direction the thumb of your right hand points when you wrap your fingers around in the direction the object is turning).

The units of angular momentum are I multiplied by the units of $\boldsymbol{\omega}$, or $kg{\cdot}m^2/s$ in the MKS system.

The important idea about angular momentum, much as with linear momentum, is that it's conserved.

Conserving angular momentum

The *principle of conservation of angular momentum* states that angular momentum is conserved if no net torques are involved.

This principle comes in handy in all sorts of problems, such as when two ice skaters start off holding each other close while spinning but then end up at arm's length. Given their initial angular velocity, you can find their final angular velocity, because angular momentum is conserved:

$$I_1\omega_1 = I_2\omega_2$$

If you can find the initial moment of inertia and the final moment of inertia, you're set. But you also come across less obvious cases where the principle of conservation of angular momentum helps out. For example, satellites don't have to travel in circular orbits; they can travel in ellipses. And when they do, the math can get a lot more complicated. Lucky for you, the principle of conservation of angular momentum can make the problems simple.

Satellite orbits: A conservation-of-angular-momentum example

Say that NASA planned to put a satellite into a circular orbit around Pluto for studies, but the situation got a little out of hand and the satellite ended up with an elliptical orbit. At its nearest point to Pluto, 6.0×10^6 meters, the satellite zips along at 9,000 meters per second.

The satellite's farthest point from Pluto is 2.0×10^7 meters. What's its speed at that point? The answer is tough to figure out unless you can come up with an angle here, and that angle is angular momentum.

Angular momentum is conserved because there are no external torques the satellite must deal with (gravity always acts perpendicular to the orbital radius). Because angular momentum is conserved, you can say that

$$I_1\omega_1 = I_2\omega_2$$

Because the satellite is so small compared to the radius of its orbit at any location, you can consider the satellite a point mass. Therefore, the moment of inertia, I, equals mr^2 (refer to the earlier section "Factoring in the moment of inertia"). The magnitude of the angular velocity equals v/r, so you can express the conservation of angular momentum in terms of the velocity like so:

$$I_1\omega_1 = I_2\omega_2$$

$$mr_1v_1 = mr_2v_2$$

You can put v_2 on one side of the equation by dividing by mr_2:

$$v_2 = \frac{r_1v_1}{r_2}$$

You have your solution; no fancy math involved at all, because you can rely on the principle of conservation of angular momentum to do the work for you. All you need to do is plug in the numbers:

$$v_2 = \frac{r_1v_1}{r_2} = \frac{(6.0\times10^6 \text{ m})(9{,}000 \text{ m/s})}{2.0\times10^7 \text{ m}} = 2{,}700 \text{ m/s}$$

At its closest point to Pluto, the satellite will be screaming along at 9,000 meters per second, and at its farthest point, it will be moving at 2,700 meters per second. Easy enough to figure out, as long as you have the principle of conservation of angular momentum under your belt.

Chapter 13

Springs 'n' Things: Simple Harmonic Motion

In This Chapter

- Understanding force when you stretch or compress springs
- Going over the basics of simple harmonic motion
- Mustering up the energy for simple harmonic motion
- Predicting a pendulum's motion and period

In this chapter, I shake things up with a new kind of motion: periodic motion, which occurs when objects are bouncing around on springs or bungee cords or are swooping on the end of a pendulum. This chapter is all about describing their motion. Not only can you describe their motion in detail, but you can also predict how much energy bunched-up springs have, how long a pendulum will take to swing back and forth, and more.

Bouncing Back with Hooke's Law

Objects that can stretch but return to their original shapes are called *elastic.* Elasticity is a valuable property, because it allows you to use objects such as springs for all kinds of applications: as shock absorbers in lunar landing modules, as timekeepers in clocks and watches, and even as hammers of justice in mousetraps.

In this section, I introduce Hooke's law, which relates forces to how much a spring is stretched or compressed.

Stretching and compressing springs

Robert Hooke, a physicist from England, undertook the study of elastic materials in the 1600s. He discovered a new law, not surprisingly called Hooke's law, which states that stretching or compressing an elastic material requires a force that's directly proportional to the amount of stretching or compressing you do. For example, to stretch a spring a distance x, you need to apply a force that's directly proportional to x:

$$F_a = kx$$

Here, F_a and x are the components of the applied force and displacement along the direction of the spring, such that

- Positive values correspond to stretching
- Negative values correspond to compression

The constant k is called the *spring constant,* and its units are newtons per meter (N/m).

Pushing or pulling back: The spring's restoring force

In accordance with Newton's third law, if an object applies a force to a spring, then the spring applies an equal and opposite force to the object. Hooke's law gives the force a spring exerts on an object attached to it with the following equation:

$$F = -kx$$

where the minus sign shows that this force is in the opposite direction of the force that's stretching or compressing the spring (see the preceding section for info on forces on springs).

The force exerted by a spring is called a *restoring force;* it always acts to restore the spring toward equilibrium. In Hooke's law, the negative sign on the spring's force means that the force exerted by the spring opposes the spring's displacement.

Figure 13-1 shows a ball attached to a spring. You can see that if the spring isn't stretched or compressed, it exerts no force on the ball. If you push the spring, however, it pushes back, and if you pull the spring, it pulls back.

Hooke's law is valid as long as the elastic material you're dealing with stays elastic — that is, it stays within its *elastic limit.* If you pull a spring too far, it loses its stretchy ability. As long as a spring stays within its elastic limit, you can say that $F = -kx$. When a spring stays within its elastic limit and obeys Hooke's law, the spring is called an *ideal spring.*

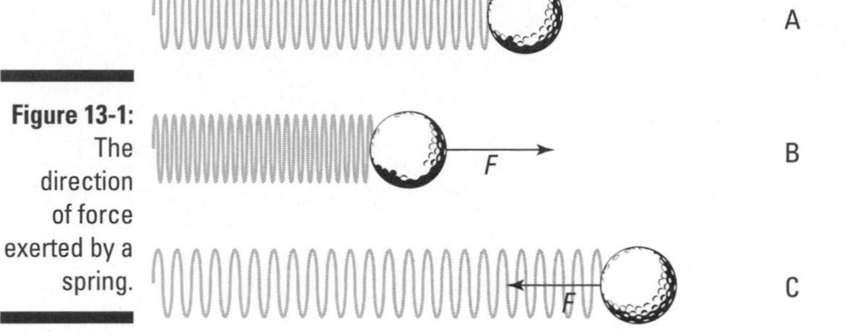

Figure 13-1: The direction of force exerted by a spring.

Suppose that a group of car designers knocks on your door and asks whether you can help design a suspension system. "Sure," you say. They inform you that the car will have a mass of 1,000 kilograms, and you have four shock absorbers, each 0.5 meters long, to work with. How strong do the springs have to be? Assuming these shock absorbers use springs, each one has to support a mass of at least 250 kilograms, which weighs the following:

$$F = mg = (250\text{ kg})(9.8\text{ m/s}^2) = 2{,}450\text{ N}$$

where F equals force, m equals the mass of the object, and g equals the acceleration due to gravity, 9.8 meters per second^2. The spring in the shock absorber will, at a minimum, have to give you 2,450 newtons of force at the maximum compression of 0.5 meters. What does this mean the spring constant should be? Hooke's law says

$$F = -kx$$

Looking only at the magnitudes and therefore omitting the negative sign (look for its return in the following section), you get

$$k = \frac{F}{x}$$

Time to plug in the numbers:

$$k = \frac{F}{x} = \frac{2{,}450\text{ N}}{0.5\text{ m}} = 4{,}900\text{ N/m}$$

The springs used in the shock absorbers must have spring constants of at least 4,900 newtons per meter. The car designers rush out, ecstatic, but you call after them, "Don't forget, you need to at least double that if you actually want your car to be able to handle potholes."

Getting Around to Simple Harmonic Motion

An *oscillatory motion* is one that undergoes repeated cycles. When the net force acting on an object is elastic, the object undergoes a simple oscillatory motion called *simple harmonic motion*. The force that tries to restore the object to its resting position is proportional to the displacement of the object. In other words, it obeys Hooke's law.

Elastic forces suggest that the motion will just keep repeating (that isn't really true, however; even objects on springs quiet down after a while as friction and heat loss in the spring take their toll). This section delves into simple harmonic motion and shows you how it relates to circular motion. Here, you graph motion with the sine wave and explore familiar concepts such as position, velocity, and acceleration.

Around equilibrium: Examining horizontal and vertical springs

Take a look at the golf ball in Figure 13-1. The ball is attached to a spring on a frictionless horizontal surface. Say that you push the ball, compressing the spring, and then you let go; the ball shoots out, stretching the spring. After the stretch, the spring pulls back and once again passes the equilibrium point (where no force acts on the ball), shooting backward past it. This happens because the ball has inertia (see Chapter 5), and when the ball is moving, bringing it to a stop takes some force. Here are the various stages the ball goes through, matching the letters in Figure 13-1 (and assuming no friction):

- **Point A:** The ball is at equilibrium, and no force is acting on it. This point, where the spring isn't stretched or compressed, is called the *equilibrium point.*
- **Point B:** The ball pushes against the spring, and the spring retaliates with force F opposing that pushing.
- **Point C:** The spring releases, and the ball springs to an equal distance on the other side of the equilibrium point. At this point, the ball isn't moving, but a force acts on it, F, so it starts going back the other direction.

The ball passes through the equilibrium point on its way back to Point B. At the equilibrium point, the spring doesn't exert any force on the ball, but the ball is traveling at its maximum speed. Here's what happens when the golf ball bounces back and forth; you push the ball to Point B, and it goes through Point A, moves to Point C, shoots back to A, moves to B, and so on: B-A-C-A-

B-A-C-A, and so on. Point A is the equilibrium point, and both Points B and C are equidistant from Point A.

What if the ball were to hang in the air on the end of a spring, as Figure 13-2 shows? In this case, the ball oscillates up and down. Like the ball on a surface in Figure 13-1, the ball hanging on the end of a spring oscillates around the equilibrium position; this time, however, the equilibrium position isn't the point where the spring isn't stretched.

The *equilibrium* position is defined as the position at which no net force acts on the ball. In other words, the equilibrium position is the point where the ball can simply sit at rest. When the spring is vertical, the weight of the ball downward matches the pull of the spring upward. If the x position of the ball corresponds to the equilibrium point, x_i, the weight of the ball, mg, must match the force exerted by the spring. Because $F = kx_i$, you can write the following:

$$mg = kx_i$$

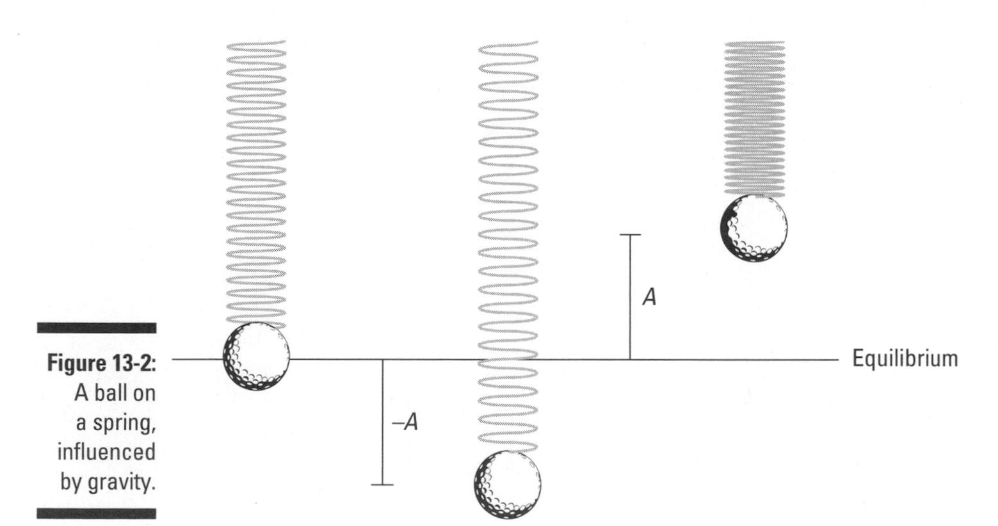

Figure 13-2: A ball on a spring, influenced by gravity.

Solving for x_i gives you the distance the spring stretches because of the ball's weight:

$$x_i = \frac{mg}{k}$$

When you pull the ball down or lift it up and then let go, it oscillates around the equilibrium position, as Figure 13-2 shows. If the spring is elastic, the ball undergoes simple harmonic motion vertically around the equilibrium position; the ball goes up a distance A and down a distance $-A$ around that position (in real life, the ball would eventually come to rest at the equilibrium position, because a frictional force would dampen this motion).

The distance A, or how high the object springs up, is an important one when describing simple harmonic motion; it's called the *amplitude*. The amplitude is simply the maximum extent of the oscillation, or the size of the oscillation.

Catching the wave: A sine of simple harmonic motion

Calculating simple harmonic motion can require time and patience when you have to figure out how the motion of an object changes over time. Imagine that one day you come up with a brilliant idea for an experimental apparatus. You decide to shine a spotlight on a ball bouncing on a spring, casting a shadow on a moving piece of photographic film. Because the film is moving, you get a record of the ball's motion as time goes on. You turn the apparatus on and let it do its thing. See the results in Figure 13-3.

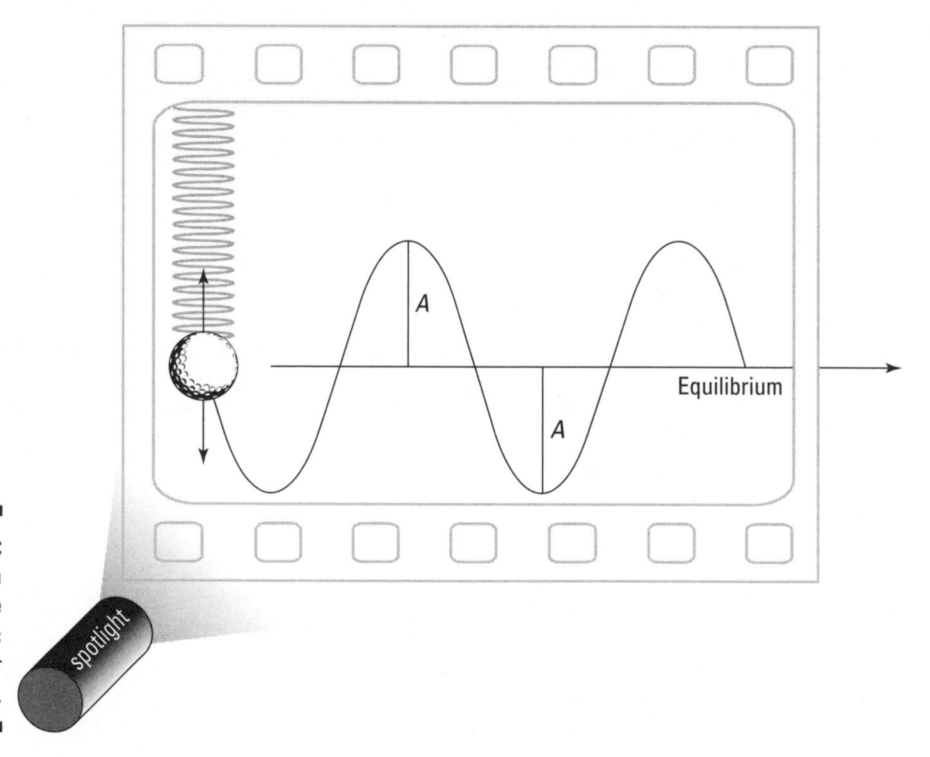

Figure 13-3: Tracking a ball's simple harmonic motion over time.

The ball oscillates around the equilibrium position, up and down, reaching amplitude A at its lowest and highest points. But take a look at the ball's track: You can tell where the ball is moving fastest because that's where the curve has the steepest slope. The ball goes fastest near the equilibrium point because of the acceleration caused by the spring force, which has been applied since the turning point. At the top and bottom, the ball is subject to plenty of force, so it slows down and reverses its motion.

The track of the ball is best modeled with a *sine wave,* which means that its track is a sine wave of amplitude A. (***Note:*** You can also use a cosine wave, because the shape is the same. The only difference is that when a sine wave is at its peak, the cosine wave is at zero, and vice versa.)

You can get a clear picture of the sine wave if you plot the sine function on an xy graph like this:

$$y = \sin x$$

In the rest of this section, I show you how the sine wave relates circular motion to simple harmonic motion.

Understanding sine waves with a reference circle

Take a look at the sine wave in a circular way. If you attach a ball to a rotating disk (see Figure 13-4) and you shine a spotlight on it, you get the same result as when you have the ball hanging from the spring (Figure 13-3): a sine wave.

The rotating disk, which you can see in Figure 13-5, is often called a *reference circle.* You can see how the vertical component of circular motion relates to the sinusoidal (sine-like) wave of simple harmonic motion. Reference circles can tell you a lot about simple harmonic motion.

As the disk turns, the angle, θ, increases in time. What does the track of the ball look like as the film moves to the right? Using a little trig, you can resolve the ball's motion along the y-axis; all you need is the vertical (y) component of the ball's position. At any one time, the ball's y position is the following:

$$y = A \sin \theta$$

The vertical displacement varies from positive A to negative A in amplitude. In fact, you can say that you already know how θ is going to change in time because $\theta = \omega t$, where ω is the single component of the angular velocity — that is, the angular velocity along the axis of rotation of the disk — and t is time:

$$y = A \sin(\omega t)$$

You can now explain the track of the ball as time goes on, given that the disk is rotating with angular velocity ω.

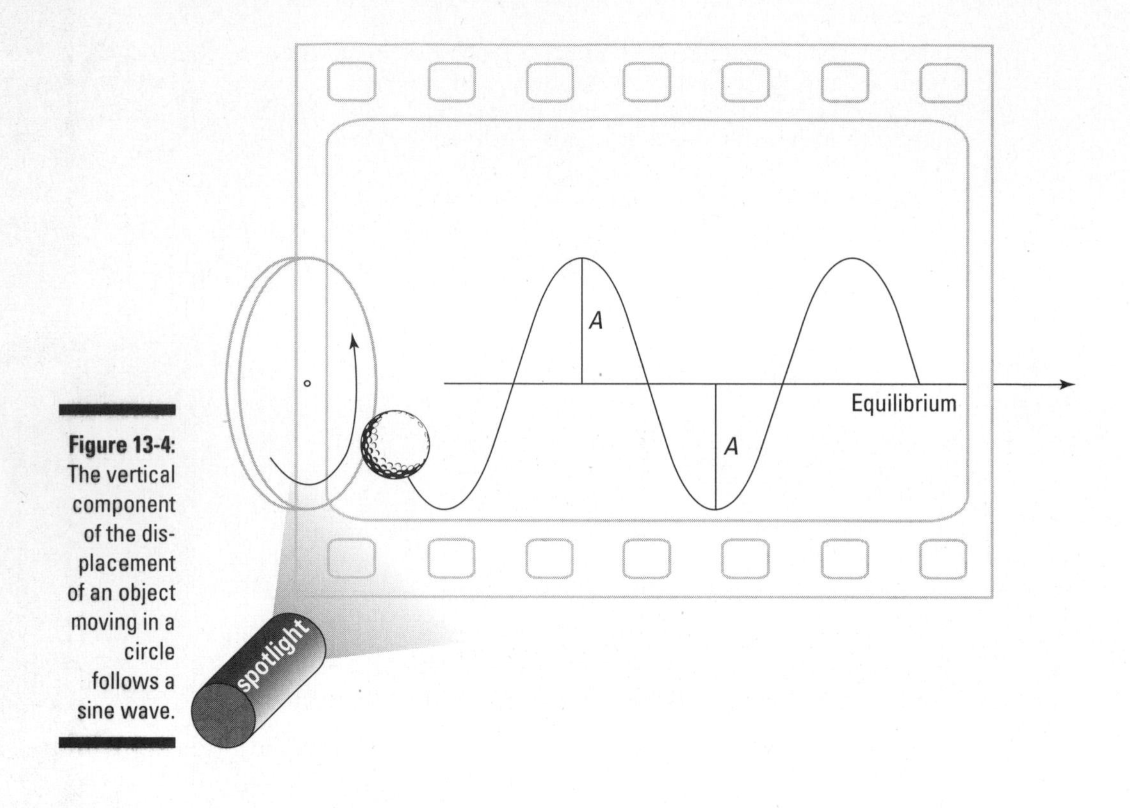

Figure 13-4: The vertical component of the displacement of an object moving in a circle follows a sine wave.

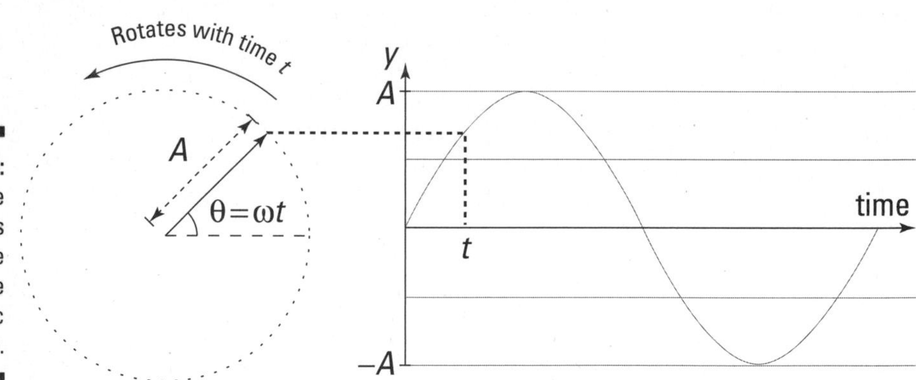

Figure 13-5: A reference circle helps you analyze simple harmonic motion.

Getting periodic

Each time an object moves around a full circle, it completes a *cycle*. The time the object takes to complete the cycle is called the *period,* and it's generally measured in seconds. The letter used for period is T.

Look at Figure 13-5 in terms of the y motion on the film. During one cycle, the ball moves from $y = A$ to $-A$ and then back to A. When the ball goes from any point on the sine wave and passes through one whole wave (including one peak and one trough) back to the next equivalent point on the sine wave later in time, it completes a cycle. The time the ball takes to move from a certain position back to that same position while moving in the same direction is its period.

How can you relate the period to something more familiar? When an object moves in a full circle, completing a cycle, the object goes 2π radians. It travels that many radians in T seconds, so its angular speed, ω (see Chapter 11), is

$$\omega = \frac{2\pi}{T}$$

Multiplying both sides by T and dividing by ω allows you to solve for the period. Now you can relate the period and the angular speed:

$$T = \frac{2\pi}{\omega}$$

Sometimes you speak in terms of the frequency of periodic motion, not the period. The *frequency* is the number of cycles that are completed per second.

For instance, if the disk from Figure 13-4 rotates at 1,000 full turns per second, the frequency, f, would be 1,000 cycles per second. Cycles per second are also called *hertz,* abbreviated Hz, so this frequency would be 1,000 hertz.

So how do you connect frequency, f, to period, T? T is the amount of time one cycle takes, so here's how you can define frequency:

$$f = \frac{1}{T}$$

Because $\omega = 2\pi/T$ and $Tf = 1$, you can rewrite the angular-velocity equation in terms of frequency:

$$\omega = \left(\frac{2\pi}{T}\right)Tf = 2\pi f$$

In simple harmonic motion, the angular velocity, ω, is often referred to as *angular frequency*. Don't confuse the wave's frequency, f, with the angular frequency.

Remembering not to speed away without the velocity

Take a look at Figure 13-5, where a ball is rotating on a disk. In the section "Understanding sine waves with a reference circle," earlier in this chapter, you figure out that

$$y - A\sin(\omega t)$$

where y stands for the y coordinate and A stands for the amplitude of the motion. At any point y, the ball also has a certain velocity, which varies in time also. So how can you describe the velocity mathematically? Well, you can relate tangential velocity to angular velocity like this (see Chapter 11):

$$v = r\omega$$

where r represents the radius. Because the radius of the circle is equal to the amplitude of wave it corresponds to, $r = A$. Therefore, you get the following equation:

$$v = A\omega$$

Does this equation get you anywhere? Sure, because the ball's shadow on the film gives you simple harmonic motion. The velocity vector (see Chapter 4) always points tangential to the circle — perpendicular to the radius — so you get the following for the y component of the velocity at any one time:

$$v_y = A\omega \cos \theta$$

And because the ball is on a rotating disk, you know that $\theta = \omega t$, so

$$v_y = A\omega \cos(\omega t)$$

This equation describes the velocity of any object in simple harmonic motion. Note that the velocity changes in time — from $-A\omega$ to 0 and then to $A\omega$ and back again to 0. So the maximum velocity, which happens at the equilibrium point, has a magnitude of $A\omega$. Among other things, this equation says that, for a given angular velocity, the maximum velocity (v) is directly proportional to the amplitude (A) of the motion: Simple harmonic motion of greater amplitude has a larger maximum velocity, and vice versa.

For example, say that you're on a physics expedition watching a daredevil team do some bungee jumping. You notice that the team members are starting by finding the equilibrium point of their new bungee cords when a jumper is dangling from it but not bouncing, so you measure that point.

The team decides to release their leader from a few meters above the equilibrium point, and you watch as he flashes past the point and then bounces back at a speed of 4.0 meters per second at the equilibrium point. Ignoring all caution, the team lifts its leader to a distance 10 times greater away from the equilibrium point and lets go again. This time you hear a distant scream as the costumed figure hurtles up and down. What's his maximum speed?

You know that, on the first run, he was going 4.0 meters per second at the equilibrium point, the point where he achieved maximum speed; you know that he started with an amplitude 10 times greater on the second try; and you know that the maximum velocity is proportional to the amplitude. Therefore, assuming that the frequency of his bounce is the same, he'll be going 40.0 meters per second at the equilibrium point — pretty speedy.

Including the acceleration

You can find the displacement of an object undergoing simple harmonic motion with the equation $y = A \sin(\omega t)$, and you can find the object's velocity with the equation $v = A\omega \cos(\omega t)$. But you have another factor to account for when describing an object in simple harmonic motion: its acceleration at any particular point. How do you figure it out? No sweat. When an object is going around in a circle, the acceleration is the centripetal acceleration (see Chapter 11), which is

$$a = r\omega^2$$

where r is the radius and ω is the single component of angular velocity (that is, the angular velocity in the direction of the [constant] axis of rotation). And because $r = A$ — the amplitude — you get the following equation:

$$a = A\omega^2$$

This equation represents the relationship between centripetal acceleration, a, and angular velocity, ω. To go from a reference circle (see the earlier section "Understanding sine waves with a reference circle") to simple harmonic motion, you take the component of the acceleration in one dimension — the y direction here — which looks like this:

$$a = -A\omega^2 \sin\theta$$

The negative sign indicates that the y component of the acceleration is always directed opposite the displacement (the ball always accelerates toward the equilibrium point). And because $\theta = \omega t$, where t represents time, you get the following equation for acceleration:

$$a = -A\omega^2 \sin(\omega t)$$

Now you have the equation to find the acceleration of an object at any point while it's moving in simple harmonic motion.

For example, say that your phone rings, and you pick it up. You hear "Hello?" from the earpiece.

"Hmm," you think. "I wonder what the maximum acceleration of the diaphragm in the phone is." The diaphragm (a metal disk that acts like an eardrum) in your phone undergoes a motion very similar to simple harmonic motion, so calculating its acceleration isn't any problem. Measuring carefully, you note that the amplitude of the diaphragm's motion is about 1.0×10^{-4} meters. So far, so good. Human speech is in the 1.0-kilohertz (1,000 hertz) frequency range, so you have the frequency, ω. And you know that the maximum acceleration equals the following:

$$a_{max} = A\omega^2$$

Also, $\omega = 2\pi f$, where f represents frequency. Replace ω with $2\pi f$, and you can plug in the amplitude and frequency to find your answer:

$$a_{max} = A(2\pi f)^2 = (1.0 \times 10^{-4}\text{ m})[2\pi(1{,}000/\text{s})]^2 \approx 3{,}940\text{ m/s}^2$$

You get a value of 3,940 meters per second^2. That seems like a large acceleration, and indeed it is; it's about 402 times the magnitude of the acceleration due to gravity! "Wow," you say. "That's an incredible acceleration to pack into such a small piece of hardware."

"What?" says the impatient person on the phone. "Are you doing physics again?"

Finding the angular frequency of a mass on a spring

If you take the information you know about Hooke's law for springs (see the earlier section "Bouncing Back with Hooke's Law") and apply it to what you know about simple harmonic motion (see "Getting Around to Simple Harmonic Motion"), you can find the angular frequencies of masses on springs, along with the frequencies and periods of oscillations. And because you can relate angular frequency and the masses on springs, you can find the displacement, velocity, and acceleration of the masses.

Hooke's law says that

$$F = -kx$$

where F is the force exerted by the string, k is the spring constant, and x is displacement from equilibrium. Because of Isaac Newton (see Chapter 5), you know that force also equals mass times acceleration:

$$F = ma$$

These force equations are in terms of displacement and acceleration, which you see in simple harmonic motion in the following forms (see the preceding section):

- $x = A\sin(\omega t)$
- $a = -A\omega^2\sin(\omega t)$

Inserting these two equations into the force equations gives you the following:

$$ma = -kx$$

$$m[-A\omega^2\sin(\omega t)] = -kA\sin(\omega t)$$

Divide both sides by $-A\sin(\omega t)$, and this equation breaks down to

$$m\omega^2 = k$$

Rearranging to put ω on one side of the equation gives you the formula for angular frequency:

$$\omega = \sqrt{\frac{k}{m}}$$

You can now find the angular frequency (angular velocity) of a mass on a spring, as it relates to the spring constant and the mass. You can also tie the angular frequency to the frequency and period of oscillation (see "Getting periodic") by using the following equation:

$$\omega = \frac{2\pi}{T} = 2\pi f$$

With this equation and the earlier angular-frequency formula, you can write the formulas for frequency and period in terms of k and m:

- $f = \frac{1}{2\pi}\sqrt{\frac{k}{m}}$
- $T = 2\pi\sqrt{\frac{m}{k}}$

Say that the spring in Figure 13-1 has a spring constant, k, of 15 newtons per meter and that you attach a 45-gram ball to the spring. What's the period of oscillation? After you convert from grams to kilograms, all you have to do is plug in the numbers:

$$\begin{aligned} T &= 2\pi\sqrt{\frac{m}{k}} \\ &= 2\pi\sqrt{\frac{0.045\text{ kg}}{15\text{ N/m}}} \approx 0.34\text{ s} \end{aligned}$$

The period of the oscillation is 0.34 seconds. How many bounces will you get per second? The number of bounces represents the frequency, which you find this way:

$$f = \frac{1}{T} = \frac{1}{0.34\text{ s}} \approx 2.9\text{ Hz}$$

You get nearly 3 oscillations per second.

Because you can relate the angular frequency, ω, to the spring constant and the mass on the end of the spring, you can predict the displacement, velocity, and acceleration of the mass, using the following equations for simple harmonic motion (see the section "Catching the wave: A sine of simple harmonic motion" earlier in this chapter):

- $y = A\sin(\omega t)$
- $v = A\omega\cos(\omega t)$
- $a = -A\omega^2\sin(\omega t)$

Using the example of the spring in Figure 13-1 — with a spring constant of 15 newtons per meter and a 45-gram ball attached — you know that the angular frequency is the following:

$$\omega = \sqrt{\frac{15\text{ N/m}}{0.045\text{ kg}}} \approx 18\text{ s}^{-1}$$

You may like to check how the units work out. Remember that $1\text{ N} = 1\text{ kg}\cdot\text{m/s}^2$, so the units you get from the preceding equation for the angular velocity work out to be

$$\sqrt{\frac{\left(\text{kg}\cdot\text{m/s}^2\right)/\text{m}}{\text{kg}}} = \sqrt{\frac{\text{kg}/\text{s}^2}{\text{kg}}} = \sqrt{\frac{1}{\text{s}^2}} = \text{s}^{-1}$$

Say, for example, that you pull the ball 10.0 centimeters before releasing it (making the amplitude 10.0 centimeters). In this case, you find that

$$x = (0.10\text{ m})\sin[(18\text{ s}^{-1})t]$$

$$v = (0.10\text{ m})(18\text{ s}^{-1})\sin[(18\text{ s}^{-1})t]$$

$$a = -(0.10\text{ m})(18\text{ s}^{-1})^2\cos[(18\text{ s}^{-1})t]$$

Factoring Energy into Simple Harmonic Motion

Along with the actual motion that takes place in simple harmonic motion, you can examine the energy involved. For example, how much energy is stored in a spring when you compress or stretch it? The work you do compressing or stretching the spring must go into the energy stored in the spring. That energy is called *elastic potential energy* and is equal to the force, *F*, times the distance, *s*:

$$W = Fs$$

As you stretch or compress a spring, the force varies, but it varies in a linear way (because in Hooke's law, force is proportional to the displacement). Therefore, you can write the equation in terms of the average force, $\overline{F}$:

$$W = \overline{F}s$$

The distance (or displacement), s, is just the difference in position, $x_f - x_i$, and the average force is $(1/2)(F_f + F_i)$. Therefore, you can rewrite the equation as follows:

$$W = \left[\frac{1}{2}(F_f + F_i)\right](x_f - x_i)$$

Hooke's law says that $F = -kx$. Therefore, you can substitute $-kx_f$ and $-kx_i$ for F_f and F_i:

$$W = \left[-\frac{1}{2}(kx_f + kx_i)\right](x_f - x_i)$$

Distributing and simplifying the equation gives you the equation for work in terms of the spring constant and position:

$$W = \frac{1}{2}kx_i^2 - \frac{1}{2}kx_f^2$$

The work done on the spring changes the potential energy stored in the spring. Here's how you give that potential energy, or the elastic potential energy:

$$PE = \frac{1}{2}kx^2$$

For example, suppose a spring is elastic and has a spring constant, k, of 1.0×10^{-2} newtons per meter, and you compress the spring by 10.0 centimeters. You store the following amount of energy in it:

$$PE = \frac{1}{2}kx^2 = \frac{1}{2}(1.0 \times 10^{-2} \text{ N/m})(0.10 \text{ m})^2 = 5.0 \times 10^{-5} \text{ J}$$

You can also note that when you let the spring go with a mass on the end of it, the mechanical energy (the sum of potential and kinetic energy) is conserved:

$$PE_1 + KE_1 = PE_2 + KE_2$$

When you compress the spring 10.0 centimeters, you know that you have 5.0×10^{-5} joules of energy stored up. When the moving mass reaches the equilibrium point and no force from the spring is acting on the mass, you have maximum velocity and therefore maximum kinetic energy — at that point, the kinetic energy is 5.0×10^{-5} joules, by the conservation of mechanical energy (see Chapter 9 for more on this topic).

Swinging with Pendulums

Other objects besides springs, such as pendulums, move in simple harmonic motion. In Figure 13-6, a ball tied to a string swings back and forth.

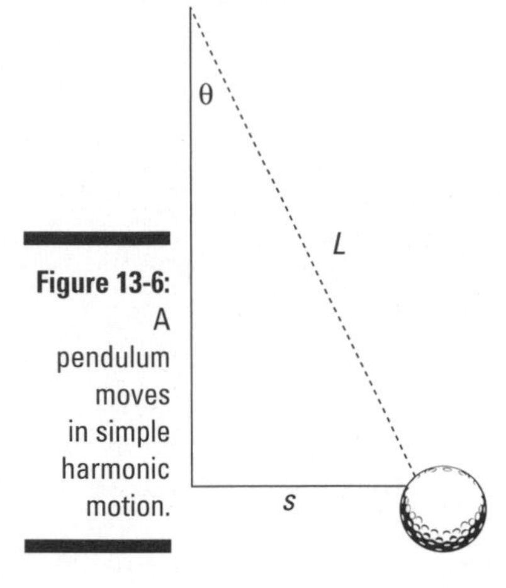

Figure 13-6: A pendulum moves in simple harmonic motion.

The torque, τ, that comes from gravity is the weight of the ball (which is a force of magnitude mg directed downward — hence the minus sign) multiplied by the lever arm, s (for more on lever arms and torque, see Chapter 11):

$$\tau = -mgs$$

Here's where you make an approximation. For small angles θ, the distance s approximately equals $L\theta$, where L is the length of the pendulum string:

$$\tau = -mgL\theta$$

This equation resembles Hooke's law, $F = -kx$, if you treat mgL as you would a spring constant. In Chapter 12, I show you the relation between torque and angular acceleration, and you see that the angular variables obey the same equations as their linear equivalents. Therefore, the calculation goes in just the same way as for the spring. In the angular variables:

$$\Sigma\tau = I\alpha$$

Just as in the case of the spring, the pendulum undergoes simple harmonic motion, with

- $\theta = A \sin(\omega t)$
- $\alpha = -A\omega^2 \sin(\omega t)$

Plug in the torque of the pendulum and the values of α and θ to get

$$I\alpha = -mgL\theta$$
$$I[-A\omega^2 \sin(\omega t)] = -mgLA \sin(\omega t)$$

Then solve for ω to get this:

$$\omega = \sqrt{\frac{mgL}{I}}$$

The moment of inertia equals mr^2 for a point mass (see Chapter 12), which you can use here, assuming that the ball is small compared to the pendulum string. For the pendulum, the radius r is the length of the string, L. This gives you the following equation:

$$\omega = \sqrt{\frac{mgL}{mL^2}}$$
$$= \sqrt{\frac{g}{L}}$$

Now you can plug this angular velocity into the equations for simple harmonic motion. You can also find the period of a pendulum with the following equation:

$$\omega = \frac{2\pi}{T}$$

where T represents period. If you substitute the preceding form for ω, you get this:

$$\sqrt{\frac{g}{L}} = \frac{2\pi}{T}$$

Then rearrange to find the period:

$$T = 2\pi\sqrt{\frac{L}{g}}$$

Note that this period is actually independent of the mass on the pendulum!

Part IV
Laying Down the Laws of Thermodynamics

In this part . . .

How much boiling water do you need to melt a 200-pound block of ice? Why would you freeze in space? Why does metal feel cold to the touch? What is an ideal gas? The answers all boil (or freeze) down to *thermodynamics,* which is the physics of thermal energy and heat flow. You find the answers to your questions in this part in the form of useful equations and explanations.

Chapter 14

Turning Up the Heat with Thermodynamics

In This Chapter

- Taking temperature in Fahrenheit, Celsius, and Kelvin
- Examining temperature change with thermal expansion
- Following along with heat flow
- Accounting for specific heat capacity
- Meeting the requirements for phase change

The concepts of heat and temperature are part of your daily life. Understanding the laws that govern the temperatures of things, how heat flows between them, and how the material and thermal properties depend on each other hasn't only furthered physicists' appreciation of the world and its workings; it has also led to technological and engineering advances. A structurally sound bridge, for example, depends on understanding the thermal expansion of any of the bridge's metal elements. The motor car works because of the thermal energy released from the combustion of gasoline and air. These, and more, are possible only with an understanding of the relationship between materials and their thermal properties.

This chapter explores heat and temperature. Physics gives you plenty of power to predict what goes on when things heat up or cool down. I discuss temperature scales, linear expansion, volume expansion, and how much of liquid at one temperature will change the temperature of another when they're put together.

Measuring Temperature

Temperature is a measure of molecular movement — how fast and how much the molecules of whatever substance you're measuring are moving. You always start a calculation or observation in physics by making measurements, and when you're discussing temperature, you have several scales at your disposal: most notably, Fahrenheit, Celsius, and Kelvin.

Fahrenheit and Celsius: Working in degrees

In the United States, the most common temperature scale is the *Fahrenheit scale,* which measures temperature in degrees. For example, the blood temperature of a healthy human being is 98.6°F — the *F* means you're using the Fahrenheit scale. In Fahrenheit's system, pure water freezes at 32°F and boils at 212°F.

However, the Fahrenheit system wasn't very reproducible in its early days, so scientists developed another system — the *Celsius scale* (formerly called the *centigrade* system). Using this system, pure water freezes at 0°C and boils at 100°C. Here's how you tie the two systems of temperature measurement together (these measurements are at sea level; they change as you go up in altitude):

- **Freezing water:** 32°F = 0°C
- **Boiling water:** 212°F = 100°C

If you do the math, you find 180°F between the points of freezing and boiling in the Fahrenheit system and 100°C in the Celsius system, so the conversion ratio is 180/100 = 18/10 = 9/5. And don't forget that the measurements are also offset by 32 degrees (the 0-degrees point of the Celsius scale corresponds to the 32-degrees point of the Fahrenheit scale). Putting these ideas together lets you convert from Celsius to Fahrenheit or from Fahrenheit to Celsius pretty easily; just remember these equations:

- **Fahrenheit to Celsius:** $C = \frac{5}{9}(F - 32)$
- **Celsius to Fahrenheit:** $F = \left(\frac{9}{5}\right)C + 32$

For example, the blood temperature of a healthy human being is 98.6°F. What does this equal in Celsius? Just plug in the numbers:

$$C = \frac{5}{9}(F - 32)$$
$$= \frac{5}{9}(98.6 - 32) \approx 37.0°C$$

Zeroing in on the Kelvin scale

In the 19th century, William Thompson created a third temperature system, one now in common use in physics — the Kelvin system (Thompson later became Lord Kelvin). The Kelvin system has become so central to physics that the Fahrenheit and Celsius systems are defined in terms of the Kelvin system — a system based on the concept of absolute zero.

Analyzing absolute zero

Molecules move more and more slowly as the temperature lowers. At *absolute zero,* the molecules almost stop, which means you can't cool them anymore. (The molecules only "almost" stop because when you get down to the scale of molecules, you're in the realm of quantum mechanics. When the molecules have as little energy as possible, they still have *zero-point energy.*) No refrigeration system in the world — or in the universe — can go any lower than absolute zero.

The Kelvin system uses absolute zero as its zero point, which makes sense. What's a little odd is that you don't measure temperature in this scale in degrees; you measure it in *kelvins*. A temperature of 100 is 100 kelvins (not 100 degrees Kelvin) in the Kelvin scale. This system has become so widely adopted that the official MKS unit of temperature is the kelvin (in practice, however, you see °C used more often in introductory physics).

Making kelvin conversions

Each kelvin is the same size as a Celsius degree, which makes converting between Celsius degrees and kelvins easy. On the Celsius scale, absolute zero is –273.15°C. This temperature corresponds to 0 kelvins, which you also write as 0 K (not, please note, 0°K).

To convert between the Celsius and Kelvin scales, use the following formulas:

- **Celsius to Kelvin:** $K = C + 273.15$
- **Kelvin to Celsius:** $C = K - 273.15$

And to convert from kelvins to Fahrenheit, you can use this formula:

$$\begin{aligned} F &= \left(\frac{9}{5}\right)(K - 273.15) + 32 \\ &= \left(\frac{9}{5}\right)K - 459.67 \end{aligned}$$

(Or you can convert kelvins to degrees Celsius and then use the conversion formulas in the earlier section "Fahrenheit and Celsius: Working in degrees.")

At what temperature does water boil in kelvins? Well, pure water boils at 100°C at sea level, so plug your numbers into the formula:

$$\begin{aligned} K &= C + 273.15 \\ &= 100 + 273.15 = 373.15 \text{ K} \end{aligned}$$

Water boils at 373.15 kelvins. Helium turns to liquid at 4.2 kelvins; what's that in degrees Celsius? Use the formula:

$$\begin{aligned} C &= K - 273.15 \\ &= 4.2 - 273.15 = -268.95°\text{C} \end{aligned}$$

Helium liquefies at –268.95°C. Pretty chilly.

The Heat Is On: Thermal Expansion

Some screw-top jars can be tough to open, which is maddening when you really want some pickles. Maybe you remember seeing your mom run stubborn jar lids under hot water when you were a kid. She did this because heat makes the lid expand, which usually makes the job of turning it much easier.

On a molecular level, thermal expansion happens because when you heat objects, the molecules bounce around faster, which leads to a physical expansion. (Note that this relationship between heating and expanding isn't true for all materials, however. For example, water becomes denser as you raise its temperature from 0°C to 4°C.)

In this section, I first cover the linear expansion of solids — how solid objects lengthen when temperature rises. I then discuss thermal expansion in 3-D so you can observe volume changes in both solids and liquids. (For info on thermal expansion in gases, flip to Chapter 16.)

Linear expansion: Getting longer

When you talk about the expansion of a solid in any one dimension under the influence of heat, you're talking about *linear expansion.* Figure 14-1 shows an image of this phenomenon.

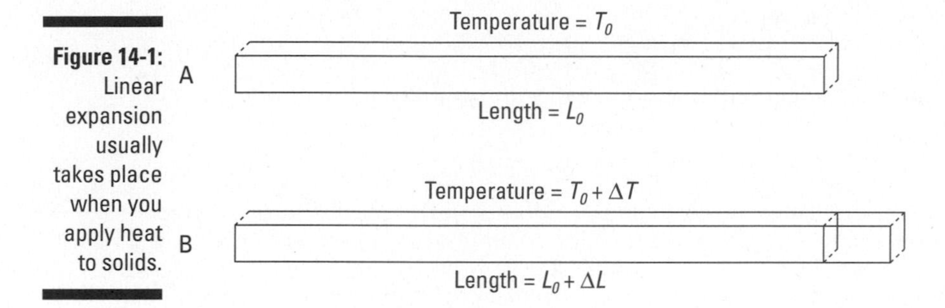

Figure 14-1: Linear expansion usually takes place when you apply heat to solids.

Relating temperature changes to changes in length

Under thermal expansion, a solid object's change in length, ΔL, is proportional to the change in temperature, ΔT. You can show this relationship mathematically.

Note: Even though initial values are represented by a subscript *i* in other chapters (L_i, for example), I use a subscript *0* (L_0, for example), which is what you're more likely to see in other texts for these kinds of equations.

First, suppose you raise the temperature of an object a small amount:

$$T = T_0 + \Delta T$$

where T represents the final temperature, T_0 represents the original temperature, and ΔT represents the change in temperature. The change of temperature results in an expansion in any linear direction of

$$L = L_0 + \Delta L$$

where L represents the final length of the solid, L_0 represents its original length, and ΔL represents the change in length.

When you heat a solid, the solid expands by a few percent, and that percentage is proportional to the change in temperature. In other words, $\Delta L/L_0$ (the fraction by which the solid expands) is proportional to ΔT (the change in temperature).

The constant of proportionality, which helps tell you exactly how much an object will expand, depends on which material you're working with. The constant of proportionality is the *coefficient of linear expansion,* which you give the symbol α. You can write this relationship as an equation this way:

$$\frac{\Delta L}{L_0} = \alpha \Delta T$$

Here's the linear-expansion equation in standard form, solved for ΔL:

$$\Delta L = \alpha L_0 \Delta T$$

People usually measure α, the coefficient of linear expansion, in units of 1/°C (that is, in $°C^{-1}$). However, because the units of Celsius and Kelvin are the same size, a difference in temperature measured in degrees Celsius is of the same magnitude when measured in kelvins. Therefore, to convert the coefficient of linear expansion from degrees Celsius to kelvins, you only have to swap the symbols.

Physics problems provide these coefficients when you need them to solve the problem. But just in case, here's a useful website that lists many of the coefficients: `www.engineeringtoolbox.com/linear-expansion-coefficients-d_95.html`.

Workin' on the railroad: A linear expansion example

Plenty of construction projects take linear expansion into account. You often see bridges with "expansion joints" connecting the bridge to the road surface. As temperatures rise, these joints allow the bridge materials to expand without buckling.

Here's a construction-based example. Say that you're called in to check out a new railroad. You look closely at the 10.0-meter-long rails, noticing that they're only 1.0 millimeter apart at the ends. "How much hotter does it get around these parts during the summer?" you ask.

"Hotter?" the chief designer guffaws. "You afraid the rails will *melt?*"

Everyone snickers at your ignorance as you check your almanac, which tells you that you can expect the rails to get 50°C hotter during a normal summer. The coefficient of linear expansion for the steel that the rails are made from is approximately 1.2×10^{-5}°C^{-1}. So how much will the typical rail expand during the hot part of summer? You know that

$$\Delta L = \alpha L_0 \Delta T$$

Plugging in the numbers gives you the expansion:

$$\Delta L = \alpha L_0 \Delta T = (1.2 \times 10^{-5}°C^{-1})(10.0\ m)(50°C^{-1}) = 6.0 \times 10^{-3}\ m$$

In other words, you can expect the rails to expand 6.0×10^{-3} meters, or 6.0 millimeters, in the summer. However, the rails are only 1.0 millimeter apart. The railroad company is in trouble.

You look at the chief designer and say, "You and I are about to have a nice, long talk about physics."

Volume expansion: Taking up more space

Linear expansion, as the name indicates, takes place in one dimension, but the world comes with three dimensions. If an object undergoes a small temperature change of just a few degrees, you can say that the volume of the solid or liquid will change in a way proportionate to the temperature change. As long as the temperature differences involved are small, the fraction by which the solid expands, $\Delta V/V_0$, is proportional to the change in temperature, ΔT (where ΔV represents the change in volume and V_0 represents the original volume).

With volume expansion, the constant involved is called the *coefficient of volume expansion.* This constant is given by the symbol β, and like α, it's often measured in °C^{-1}. Using β, here's how you can express the equation for volume expansion:

$$\frac{\Delta V}{V_0} = \beta \Delta T$$

When you solve for ΔV, you get the volume-expansion equation in standard form:

$$\Delta V = \beta V_0 \Delta T$$

You've created the analog (or equivalent) of the equation $\Delta L = \alpha L_0 \Delta T$ for linear expansion (see the earlier section "Linear expansion: Getting longer").

If the lengths and temperature changes are small, you find that $\beta = 3\alpha$ for most solids. This makes sense, because you go from one dimension to three. For example, for steel, α is 1.2×10^{-5}°C^{-1} and β is 3.6×10^{-5}°C^{-1}. Liquids also undergo linear volume expansion, but the preceding relation between β and α does not apply generally.

Tanker trucks: Looking at expanding liquids

Say you're at the gasoline refinery when you notice that workers are filling all the 5,000-gallon tanker trucks to the very brim before driving off on a hot summer day. "Uh oh," you think as you get your calculator out. For gasoline, $\beta = 9.5 \times 10^{-4}$°C^{-1}, and you figure that it's 10.0°C warmer in the sunshine than in the building, so here's how much the volume of gasoline will increase:

$$\Delta V = \beta V_0 \Delta T = (9.5 \times 10^{-4}°\text{C}^{-1})(5{,}000 \text{ gal})(10.0°\text{C}) = 47.5 \text{ gal}$$

Not good news for the refinery — those 5,000-gallon tankers of gasoline that are filled to the brim have to carry 5,047.5 gallons of gasoline after they go out in the sunshine. The gas tanks may also expand, but the β of steel is much less than the β of gasoline. Should you tell the refinery workers? Or should you ask for a bigger fee first?

First you negotiate your whopper fee, and then you go explain the problem to the foreman. "Holy smokes!" he cries. "We'd have gasoline pouring out of the caps on the top of our trucks." He stops the trucks and gets some gasoline taken out of each before they're sent on their way.

Radiators: Seeing expanding liquids and containers

The foreman of a gasoline refinery notices that his workers are filling the radiators of the trucks full up to the brim. "Holy smokes!" he cries. "What about volume expansion? The coolant will be pouring out of those radiators when they get hot." True, the coolant will expand. But doesn't everyone fill radiators to the brim?

Most cars have a plastic overflow reservoir that catches the overflow as it happens. So are the gasoline company's radiators safe? Each radiator holds 15 quarts of coolant, which is 1.4×10^{-2} cubic meters, and has a 1-quart coolant reservoir, which is 9.5×10^{-4} cubic meters. Will a radiator overflow more than the reservoir can handle?

You get out your clipboard. Okay, the radiator takes 15 quarts (1.4×10^{-2} m^3) of coolant, and you happen to know that the β for the coolant is $\beta = 4.1 \times 10^{-4}$°C^{-1}. You want to be precise this time and take into account the expansion of the radiator as well. The radiator is made of copper (with a thin outer layer of aluminum, which you can neglect in this example), so $\beta = 5.1 \times 10^{-5}°C^{-1}$. If the radiator starts at 20°C and heats up to its working temperature of 92°C, will a 1-quart (9.5×10^{-4} m^3) coolant reservoir be enough to catch the overflow?

The foreman watches tensely as you begin your calculations. Here's the formula for the expansion of the coolant:

$$\Delta V_c = \beta_c V_{0c} \Delta T$$

You know that $\beta_c = 4.1 \times 10^{-4}°C^{-1}$, $V_{0c} = 1.4 \times 10^{-2}$ m^3, and $\Delta T = 92°C - 20°C = 72°C$ in this example. Plug in these numbers and solve:

$$\Delta V_c = \beta_c V_{0c} \Delta T = 4.2 \times 10^{-4} \text{ m}^3$$

Okay, so the coolant will expand by 4.2×10^{-4} m^3, which is equal to 0.44 quarts. But the radiator will also expand, meaning it can hold more coolant. This time, you take that expansion into account to get a more accurate answer.

Because the radiator is made of copper, it'll expand as though it were made of solid copper, which makes the math easier. Here's the change in volume for the radiator:

$$\Delta V_r = \beta_r V_{0r} \Delta T$$

Here, $\beta_r = 5.1 \times 10^{-5}°C^{-1}$, $\Delta V_{0r} = 1.4 \times 10^{-2}$ m^3, and $\Delta T = 92°C - 20°C = 72°C$. Plug in your numbers and solve:

$$\begin{aligned} \Delta V_r &= \beta_r V_{0r} \Delta T \\ &= (5.1 \times 10^{-5}°C^{-1})(1.4 \times 10^{-2} \text{ m}^3)(72°C) \\ &\approx 5.2 \times 10^{-5} \text{ m}^3 \end{aligned}$$

The total overflow is equal to the coolant expansion minus the amount the radiator expands, so plug in your numbers:

$$\begin{aligned} \Delta V &= \Delta V_c - \Delta V_r \\ &= 4.2 \times 10^{-4} \text{ m}^3 - 5.2 \times 10^{-5} \text{ m} \\ &\approx 3.7 \times 10^{-4} \text{ m}^3 \end{aligned}$$

So each radiator will overflow a little more than a third of a quart, and the overflow reservoir is 1 quart in volume. You turn to the foreman and say, "The radiators are fine, with a good safety margin."

"Whew," says the foreman.

Heat: Going with the Flow (Of Thermal Energy)

What, really, is heat? When you touch a hot object, heat flows from the object to you, and your nerves record that fact. When you touch a cold object, heat flows from you to that object, and again, your nerves keep track of what's happening. Your nerves record why objects feel hot or cold — because heat flows from them to you or from you to them.

To understand heat, you need to understand thermal energy. *Thermal energy* is the energy that a body has in the vibrations of its molecules — the energy stored in the internal molecular motion of an object. The temperature of a body usually increases with its thermal energy.

When two bodies are brought into thermal contact, thermal energy is free to be exchanged between them. If no thermal energy flows between them, they are in *thermal equilibrium.* In other words, they are in a kind of balance. Two objects in thermal equilibrium are said to have the same *temperature*. If thermal energy *does* flow between them — an object at a higher temperature is in thermal contact with an object at a lower temperature and the thermal energy flows from the hotter body to the cooler one — they are not in thermal equilibrium.

In physics terms, *heat* is thermal energy that flows from objects of higher temperatures to objects of lower temperatures. The unit of this energy in the MKS system is the *joule* (J) — the same unit you use for other forms of energy and work (see Chapter 9).

One *calorie* is defined as the amount of heat needed to raise the temperature of 1.0 gram of water 1.0°C, so 1 calorie = 4.186 joules. Nutritionists use the food-energy term *Calorie* (capital *C*) to stand for 1,000 calories — 1.0 kilocalorie (kcal), so 1.0 Calorie = 4,186 joules. Engineers use another unit of measurement as well: the British thermal unit (Btu). One Btu is the amount of heat needed to raise 1 pound of water 1.0°F. To convert, you can use the relation 1 British thermal unit = 1,055 joules.

This section covers heat and how the change in energy affects temperature. I also discuss phase changes, special cases in which a substance can absorb heat without changing temperature.

Getting specific with temperature changes

At a given temperature, different materials can hold different amounts of thermal energy. For instance, if you warm up a potato, it can hold its heat longer (as your tongue can testify) than a lighter material such as cotton candy. Why? Because the potato stores more thermal energy for a given change in temperature; therefore, more heat has to flow to cool the potato than is needed to cool the cotton candy. The measure of how much heat an object of a given mass can hold at a given temperature is called its *specific heat capacity*.

Suppose you see someone making a pot of coffee. You measure exactly 1.0 kilogram of brewed coffee in the pot, and then you get down to the real measurements. You find out that you need 4,186 joules of heat energy to raise the temperature of the coffee by 1°C, but you need only 840 joules to raise 1.0 kilogram of glass by 1°C; the coffee and glass have different specific heat capacities. The energy goes into the substance being heated, which stores the energy as internal energy until it leaks out again. (***Note:*** If you need 4,186 joules to raise 1.0 kilogram of coffee by 1°C, you need double that, 8,372 joules, to raise 2.0 kilograms of coffee by 1°C or to raise 1.0 kilogram of coffee by 2°C.)

The following equation relates the amount of heat needed to raise an object's temperature to the change in temperature and the amount of mass involved:

$$Q = cm\Delta T$$

Here, Q is the amount of heat energy involved (measured in joules if you're using the MKS system), m is the mass, ΔT is the change in temperature, and c is a constant called the *specific heat capacity,* which is measured in joules per kilogram-degree Celsius, or J/(kg·°C). In Chapter 16, you can find a calculation of the specific heat capacity for the special case of an ideal gas, but usually physicists calculate specific heat capacity through experiment, so most problems give you c or refer you to a table of specific-heat values for various materials.

You can use the heat equation to find out how temperature changes when you mix liquids of different temperatures. Suppose you have 45 grams of coffee in your cup, but it cooled while you were figuring out the coffee's specific heat. You call over your host. The coffee is 45°C, but you like it at 65°C. The host gets up to pour some more. "Just a minute," you say. "The coffee in the pot is 95°C. Wait until I calculate exactly how much you need to pour."

The following equation represents the heat lost by the new mass of coffee, m_1:

$$\Delta Q_1 = cm_1(T - T_{1,0})$$

And here's the heat gained by the existing coffee, mass m_2:

$$\Delta Q_2 = cm_2(T - T_{2,0})$$

Assuming you have a superinsulating coffee mug, no energy leaves the system to the outside, and because energy cannot be created or destroyed, energy is conserved within such a closed system; therefore, the heat lost by the new coffee is the heat that the existing coffee gains, so

$$\Delta Q_1 = -\Delta Q_2$$

Therefore, you can say the following:

$$cm_1(T - T_{1,0}) = -cm_2(T - T_{2,0})$$

Dividing both sides by the specific heat capacity, *c*, and plugging in the numbers gives you the following:

$$\begin{aligned} m_1\left(T - T_{1,0}\right) &= m_2\left(T - T_{2,0}\right) \\ m_1\left(65°\text{C} - 95°\text{C}\right) &= -\left(0.045\text{ kg}\right)\left(65°\text{C} - 45°\text{C}\right) \\ m_1 &= \frac{-\left(0.045\text{ kg}\right)\left(65°\text{C} - 45°\text{C}\right)}{\left(65°\text{C} - 95°\text{C}\right)} \\ m_1 &= 0.03\text{ kg} \end{aligned}$$

You need 0.03 kilograms, or 30 grams. Satisfied, you put away your calculator and say, "Give me exactly 30 grams of that coffee."

Just a new phase: Adding heat without changing temperature

Phase changes occur when materials change state, going from liquid to solid (as when water freezes), solid to liquid (as when rocks melt into lava), liquid to gas (as when you boil water for tea), and so on. When the material in question changes to a new state — liquid, solid, or gas (you can also factor in a fourth state: plasma, a superheated gas-like state) — some heat goes into or comes out of the process without changing the temperature.

You can even have solids that turn directly into gas. As dry ice (frozen dioxide gas) gets warmer, it turns into carbon dioxide gas. This process is called *sublimation.*

Imagine you're calmly drinking your lemonade at an outdoor garden party. You grab some ice to cool your lemonade, and the mixture in your glass is now half ice, half lemonade (which you can assume has the same specific heat as water), with a temperature of exactly 0°C.

As you hold the glass and watch the action, the ice begins to melt — but the contents of the glass don't change temperature. Why? The heat (thermal energy) going into the glass from the outside air is melting the ice, not warming the mixture up. So does this make the equation for heat energy ($Q = cm\Delta T$) useless? Not at all — it just means that the equation doesn't apply for a phase change.

In this section, you see how heat affects temperature before, during, and after phase changes.

Breaking the ice with phase-change graphs

If you graph the heat added to a system versus the system's temperature, the graph usually slopes upward; adding heat increases temperature. However, the graph levels out during phase changes, because on a molecular level, making a substance change state requires energy. After all the material has changed state, the temperature can rise again.

Imagine that someone has taken a bag of ice and thoughtlessly put it on the stove. Before it hit the stove, the ice was at a temperature below freezing (–5°C), but being on the stove is about to change that. You can see the change taking place in graph form in Figure 14-2.

As long as no phase change takes place, the equation $Q = cm\Delta T$ holds (the specific heat capacity of ice is around 2.0×10^{-3} J/kg·°C), which means that the temperature of the ice will increase linearly as you add more heat to it, as you see in the graph.

However, when the ice reaches 0°C, the ice is getting too warm to hold its solid state, and it begins to melt, undergoing a phase change. When you melt ice, breaking up the crystalline ice structure requires energy, and the energy needed to melt the ice is supplied as heat. That's why the graph in Figure 14-2 levels off in the middle — the ice is melting. You need heat to make the ice change phase to water, so even though the stove adds heat, the temperature of the ice doesn't change as it melts.

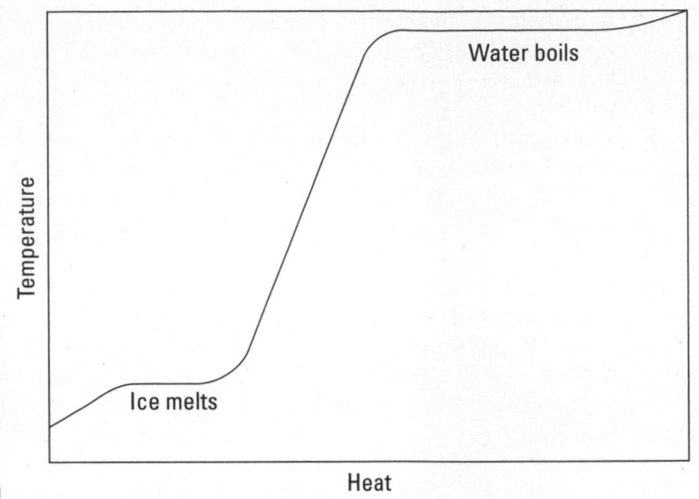

Figure 14-2: Phase changes of water.

As you watch the bag of ice on the stove, however, you note that all the ice eventually melts into water. Because the stove is still adding heat, the temperature begins to rise, which you see in Figure 14-2. The stove adds more and more heat to the water, and in time, the water starts to bubble. "Aha," you think. "Another phase change." And you're right: The water is boiling and becoming steam. The bag holding the ice seems pretty resilient, and it expands while the water turns to steam.

You measure the temperature of the water. Fascinating — although the water boils, turning into steam, the temperature doesn't change. Once again, you need to add heat to incite a phase change — this time from water to steam. You can see in Figure 14-2 that as you add heat, the water boils, but the temperature of that water doesn't change.

What's going to happen next, as the bag swells to an enormous volume? You never get to find out, because the bag finally explodes. You pick up a few shreds of the bag and examine them closely. How can you account for the heat that's needed to change the state of an object? How can you add something to the equation for heat energy to take into account phase changes? That's where the idea of latent heat comes in.

Understanding latent heat

Latent heat is the heat per kilogram that you have to add or remove to make an object change its state; in other words, latent heat is the heat needed to make a phase change happen. Its units are joules per kilogram (J/kg) in the MKS system.

Physicists recognize three types of latent heat, corresponding to the changes of phase between solid, liquid, and gas:

- **The latent heat of fusion, L_f:** The heat per kilogram needed to make the change between the solid and liquid phases, as when water turns to ice or ice turns to water
- **The latent heat of vaporization, L_v:** The heat per kilogram needed to make the change between the liquid and gas phases, as when water boils or when steam condenses into water
- **The latent heat of sublimation, L_s:** The heat per kilogram needed to make the change between the solid and gas phases, as when dry ice evaporates

Water's latent heat of fusion of water, L_f, is 3.35×10^5 J/kg, and its latent heat of vaporization, L_v, is 2.26×10^6 J/kg. In other words, you need 3.35×10^5 joules to melt 1 kilogram of ice at 0°C (just to melt it, not to change its temperature). And you need 2.26×10^6 joules to boil 1 kilogram of water into steam.

Here's the formula for heat transfer during phase changes, where ΔQ is the change in heat, m is the mass, and L is the latent heat:

$$\Delta Q = mL$$

Here, L takes the place of the ΔT (change in temperature) and c (specific heat capacity) terms in the temperature-change formula.

Suppose you're in a restaurant with a glass of 100.0 grams of water at room temperature, 25°C, but you'd prefer ice water at 0°C. How much ice would you need? You can find the answer using the heat formulas for both change in temperature and phase change.

You get out your clipboard, reasoning that the heat absorbed by the melting ice must equal the heat lost by the water you want to cool. Here's the heat lost by the water you're cooling:

$$\Delta Q_{water} = cm\Delta T = cm(T - T_0)$$

where ΔQ_{water} is the heat lost by the water, c is the specific heat capacity of water, m is the mass of the water, ΔT is the change in temperature of the water, T is the final temperature, and T_0 is the initial temperature.

Plugging in the numbers tells you how much heat the water needs to lose:

$$\Delta Q_{water} = cm(T - T_0)$$
$$= (4{,}186 \text{ J kg}^{-1} \text{ K}^{-1})(0.100 \text{ kg})(0 \text{ K} - 25 \text{ K}) \approx -1.04 \times 10^4 \text{ J}$$

Therefore, the water needs to lose 1.04×10^4 joules of heat.

So how much ice would that amount of heat melt? That is, how much ice at 0°C would you need to add to cool the water to 0°C? That would be the following amount, where L_f is the latent heat of fusion for ice:

$$\Delta Q_{ice} = m_{ice} L_f$$

For ice, L_m is 3.35×10^5 J/kg, so you get this answer:

$$\Delta Q_{ice} = m_{ice}(3.35 \times 10^5 \text{ J/kg})$$

You know this has to be equal to the heat lost by the water, so you can set this equal and opposite to ΔQ_{water}, or -1.04×10^4 joules:

$$\Delta Q_{ice} = -\Delta Q_{water}$$
$$m_{ice}(3.35 \times 10^5 \text{ J/kg}) = -(-1.04 \times 10^4 \text{ J})$$

In other words,

$$m_{ice} = \frac{1.04 \times 10^4 \text{ J}}{3.35 \times 10^5 \text{ J/kg}} = 3.10 \times 10^{-2} \text{ kg}$$

So you need 3.10×10^{-2} kilograms, or 31.0 grams of ice.

"Pardon me," you say to the waiter. "Please bring me exactly 31.0 grams of ice at precisely 0°C."

Chapter 15

Here, Take My Coat: How Heat Is Transferred

In This Chapter

- Examining natural and forced convection
- Transferring heat through conduction
- Shining the light on radiation

Heat is the flow of thermal energy from one point to another (see Chapter 14). You witness the transfer of heat every day. You cook some pasta, and you see currents of water cycling the noodles in the pan. You pick up the pan without a hand towel, and you burn your hand. You look to the sky on a summer day, and you feel your face warming up. You give your coat to your date, and you watch his or her feelings for you warm up (through radiation, of course!).

In this chapter, I discuss the three primary ways in which heat can be transferred. You find out how to predict how quickly pot handles get hot, see why heat rises, and discover how the sun warms the Earth.

Convection: Letting the Heat Flow

Convection is a means of transferring thermal energy (heat) in a fluid. In *convection,* the flowing fluid carries energy along, mixing with the rest of the fluid and thereby transferring the thermal energy. Through this mixing, thermal energy moves from a higher-temperature region to a lower-temperature region.

Convection occurs in both liquids and gases, because both liquids and gases are fluids. *Buoyancy,* which is the upward force on the part of the fluid that's less dense than the surrounding fluid, often drives the fluid's motion. Fluids expand when you add heat, changing the fluid's density (refer to Chapter 14 for info on thermal expansion). Cooler, denser regions of fluid tend to sink as warmer, less dense regions rise, causing the fluid to flow.

Figure 15-1 shows a cross-section of a pan of water coming to the boil. The water at the bottom heats up, expands slightly, and then rises in the pan by buoyant forces. The warm fluid carries the thermal energy from the bottom of the pan to the top.

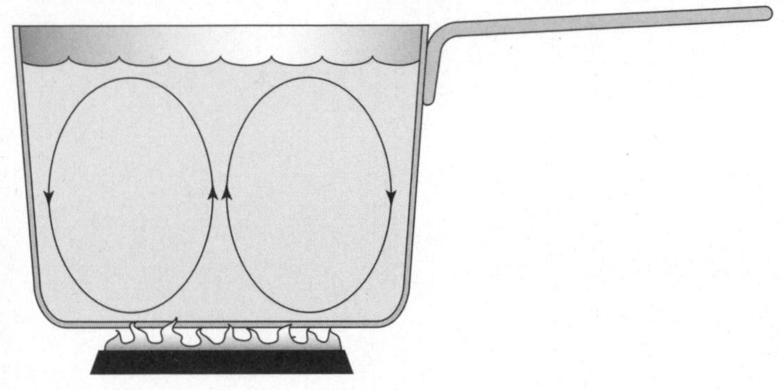

Figure 15-1: You can see convection in action by boiling a pot of water.

Convection may be natural or forced, and the following subsections give you the story on both.

Hot fluid rises: Putting fluid in motion with natural convection

You may have heard the maxim "heat rises," which is all about convection. However, a more accurate statement is that "hot fluid rises." In substances where convection is free to take place — that is, in gases and liquids — hotter material naturally ends up on top and cooler material ends up on the bottom because of buoyancy.

If your house has two floors, you often end up with the bottom floor being cooler than the top floor. The warmer air rises by buoyancy, which drives the convection. Physicists refer to this type of convection as *natural convection* because it isn't externally driven.

To understand how natural convection works, look at the microscopic picture. Any substance is made up of molecules, tiny particles that zip around at varying speeds. When a gas or liquid becomes hot, its molecules move faster. If you have a heating element that contacts the bottom of the substance — such as a wood stove at the bottom of a room or a stove element that's heating a kettle of water — the molecules near the heating element become hot. Hotter molecules have more kinetic energy and so can zip around faster and hit other molecules harder.

Because they move faster and hit harder, hotter molecules make the substance in their immediate area less dense. That is, they have more energy with which to push other molecules out of the way. The molecules that have been hit also have more energy to push other molecules out of the way, so the substance in the immediate vicinity of the heating element becomes less dense.

A unit volume of material that's less dense weighs less than a unit volume of the surrounding material, and if that material is a gas or liquid, the less dense stuff rises. Because the denser material has more mass per volume, it sinks under the influence of gravity.

Anyone who has ever flown in a plane is familiar with natural convection in the form of turbulence. Turbulence is caused by the sun's heating of the Earth, which in turn heats the air above it. The hot air rises through the atmosphere, and airplanes, going along on their merry way, fly through these rising columns of rising hot air. If you look out the window of the plane, you may see birds riding these columns, called *thermals,* as well. If you see a bird rising but not flapping its wings, it's most likely hitching a ride on a thermal.

Controlling the flow with forced convection

With natural convection, you rely on the fact that hot fluid rises to transfer heat. But sometimes natural convection is the opposite of what you want. With *forced convection,* you control the movement of the warm or cool fluid, often using a fan or pump.

For example, take a room on a cold winter's day. Because heat rises, the hotter air in the room drifts up to the ceiling, while the cooler air in the room settles near the bottom of the room, where you are. So in time, all the hot air in the room collects near the ceiling, and all the cold air collects near the floor. Although you were originally quite cozy, you may now be getting pretty cold — all as a result of natural convection.

What can you do? You can turn on your room's ceiling fan in reverse! Ceiling fans force the air to circulate, so the hot air near the top of the room moves downward. The warmer air at the top of the room now ends up at the bottom of the room again, where you are. Just make sure you choose a low speed so you don't create a breeze.

You find forced convection all around you. The fans in a desktop computer, for example, cause forced convection (and the lack of room for a fan in laptop computers has caused plenty of overheating problems). Refrigerators use fans to blow away heat, again relying on forced convection.

Natural convection is the dominant way heat is transferred through the interior of any standard oven (microwave ovens, on the other hand, use electromagnetic radiation to jostle the water molecules in the food — see Chapter 19 for details). In traditional ovens, the buoyancy of the hot air distributes the heat. Ovens specifically labeled *convection ovens* use a fan to increase the heat transfer by convection. The fan drives the air inside the oven to move more and thereby distribute the heat faster.

Here's a last example, this time of forced convection happening twice in the same system. Cars generate a lot of heat when they're running. To keep the engine cool, a pump circulates coolant throughout the engine. The liquid coolant transfers heat away from the engine to the radiator to keep the car from overheating. And the radiator itself is another example of forced convection, moving the air not with a cooling fan but with the motion of the car itself: the car drives air through the radiator, cooling it as the car moves. When the car isn't moving, the engine produces less heat, so there's less need to dissipate heat from the radiator.

Too Hot to Handle: Getting in Touch with Conduction

Conduction transfers heat through material directly, through contact. Take a look at the metal pot in Figure 15-2 and its metal handle; the pot has been boiling for 15 minutes. Would you want to lift it off the fire by grabbing the handle without an oven mitt? Probably not. The handle is hot because of conduction of heat through the metal handle.

On the molecular level, the molecules near the heat source are heated and begin vibrating faster. They bounce off nearby molecules and cause them to vibrate faster. That increased bouncing is what heats a substance.

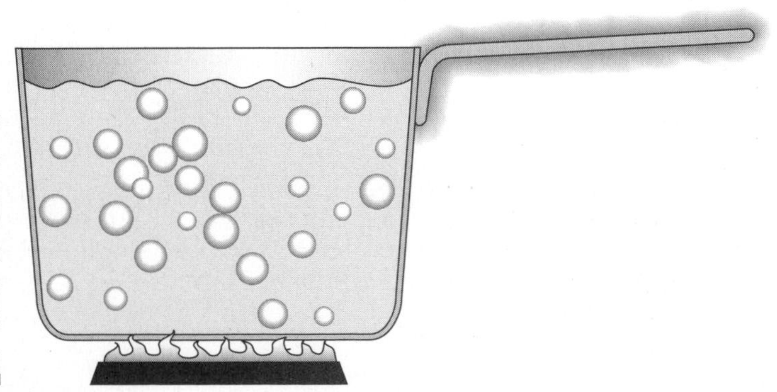

Figure 15-2: Conduction heats the pot that holds the boiling water.

How the elephant got its ears: A physics lesson in body design

As bodies become larger, their volume grows faster than their surface area. Cooling a larger body becomes more difficult because for every unit of volume of the body, there's less surface area through which the heat can escape. This idea also applies to animals, and it partly explains why the elephant needs such large ears. Because an elephant has such a large body, it has lots of heat to conduct through its body and then from its skin to the air; but relative to its large volume, the elephant doesn't have much surface area through which to conduct the heat. So the elephant sports two great big ears, with a large surface area, through which to conduct away its heat.

Some materials, such as most metals, conduct heat better than others, such as porcelain, wood, or glass. The way substances conduct heat depends a great degree on their molecular structures, so different substances react differently.

Finding the conduction equation

You have to take different properties of objects into account when you want to examine the conduction that takes place. If you have a bar of steel, for example, you have to consider the bar's area and length, along with the temperature at different parts of the bar.

Take a look at Figure 15-3, where a bar of steel is being heated on one end and the heat is traveling by conduction toward the other side. Can you find out the thermal energy transferred? No problem.

Here are the factors that affect the rate of conduction:

- **Temperature difference:** The greater the difference in temperature between the two ends of the bar, the greater the rate of thermal energy transfer, so more heat is transferred. The heat, Q, is proportional to the difference in temperature, ΔT:

  $$Q \propto \Delta T$$

- **Cross-sectional area:** A bar twice as wide conducts twice the amount of heat. In general, the amount of heat conducted, Q, is proportional to the cross-sectional area, A, like this:

  $$Q \propto A$$

- **Length (distance heat must travel):** The longer the bar, the less heat that will make it all the way through. Therefore, the conducted heat is inversely proportional to the length of the bar, *l*:

  $$Q \propto \frac{1}{l}$$

- **Time:** The amount of heat transferred, *Q*, depends on the amount of time that passes, *t* — twice the time, twice the heat. Here's how you express this idea mathematically:

  $$Q \propto t$$

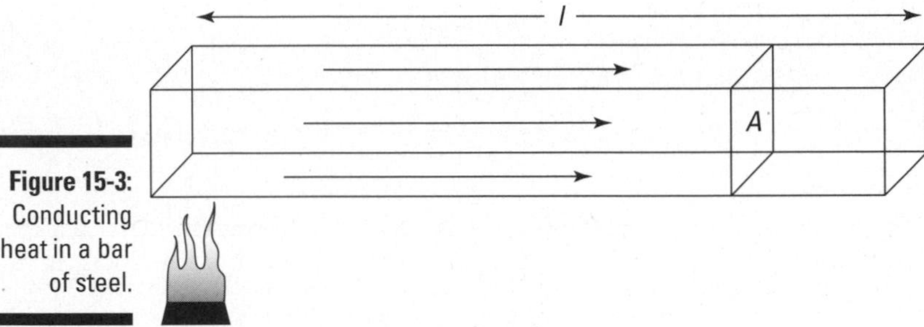

Figure 15-3: Conducting heat in a bar of steel.

Now you can put the variables together, using *k* as a constant of proportionality that's yet to be determined.

Here's the equation for heat transfer by conduction through a material:

$$Q = \frac{kA\Delta Tt}{l}$$

This equation represents the amount of heat transferred by conduction in a given amount of time, *t*, down a length *l*, where the cross-sectional area is *A*. Here, *k* is the material's *thermal conductivity,* measured in joules per second-meters-degrees Celsius, or J/s·m·°C.

Working with thermal conductivity

Different materials (such as glass, steel, copper, and bubble gum) conduct heat at different rates, so the thermal conductivity constant depends on the material in question. Lucky for you, physicists have measured the constants for various materials already. Check out some of the values in Table 15-1.

Table 15-1 Thermal Conductivities for Various Materials

Material	*Thermal Conductivity (J/s·m·°C)*
Diamond	1,600
Silver	420
Copper	390
Brass	110
Lead	35
Steel	14.0
Glass	0.80
Water	0.60
Body fat	0.20
Wood	0.15
Wool	0.04
Air	0.0256
Styrofoam	0.01

The thermal conductivity of the steel part of a pot handle is 14.0 J/s·m·°C (see Table 15-1). Take a look at Figure 15-3. Suppose the handle is 15 centimeters long, with a cross-sectional area of 2.0 square centimeters (2.0×10^{-4} m^2). If the fire at one end is 600°C, how much heat would be pumped into your hand in 1 second if you grabbed the handle? The equation for heat transfer by conduction is

$$Q = \frac{kA\Delta Tt}{l}$$

If you assume that the end of the cool end of the handle starts at about room temperature, 25°C, you get the following amount of heat transferred in a time t:

$$Q = \frac{kA\Delta Tt}{l} = \frac{(14 \text{ J/s}\cdot\text{m}\cdot°\text{C})(2.0\times10^{-4} \text{ m}^2)(600°\text{C} - 25°\text{C})t}{0.15 \text{ m}}$$
$$= (10.7 \text{ J/s})t$$

You can see that in 1 second, 10.7 joules of thermal energy would enter your hand.

If 10.7 joules of heat is being transferred to the end of the handle each second, then the heat transfer is 10.7 joules per second, or 10.7 watts. As the seconds go by, the joules of heat add up, making the handle hotter and hotter. Note that the conduction rate of 10.7 watts will decrease with time because the end of the handle heats up, giving you a smaller value for ΔT.

Camping with the Johnsons: A conduction example

The vacationing Johnson family wants to know whether they have enough ice in their ice chest to last for 12 hours while they're out camping, so they ask you, the famous consulting physicist. A quick glance at your outdoor thermometer tells you that the outside temperature is 35°C. You measure the walls of the Styrofoam ice chest to be 2.0 centimeters thick. The total surface area of the ice chest is 0.66 square meters.

The final measurement you note on your clipboard is that the Johnsons have loaded the ice chest with exactly 1.5 kilograms of ice at 0°C. So how long will 1.5 kilograms of ice take to melt in that chest?

Because, in this situation, you have an amount of heat conducting through a material of known surface area and thickness, you can start with the conduction equation:

$$Q = \frac{kA\Delta Tt}{l}$$

You want to know the time, so solve the equation for t:

$$t = \frac{Ql}{kA\Delta T}$$

Now think about which values you already know. The heat has to travel the width of the cooler to escape, so $l = 2.0 \text{ cm} = 0.020 \text{ m}$. Styrofoam's thermal conductivity is $k = 0.010 \text{ J/s·m·°C}$, and the ice chest's surface area is $A = 0.66 \text{ m}^2$. The difference between inside and outside temperature is $\Delta T = 35°\text{C} - 0°\text{C} = 35°\text{C}$. Now all you need is Q, the amount of heat needed to melt the ice.

You can use water's latent heat of fusion to figure out how much heat you need to change the ice from a solid to liquid state (see Chapter 14 for details on heat and phase changes). In general, you need the following amount of heat to melt ice at 0°C:

$$Q = mL$$

where Q is the amount of heat needed, m is the mass of ice, and L is water's latent heat of fusion, 3.35×10^5 J/kg. Plugging in the numbers gives you the amount of heat you need:

$$Q = mL = (1.5 \text{ kg})(3.35 \times 10^5 \text{ J/kg}) \approx 5.0 \times 10^5 \text{ J}$$

Now you know Q, so you have enough info to use the conduction equation. Plugging the numbers in the conduction equation (solved for time) gives you the answer:

$$t = \frac{(5.0 \times 10^5\ \text{J})(0.020\ \text{m})}{(0.010\ \text{J/s}\cdot\text{m}\cdot{}^\circ\text{C})(0.66\ \text{m}^2)(35^\circ\text{C})} \approx 44{,}000\ \text{s}$$

You tell the Johnsons, "It will take 44,000 seconds for your ice to melt."

"How long is that?" they ask.

Well, you think, 60 seconds are in a minute, and 60 minutes are in an hour, so you do a few more calculations:

$$t = (44{,}000\ \cancel{\text{s}})\left(\frac{1\ \cancel{\text{min}}}{60\ \cancel{\text{s}}}\right)\left(\frac{1\ \text{hr}}{60\ \cancel{\text{min}}}\right) \approx 12\ \text{hr}$$

"Your ice will last for 12 hours," you tell them, handing them your bill.

Considering conductors and insulators

Materials with high *thermal conductivity,* such as copper, conduct heat well. For example, you may have seen copper wire in indoor/outdoor thermometers. The wires conduct heat in from outside so that the thermometer can measure the outside temperature. Diamond is a far better conductor of heat than copper, as you can see in Table 15-1 (but building indoor/outdoor thermometers with diamonds would be a little pricey).

On the other end of the scale, some materials act as heat *insulators* because their thermal conductivity is so low. For example, body fat has a low thermal conductivity, as you can see in Table 15-1, and so it's a natural insulator. Thus, body fat can help keep you warm on cold days.

Of course, in terms of conduction, the champion thermal insulator of all time is a vacuum. Conduction relies on the movement of heat through material, and if there's no material, there can't be any conduction. Nothing can conduct the heat in that case.

For that reason, people use vacuum flasks to keep foods hot or cold. These flasks have a double wall with a vacuum between the walls, so no heat can be conducted from inside to outside or outside to inside. Therefore, your soup stays hot, or your iced tea stays cold.

Cool to the touch

Why do metals feel cold to the touch when they're at room temperature? If you know something about thermal conduction, the phenomenon makes sense. Metals are such good thermal conductors that they carry away the heat from your fingers very quickly, which leads to a drop in temperature in your skin. The nerves in your skin detect this drop in temperature and send the message to your brain that you're touching something cold. Wood, on the other hand, is not a good conductor (it's an insulator), so little heat is conducted away from your fingertips. Your brain interprets this to mean that the wood is warmer than the metal, when they're actually both in thermal equilibrium with the room — that is, the same temperature!

Some thermal conductivity does exist between inside and outside in a vacuum flask. There's always some path for heat to take, such as the stopper of the flask itself. Because some heat is conducted, vacuum flasks keep hot foot hot or cold food cold only for a little while. Theoretically, if you had a capsule of food floating in a vacuum, there'd be no heat loss or gain through conduction at all — but there'd still be heat loss or gain through radiation.

Radiation: Riding the (Electromagnetic) Wave

Radiation is another way to transfer heat. You experience radiation personally whenever you get out of the shower soaking wet in the dead of winter and bask in the warmth of the heat lamp in your bathroom. Why? Because of a little physics, of course. The heat lamp, which you see in Figure 15-4, beams out heat to you and keeps you warm through radiation.

With *radiation,* electromagnetic waves carry the energy (you can find plenty of info on electromagnetic waves in *Physics II For Dummies*). Electromagnetic radiation comes from accelerating electric charges. On a molecular level, that's what happens as objects warm up — their molecules move around faster and faster and bounce off other molecules hard.

Heat energy transferred through radiation is as familiar as the light of day; in fact, it *is* the light of day. The sun is a huge thermal reactor about 93 million miles away in space, and neither conduction nor convection can produce any of the energy that arrives to Earth through the vacuum of space. The sun's energy gets to the Earth through radiation, which you can confirm on a sunny day just by standing outside and letting the sun's rays warm your face.

Figure 15-4: An incandescent light bulb radiates heat into its environment.

Mutual radiation: Giving and receiving heat

Every object around you is continually radiating, unless its temperature is at absolute zero (which is a little unlikely because you can't physically get to a temperature of absolute zero, with no molecular movement). A scoop of ice cream, for example, radiates. Even you radiate all the time, but that radiation isn't visible as light because it's in the infrared part of the spectrum. However, that light is visible to infrared scopes, as you've probably seen in the movies or on television.

You radiate heat in all directions all the time, and everything in your environment radiates heat back to you. When you have the same temperature as your surroundings, you radiate as fast and as much to your environment as it does to you. When two things are in thermal contact but no thermal energy is exchanged between them, they're in *thermal equilibrium*. If two things are in thermal equilibrium, they have the same temperature.

If your environment didn't radiate heat back to you, you'd freeze, which is why space is considered so "cold." There's nothing cold to touch in space, and you don't lose heat through conduction or convection. Rather, the environment doesn't radiate back at you, which means that the heat you radiate away is lost. You can freeze very fast from the lost heat.

When an object heats up to about 1,000 kelvins, it starts to glow red (which may explain why, even though you're radiating, you don't glow red in the visible light spectrum). As the object gets hotter, its radiation moves up in the spectrum through orange, yellow, and so on up to white hot at somewhere around 1,700 K (about 2,600°F).

Radiant heaters with coils that glow red rely on radiation to transfer heat. Convection takes place as air heats, rises, and spreads around the room (and conduction can occur if you touch the heater on a hot spot by mistake — not the most desirable of heat transfers!). But the heat transfer to you takes place mostly through radiation.

Blackbodies: Absorbing and reflecting radiation

Humans understand heat radiation and absorption in the environment intuitively. For example, on a hot day, you may avoid wearing a black t-shirt, because you know it would make you hotter. A black t-shirt absorbs light from the environment while reflecting less of it back than a white t-shirt. The white t-shirt keeps you cooler because it reflects more radiant heat back to the environment.

Some objects absorb more of the light that hits them than others. Objects that absorb all the radiant energy that strikes them are called *blackbodies.* A blackbody absorbs 100 percent of the radiant energy striking it, and if it's in equilibrium with its surroundings, it emits all the radiant energy as well.

In terms of reflection and absorption of radiation, most objects fall somewhere between mirrors, which reflect almost all light, and blackbodies, which absorb all light. The middle-of-the-road objects absorb some of the light striking them and emit it back into their surroundings. Shiny objects are shiny because they reflect most of the light, which means they don't have to emit as much heat radiantly into the room as other objects. Dark objects appear dark because they don't reflect much light, which means they have to emit more as radiant heat (usually lower down in the spectrum, where the radiation is infrared and can't be seen).

The Stefan-Boltzmann constant

How much heat does a blackbody emit when it's at a certain temperature? The amount of heat radiated is proportional to the time you allow — twice as long, twice as much heat radiated, for example. So you can write the heat relation, where t is time, as follows:

$$Q \propto t$$

And as you may expect, the amount of heat radiated is proportional to the total area doing the radiating. So you can also write the relation as follows, where A is the area doing the radiating:

$$Q \propto At$$

Temperature, T, has to be in the equation somewhere — the hotter an object, the more heat radiated. Experimentally, physicists found that the amount of heat radiated is proportional to T to the fourth power, T^4. So now you have the following relation:

$$Q \propto AtT^4$$

To show the exact relationship between heat and the other variables, you need to include a constant, which physicists measured experimentally. To find the heat emitted by a blackbody, you use the Stefan-Boltzmann constant, σ, which goes in the equation like this:

$$Q = \sigma AtT^4$$

The value of σ is 5.67×10^{-8} J/s·m^2·K^4. Note, however, that this constant works only for blackbodies that are perfect emitters.

The Stefan-Boltzmann law of radiation

Most objects aren't perfect emitters, so you have to add another constant most of the time — one that depends on the substance you're working with. The constant is called *emissivity, e.*

The *Stefan-Boltzmann law of radiation* says the following:

$$Q = e\sigma AtT^4$$

where e is an object's emissivity, σ is the Stefan-Boltzmann constant 5.67×10^{-8} J/s·m^2·K^4, A is the radiating area, t is time, and T is the temperature in kelvins.

Finding heat from the human body

A person's emissivity is about 0.98. At a body temperature of 37°C, how much heat does a person radiate each second? First, you have to factor in how much area does the radiating. If you know that the surface area of the human body is $A = 1.7$ m^2, you can find the total heat radiated by a person by plugging the numbers into the Stefan-Boltzmann law of radiation equation, making sure you convert the temperature to kelvins:

$$Q = e\sigma AtT^4$$
$$Q = (0.98)(5.67\times10^{-8}\ \text{J/s}\cdot\text{m}^2\cdot\text{K}^4)(1.07\ \text{m}^2)[(37+273.15)\text{K}]^4 t$$
$$Q \approx 550t$$

Then dividing both sides by t, you get

$$\frac{Q}{t} = 550\ \text{W}$$

You get a value of 550 joules per second, or 550 watts. That may seem high, because skin temperature isn't the same as internal body temperature, but it's in the ballpark.

Doing star calculations

Here's another example: A knock sounds on your door at around 10 p.m. Surprised, you open the door, and a number of astronomers enter. "We need you to measure the radius of Betelgeuse," they say.

"Betelgeuse, the star?" you ask. "You want me to measure the radius of a star 640 light-years from Earth?"

"If it's not too much trouble," they reply. "We heard it was a supergiant star, and we wanted to know how big it was."

You get out your telescope and find Betelgeuse. Using the set of instruments you always carry in your pocket, you use the spectrum of the star to measure its temperature (the distribution of the intensity of the light over the different wavelengths is directly related to its surface temperature because stars radiate like blackbodies). The temperature is about 2,900 kelvins and the star's power output is 4.0×10^{30} watts.

Because you know the rate at which the star is radiating energy and its surface temperature, you can use the Stefan-Boltzman law of radiation to relate the star's surface area to these known values. Then assuming the star is a sphere, you can easily work out the radius of the sphere that has that surface area.

You know that $Q = e\sigma AtT^4$, so the power is

$$\frac{Q}{t} = e\sigma AT^4$$

And you can solve for the surface area of the star, A, like this:

$$\frac{Q/t}{e\sigma T^4} = A$$

Assuming that Betelgeuse is a sphere, you can connect the surface area to the star's radius by using this formula for spheres:

$$A = 4\pi r^2$$

Solve for r:

$$r = \sqrt{\frac{A}{4\pi}}$$

Plugging in the star's surface-area expression for A, you get the following:

$$r = \sqrt{\frac{Q/t}{4\pi e\sigma T^4}}$$

Assuming that $e = 1$, plugging in numbers gives you:

$$r = \sqrt{\frac{\left(4.0\times10^{30}\text{ J/s}\right)}{4\pi(1)\left(5.67\times10^{-8}\text{ J/s}\cdot\text{m}^2\cdot\text{K}^4\right)(2{,}900\text{ K})^4}}$$
$$\approx 2.8\times10^{11}\text{ m}$$

That's a pretty big radius for a star. If the sun had that radius, the Earth would be inside it — and so would Mars.

"Two hundred eighty million kilometers," you tell the astronomers, and you hand them your bill.

Chapter 16

In the Best of All Possible Worlds: The Ideal Gas Law

In This Chapter

- Getting gassy with Avogadro's number
- Finding your ideal gas
- Mastering Boyle's and Charles's laws
- Keeping up with the molecules in ideal gases

Gas gets just about everywhere — in balloons, in the wind, in your stove, even in your lungs. In physics, an atom-by-atom (or molecule-by-molecule) knowledge of these gases is essential when you start working with heat, pressure, volume, and more.

Get out your stomach pills, because in this chapter, you get gassy! This chapter focuses on the ideal gas law, which explains the relationship among pressure, heat, volume, and the amount of a gas. But first, I introduce you to the mole, a measurement that helps you work with gases on the molecular level.

Digging into Molecules and Moles with Avogadro's Number

To look at gases on the molecular level, you need to know how many molecules you have in a certain sample. Counting the molecules is impractical, so instead, physicists use a measurement called a *mole* to relate the mass of a sample to the number of molecules it contains.

A *mole* (abbreviated *mol*) is the number of atoms in 12.0 grams of carbon isotope 12. *Carbon isotope 12* — also called carbon-12, or just carbon 12 — is the most common version of carbon. *Isotopes* of an element have the same number of protons but different numbers of neutrons. Carbon-12 has six protons and six neutrons (a total of 12 particles); however, some carbon atoms (isotopes) have a few more neutrons in them — carbon-13, for example, has seven neutrons. The average mass of a mole of a mixture of the carbon isotopes works out to be 12.011 grams.

The number of atoms in one mole (in 12.0 grams of carbon-12) has been measured as 6.022×10^{23}, which is called *Avogadro's number,* N_A.

Do you find the same number of atoms in, say, 12.0 grams of sulfur? Nope. Each sulfur atom has a different mass from each carbon atom, so even if you have the same number of grams, you have a different number of atoms.

How much more mass does an atom of sulfur have than an atom of carbon-12? If you check the periodic table of elements hanging on the wall in a science lab, you find that the *atomic mass* of sulfur is 32.06. (***Note:*** The atomic mass usually appears under the element's symbol.) But 32.06 what? It's 32.06 *atomic mass units*, u, where each atomic mass unit is 1/12 of the mass of a carbon-12 atom.

A mole of carbon-12 (6.022×10^{23} atoms of carbon-12) has a mass of 12.0 grams, and the mass of your average sulfur atom is bigger than the mass of a carbon-12 atom:

$$\text{Sulfur mass} = 32.06 \text{ u}$$

$$\text{Carbon 12 mass} = 12 \text{ u}$$

Therefore, a mole of sulfur atoms must have this mass:

$$\frac{32.06 \text{ u}}{12 \text{ u}}(12.0 \text{ g}) = 32.06 \text{ g}$$

How convenient! A mole of an element has the same mass in grams as its atomic mass in atomic units. You can read the atomic mass of any element in atomic units off any periodic table. For instance, you can find that a mole of silicon (atomic mass: 28.09 u) has a mass of 28.09 grams, a mole of sodium (atomic mass: 22.99 u) has a mass of 22.99 grams, and so on. Each of those moles contains 6.022×10^{23} atoms.

Now you can determine the number of atoms in a diamond, which is solid carbon (atomic mass: 12.01 u). A mole is 12.01 grams of diamond, so when you find out how many moles you have, you multiply that times 6.022×10^{23} atoms. Then if you like, you can work out how many atoms of carbon are in a 1 carat diamond: 1 carat is equal to 0.200 grams, so here's how many atoms you have:

$$\frac{0.200 \text{ g}}{12.01 \text{ g}}\left(6.022 \times 10^{23}\right) \approx 1.00 \times 10^{22}$$

Not every object is made up of a single kind of atom. When atoms combine, you have molecules. For example, water is made up of two hydrogen atoms for every one oxygen atom (H_2O). Instead of the atomic mass, you look for the *molecular mass,* which is also measured in atomic mass units. For example, the molecular mass of water is 18.0153 atomic mass units, so 1 mole of water molecules has a mass of 18.0153 grams.

Some physics problems provide the molecular mass; others require you to calculate molecular mass by using the atomic mass and the compound's molecular formula. That is, you add the atomic masses of the individual atoms in the molecule.

Relating Pressure, Volume, and Temperature with the Ideal Gas Law

When you start working atom-by-atom and molecule-by-molecule, you begin working with gases from a physics point of view. For example, you can relate the temperature, pressure, volume, and number of moles together for a gas. The relation I introduce in this section doesn't always hold true, but it always works for ideal gases.

Ideal gases are those gases for which the ideal gas law holds. This law is an idealized model in which the gas's particles are small compared to the average distance between them and only interact by colliding elastically. It so happens that there is no such "ideal" gas, but real gases best approximate this scenario when the pressure is low and the temperature is high. Ideal gases are very light, like helium.

Forging the ideal gas law

By using the *ideal gas law,* you can predict the pressure of an ideal gas if given how much gas you have, its temperature, and the volume you've enclosed it in. Here's how the various factors affect pressure:

- **Temperature:** Experiments show that if you keep the volume constant and heat a gas, the pressure goes up linearly, as you see in Figure 16-1. In other words, at a constant volume, where T is the temperature measured in kelvins and P is the pressure, the pressure is proportional to temperature:

  $$P \propto T$$

- **Volume:** If you let the volume vary, you also find that the pressure is inversely proportional to the volume:

  $$P \propto \frac{T}{V}$$

  For instance, if the volume of a gas doubles, its pressure is cut in half.

- **Moles:** When the volume and temperature of an ideal gas are constant, the pressure is proportional to the number of moles of gas you have — twice the amount of gas, twice the pressure (see the earlier section "Digging into Molecules with Avogadro's Number" for info on moles). If the number of moles is n, then you can say the following:

  $$P \propto \frac{nT}{V}$$

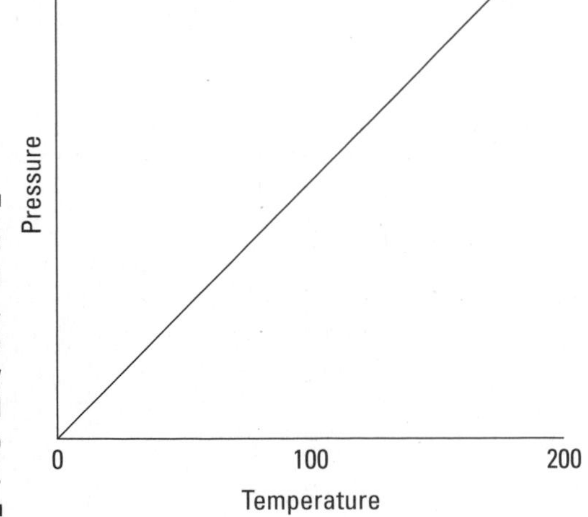

Figure 16-1: For an ideal gas, pressure is directly proportional to temperature.

Adding a constant, R — *the universal gas constant,* which has a value of 8.31 joules/mole-kelvin (J/mol·K) — gives you the *ideal gas law,* which relates pressure, volume, number of moles, and temperature:

$$PV = nRT$$

The unit of pressure is the pascal and the unit of volume is meters^3, and they combine to give the joule; when the quantity of gas, n, is measured in moles and the temperature, T, is measured in kelvins, then the units of the universal gas constant, R, are joules/mole-kelvin (J/mol·K).

You can also express the ideal gas law a little differently by using the total number of molecules, N, and Avogadro's number, N_A (see the earlier section "Digging into Molecules and Moles with Avogadro's Number"):

$$PV = nRT = \left(\frac{N}{N_A}\right)RT$$

The constant R/N_A is also called *Boltzmann's constant, k,* and it has a value of 1.38×10^{-23} J/K. Using this constant, the ideal gas law becomes

$$PV = NkT$$

Say that you're measuring a volume of 1 cubic meter filled with 600 moles of helium at room temperature, 27°C, which is very close to an ideal gas under these conditions. What's the pressure of the gas? Using this form of the ideal gas law, $PV = nRT$, you can put P on one side by dividing by V. Now plug in the numbers, making sure you convert temperature to kelvins (see Chapter 14 for details):

$$P = \frac{nRT}{V} = \frac{(600.0 \text{ mol})(8.31 \text{ J/mol}\cdot\text{K})\left[(273.15+27)\text{ K}\right]}{1.0 \text{ m}^3} \approx 1.50\times10^6 \text{ N/m}^2$$

The pressure on all the walls of the container is 1.50×10^6 N/m^2. Notice the units of pressure here — newtons per square meter. The unit is used so commonly that it has its own name in the MKS system: *pascals,* or Pa.

One pascal equals 1 newton per square meter, or 1.45×10^{-4} pounds per square inch. Atmospheric pressure is 1.013×10^5 pascals, which is 14.70 pounds per square inch. The pressure of 1 atmosphere is also given in *torr* on occasion, and 1.0 atmosphere = 760 torr.

In this example, you have a pressure of 1.50×10^6 pascals, which is about 15 atmospheres.

Working with standard temperature and pressure

You may come across a special set of conditions when talking about gases — *standard temperature and pressure,* or STP. The standard pressure is 1 atmosphere (or 1.013×10^5 Pa), and the standard temperature is 0°C (or 273.15 K).

You can use the ideal gas law to calculate that at STP, 1.0 mole of an ideal gas occupies 22.4 liters of volume (1.0 liter is 1×10^{-3} m^3). How do you get 22.4 liters? You know that $PV = nRT$, and solving for V gives you

$$V = \frac{nRT}{P}$$

Plug in STP conditions and do the math:

$$V = \frac{nRT}{P} = \frac{(1.0)(8.31 \text{ J/mol}\cdot\text{K})(273 \text{ K})}{1.013 \times 10^5 \text{ Pa}} \approx 22.4 \times 10^{-3} \text{ m}^3$$

And that's 22.4 liters, the volume 1 mole of ideal gas occupies at STP conditions.

A breathing problem: Checking your oxygen

Here's an example using the ideal gas law. There you are, walking in the park, when you notice a man sitting on a park bench and gasping. You ask what's wrong. "I don't think I'm getting enough oxygen in my lungs," he says.

You decide to check it out. Using a large plastic bag you happen to have with you, you measure his lung capacity as 5.0 liters. How many molecules of oxygen does that correspond to? You know that, to a good approximation, air is an ideal gas, so you can use the ideal gas law:

$$PV = NkT$$

Solving for N, the number of molecules, gives you the following equation:

$$N = \frac{PV}{kT}$$

The pressure in the lungs is about atmospheric pressure, so $P = 1.0 \times 10^5$ Pa. The temperature in the lungs is body temperature, so $T = 37°C$, or about 310 kelvins ($K = C + 273.15$). V is the volume of the lungs, which you've measured at 5.0 liters, or 0.0050 cubic meters. Putting this all together and doing the math gives you the answer:

$$N = \frac{PV}{kT} = \frac{\left(1.0 \times 10^5 \text{ Pa}\right)\left(0.0050 \text{ m}^3\right)}{\left(1.38 \times 10^{-23} \text{ J/K}\right)\left(310 \text{ K}\right)} \approx 1.2 \times 10^{23}$$

Being a physicist, you happen to know that the gas in the lungs is about 14 percent oxygen (a little less than in air by itself), so the number of oxygen molecules in the man's lungs equals

$$(0.14)(1.2 \times 10^{23} \text{ molecules}) \approx 1.7 \times 10^{22} \text{ molecules}$$

You turn to the man and say, "You have approximately 17 sextillion molecules of oxygen in your lungs." That's more than enough.

Boyle's and Charles's laws: Alternative expressions of the ideal gas law

You can often express the ideal gas law in different ways. For example, you can express the relationship between the pressure and volume of an ideal gas before and after one of those quantities changes at a constant temperature like this:

$$P_f V_f = P_i V_i$$

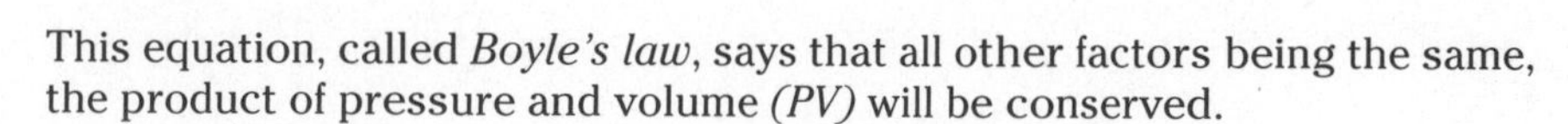
This equation, called *Boyle's law*, says that all other factors being the same, the product of pressure and volume *(PV)* will be conserved.

At constant pressure, you can say that the following relationship is true for an ideal gas:

$$\frac{V_f}{T_f} = \frac{V_i}{T_i}$$

This equation, called *Charles's law,* says that the ratio of volume to temperature *(V/T)* will be conserved for an ideal gas, all other factors being the same.

Here's an example that uses Boyle's law. You're taking a well-deserved vacation on the beach when the director of the Acme Tourism Company runs up to you and says, "We usually let our tourists scuba dive at 10.0 meters for ten minutes, but one of our tourists says he wants to stay down for a half an hour. He's going to run out of air, and we'll be sued!"

"Hmm," you say. "Let me have your data." You take some papers from the director and see that the Acme Tourism Company has scuba tanks with a volume of 0.015 cubic meters that are pressurized to 2.0×10^7 pascals. The diver breathes at a rate of 0.04 cubic meters per minute.

Because you're a physicist, you happen to know how scuba tanks work. They keep the air in the lungs at the same pressure as the surrounding water pressure (or else the lungs may collapse).

The way to tackle this problem is to take the volume of pressurized air inside the tank and work out what its volume would be if it were all released at the pressure you find at the submerged depth. You know the oxygen requirements of the body as a volume rate, so you can work out how long the pressurized air can sustain breathing from the volume the air would have at the submerged pressure. To work out the volume the pressurized air in the tank would have when released at the submerged depth, you use Boyle's law:

$$P_f V_f = P_i V_i$$

You know the pressure of the scuba tank and its volume. Now you need to know the pressure at the diver's depth to be able to calculate the available volume of air for the diver. You can find the water pressure with this equation (see Chapter 8), where P_w is the pressure of the water, ρ is the water's density, g is the acceleration due to gravity, and h is the change in depth:

$$P_w = \rho g h$$

The density of water is about 1,025 kilograms per cubic meter, and the diver descends 10.0 meters, so you have the following pressure from the water:

$$P_w = \rho g h = (1{,}025 \text{ kg/m}^3)(9.8 \text{ m/s}^2)(10.0 \text{ m}) \approx 1.0 \times 10^5 \text{ Pa}$$

To get the total pressure at the diver's depth, add the pressure of air on the surface of the water — that is, atmospheric pressure, P_a — to the water pressure:

$$P_f = P_a + P_w = (1.0 \times 10^5 \text{ Pa}) + (1.0 \times 10^5 \text{ Pa}) = 2.0 \times 10^5 \text{ Pa}$$

Now you're ready to use Boyle's equation:

$$P_f V_f = P_i V_i$$

where P_i is the pressure of the tank and V_i is the volume of the tank.

You want to find the volume of air available to the diver, so solve for V_f:

$$V_f = \frac{P_i V_i}{P_f}$$

Plugging in the numbers and doing the math gives you the volume of air available:

$$V_f = \frac{P_i V_i}{P_f} = \frac{(2.0\times 10^7 \text{ Pa})(0.015 \text{ m}^3)}{2.01\times 10^5 \text{ Pa}} \approx 1.5 \text{ m}^3$$

How long will that last? The diver breathes at a rate of 0.04 cubic meters per minute, so 1.5 cubic meters of air is enough for

$$\text{Total air time} = \frac{1.5 \text{ m}^3}{0.04 \text{ m}^3/\text{min}} \approx 38 \text{ min}$$

You tell the director that the diver will have enough air for 38 minutes. The director breathes a sigh of relief. "No lawsuit, then?"

"No lawsuit."

"But what if the diver goes down to 30.0 meters?" You figure out the new pressure from the water at 30 meters:

$$P_w = \rho g h = (1{,}025 \text{ kg/m}^3)(9.8 \text{ m/s}^2)(30.0 \text{ m}) \approx 3.0 \times 10^5 \text{ Pa}$$

And add air pressure to get

$$P_f = P_a + P_w = (1.01 \times 10^5 \text{ Pa}) + (3.0 \times 10^5 \text{ Pa}) = 4.01 \times 10^5 \text{ Pa}$$

As before, you want to find V_f, the volume of air available to the diver:

$$V_f = \frac{P_i V_i}{P_f}$$

Plugging in the numbers and doing the math gives you this result:

$$V_f = \frac{P_i V_i}{P_f} = \frac{(2.0\times 10^7 \text{ Pa})(0.015 \text{ m}^3)}{4.01\times 10^5 \text{ Pa}} \approx 0.75 \text{ m}^3$$

At a rate of 0.03 cubic meters per minute, 0.75 cubic meters of air is enough for

$$\text{Total air time} = \frac{0.75 \text{ m}^3}{0.04 \text{ m}^3/\text{min}} \approx 19 \text{ min}$$

"Uh oh," you tell the director.

"Lawsuit?" the director asks.

"Lawsuit."

Tracking Ideal Gas Molecules with the Kinetic Energy Formula

You can examine certain properties of molecules of an ideal gas as they zip around. For instance, you can calculate the average kinetic energy of each molecule with a very simple equation:

$$KE_{avg} = \frac{3}{2}kT$$

where k is Boltzmann's constant, 1.38×10^{-23} joules per kelvin (J/K), and T is the temperature in kelvins. And because you can determine the mass of each molecule if you know which gas you're dealing with (see the section "Digging into Molecules and Moles with Avogadro's Number" earlier in this chapter), you can figure out the molecules' speeds at various temperatures.

Predicting air molecule speed

Imagine you're at a picnic with friends on a beautiful spring day. You can't see the air molecules whizzing around you, but you can predict their average speeds. You get out your calculator and thermometer. You measure the air temperature at about 28°C, or 301 kelvins (see Chapter 14 for this conversion). You know that for the molecules in the air, you can measure their average kinetic energy with

$$KE_{avg} = \frac{3}{2}kT$$

Now plug in the numbers:

$$KE_{avg} = \frac{3}{2}\left(1.38 \times 10^{-23} \text{ J/K}\right)\left(301 \text{ K}\right) \approx 6.23 \times 10^{-21} \text{ J}$$

The average molecule has a kinetic energy of 6.23×10^{-21} joules. The molecules are pretty small — what speed does 6.23×10^{-21} joules correspond to?

Well, you know that $KE = (1/2)mv^2$ where m is the mass and v is the velocity (see Chapter 9). Therefore,

$$v = \sqrt{\frac{2KE}{m}}$$

Air is mostly nitrogen, and each nitrogen atom has a mass of about 14.0 u = 2.32×10^{-26} kg (you can figure that one out yourself by finding the mass of a mole of nitrogen and dividing by the number of atoms in a mole, N_A). In air, nitrogen molecules form molecules comprising two nitrogen atoms, so the mass of these molecules is 28.0 u = 4.65×10^{-26} kg. You can plug in the numbers to get

$$v = \sqrt{\frac{2KE}{m}} = \sqrt{\frac{2\left(6.23 \times 10^{-21}\text{ J}\right)}{\left(4.65 \times 10^{-26}\text{ kg}\right)}} \approx 518\text{ m/s}$$

Yow! What an image; huge numbers of the little guys crashing into you at 1,160 miles per hour! Good thing for you the molecules are so small. Imagine if each air molecule weighed a couple of pounds. Big problems.

Calculating kinetic energy in an ideal gas

Molecules have very little mass, but gases contain many, many molecules, and because they all have kinetic energy, the total kinetic energy can pile up pretty fast. How much total kinetic energy can you find in a certain amount of gas? Each molecule has this average kinetic energy:

$$KE_{avg} = \frac{3}{2}kT$$

To figure the total kinetic energy, you multiply the average kinetic energy by the number of molecules you have, which is nN_A, where n is the number of moles:

$$KE_{total} = \frac{3}{2}nN_AkT$$

N_Ak equals R, the universal gas constant (see the section "Forging the ideal gas law" earlier in this chapter), so this equation becomes the following:

$$KE_{total} = \frac{3}{2}nRT$$

If you have 6.0 moles of ideal gas at 27°C, here's how much internal energy is wrapped up in thermal movement (make sure you convert the temperature to kelvins):

$$KE_{total} = \frac{3}{2}nRT = \frac{3}{2}(6.0 \text{ mol})(8.31 \text{ J/K})\left[(273.15 + 27) \text{ K}\right] \approx 2.24 \times 10^6 \text{ J}$$

This converts to about 5 kilocalories, or *Calories* (the kind of energy unit you find on food wrappers).

Suppose you're testing out your new helium blimp. As it soars into the sky, you stop to wonder, as any physicist might, just how much internal energy there is in the helium gas that the blimp holds. The blimp holds 5,400 cubic meters of helium at a temperature of 283 kelvins. The pressure of the helium is slightly greater than atmospheric pressure, 1.1×10^5 pascals. So what is the total internal energy of the helium?

The total–kinetic energy formula tells you that $KE_{total} = (3/2)nRT$. You know T, but what's n, the number of moles? You can find the number of moles of helium with the ideal gas equation:

$$PV = nRT$$

Solving for n gives you the following:

$$n = \frac{PV}{RT}$$

Plug in the numbers and solve to find the number of moles:

$$n = \frac{PV}{RT} = \frac{(1.1 \times 10^5 \text{ Pa})(5{,}400 \text{ m}^3)}{(8.31 \text{ J/mol} \cdot \text{K})(283 \text{ K})} \approx 2.5 \times 10^5 \text{mol}$$

So you have 2.5×10^5 moles of helium. Now you're ready to use the equation for total kinetic energy:

$$KE_{total} = \frac{3}{2}nRT$$

Putting the numbers in this equation and doing the math gives you

$$KE_{total} = \frac{3}{2}(2.5 \times 10^5 \text{ mol})(8.31 \text{ J/mol} \cdot \text{K})(283 \text{ K}) \approx 8.8 \times 10^8 \text{ J}$$

So the internal energy of the helium is 8.8×10^8 joules. That's about the same energy stored in 94,000 alkaline batteries.

Chapter 17

Heat and Work: The Laws of Thermodynamics

In This Chapter

- Achieving thermal equilibrium
- Storing heat and energy under different conditions
- Revving up heat engines for efficiency
- Dropping close to absolute zero

If you've ever had an outdoor summer job, you know all about heat and work, a relationship encompassed by the term *thermodynamics.* This chapter brings together those two cherished topics, which I cover in detail in Chapter 9 (work) and Chapter 14 (heat).

Thermodynamics has laws one through three, much like Newton, but it does Newton one better: Thermodynamics also has a zeroth law. You may think that odd, because few other sets of everyday objects start off that way ("Watch out for that zeroth step — it's a doozy . . ."), but you know how physicists love their traditions.

In this chapter, I cover thermal equilibrium (the zeroth law), heat and energy conservation (the first law), heat flow (the second law), and absolute zero (the third law). Time to throw the book at thermodynamics.

Thermal Equilibrium: Getting Temperature with the Zeroth Law

Two objects are in *thermal equilibrium* if heat can pass between them but no heat is actually doing so. For example, if you and the swimming pool you're in are at the same temperature, no heat is flowing from you to it or from it to you (although the possibility is there). You're in thermal equilibrium. On the other hand, if you jump into the pool in winter, cracking through the ice covering, you won't be in thermal equilibrium with the water. And you don't want to be. (Don't try this physics experiment at home!)

To check for thermal equilibrium (especially in cases of frozen swimming pools that you're about to jump into), you should use a thermometer. You can check the temperature of the pool with the thermometer and then check your temperature. If the two temperatures agree — in other words, if you're in thermal equilibrium with the thermometer, and the thermometer is in thermal equilibrium with the pool — you're in thermal equilibrium with the pool.

The *zeroth law of thermodynamics* says that if two objects are in thermal equilibrium with a third, then they're in thermal equilibrium with each other. Then you can say that each of these objects has a thermal property that they all share — this property is called *temperature*.

Among other jobs, the zeroth law sets up the idea of temperature as an indicator of thermal equilibrium. The two objects mentioned in the zeroth law are in equilibrium with a third, giving you what you need to set up a scale such as the Kelvin scale.

Conserving Energy: The First Law of Thermodynamics

The *first law of thermodynamics* deals with energy conservation. One of the forms of energy involved is the *internal energy* that resides in the motion of the atoms and molecules (vibrations and random jostling). Another of the terms in this law is *heat,* which is a transfer of thermal energy. And finally, there is *work,* which is a transfer of mechanical energy; for example, work is done on a gas when it is compressed. The first law of thermodynamics states that these energies, together, are conserved. The initial internal energy in a system, U_i, changes to a final internal energy, U_f, when heat, Q, is absorbed or released by the system and the system does work, W, on its surroundings (or the surroundings do work on the system), such that

$$U_f - U_i = \Delta U = Q - W$$

For mechanical energy to be conserved (see Chapter 9), you have to work with systems where no energy is lost to heat — there could be no friction, for example. All that changes now. Now you can break down the total energy of a system, which includes heat, work, and the internal energy of the system.

These three quantities — heat, work, and internal energy — make up all the energy you need to consider. When you add heat, Q, to a system, and that system doesn't do work, the amount of internal energy in the system, which is given by the symbol U, changes by Q. A system can also lose energy by doing work on its surroundings, such as when an engine lifts weight at the end of a cable. When a system does work on its surroundings and gives off no waste heat, its internal energy, U, changes by W. In other words, you're in a position to think in terms of heat as energy, so when you take into account all three quantities — heat, work, and the internal energy — energy is conserved.

The first law of thermodynamics is a powerful one because it ties all the quantities together. If you know two of them, you can find the third.

Calculating with conservation of energy

The most confusing part about using $\Delta U = Q - W$ is figuring out which signs to use. The quantity Q (heat transfer) is positive when the system absorbs heat and negative when the system releases heat. The quantity W (work) is positive when the system does work on its surroundings and negative when the surroundings do work on the system.

To avoid confusion, don't try to figure out the positive or negative values of every mathematical quantity in the first law of thermodynamics; work from the idea of energy conservation instead. Think of values of work and heat flowing out of the system as negative:

- **The system absorbs heat:** $Q > 0$
- **The system releases heat:** $Q < 0$
- **The system does work on the surroundings:** $W > 0$.
- **The surroundings do work on the system:** $W < 0$.

Practicing the sign conventions

Say that a motor does 2,000 joules of work on its surroundings while releasing 3,000 joules of heat. By how much does its internal energy change? In this case, you know that the motor does 2,000 joules of work on its surroundings, so its internal energy *(U)* will decrease by 2,000 joules. And the system also releases 3,000 joules of heat while doing its work, so the internal energy of the system decreases by an additional 3,000 joules. Thinking this way makes the total change of internal energy the following:

$$\Delta U = -2{,}000 \text{ J} - 3{,}000 \text{ J} = -5{,}000 \text{ J}$$

The internal energy of the system decreases by 5,000 joules, which makes sense. On the other hand, what if the system *absorbs* 3,000 joules of heat from the surroundings while doing 2,000 joules of work on those surroundings? In this case, you have 3,000 joules of energy going in and 2,000 joules going out. The signs are now easy to understand:

$$\Delta U = -2{,}000 \text{ J [work going out]} + 3{,}000 \text{ J [heat coming in]} = 1{,}000 \text{ J}$$

In this case, the net change to the system's internal energy is +1,000 joules.

You can also see negative work when the surroundings do work on the system. Say, for example, that a system absorbs 3,000 joules at the same time that its surroundings perform 4,000 joules of work on the system. You can tell that both of these energies will flow into the system, so the system's internal energy goes up by 3,000 J + 4,000 J = 7,000 J. If you want to go by the numbers, use this equation:

$$\Delta U = Q - W$$

Then note that because the surroundings are doing work on the system, W is considered negative. Therefore, you get the following equation:

$$\Delta U = Q - W = +3{,}000 \text{ J} - (-4{,}000 \text{ J}) = 7{,}000 \text{ J}$$

Say that the system absorbs 1,600 joules of heat from the surroundings and performs 2,300 joules of work on the surroundings. What is the change in the system's internal energy? Use the equation $\Delta U = Q - W$. Here, Q is positive, because energy is absorbed by the system, and work is also positive, because work is done by the system, so you have

$$\Delta U = Q - W = +1{,}600 \text{ J} - (+2{,}300 \text{ J}) = 700 \text{ J}$$

So the internal energy of the system decreases by 700 joules.

Now say that the system absorbs 1,600 joules of heat while the surroundings do 2,300 joules of work on the system. What's the change in the internal energy of the system?

In this case, the work done by the system is negative — that is, the surroundings do work on the system. So using $\Delta U = Q - W$, you do the following calculations:

$$\begin{aligned}\Delta U &= Q - W \\ &= +1{,}600 \text{ J} - (-2{,}300 \text{ J}) \\ &= +1{,}600 \text{ J} + 2{,}300 \text{ J} \\ &= 3{,}900 \text{ J}\end{aligned}$$

So in this case, where the system both absorbs heat and work is done on it, the change in internal energy is 3,900 joules.

Trying a first-law-of-thermodynamics sample problem

The president of Acme Gas comes up to you, the world-famous physicist. "Our gases are getting lazy," the president says. "We have two processes, and we need to select the process where the gas does the most work for us. In both methods, the temperature of 6.0 moles of ideal gas is reduced from 590 kelvins to 400 kelvins. In method one, 5,500 joules of heat flow into the gas, while in method two, 1,500 joules of heat flow into the gas. So in which method does the gas do more work?"

Hmm, you think. Now's the time to use the equation $\Delta U = Q - W$. You want to find the work, so you solve for the work done by the gas:

$$W = Q - \Delta U$$

You know how much heat, Q, flows into the gas in each method because the president just told you those numbers. But what about the change in internal energy of the gas? You know that the internal kinetic energy of an ideal gas is the following (taking a tip from Chapter 16):

$$KE = \frac{3}{2}nRT$$

And because the gas is ideal, the molecules don't interact with one another, so the gas has no potential energy; therefore, the total internal energy of the gas is simply the kinetic energy:

$$U = KE = \frac{3}{2}nRT$$

That means that the internal energy of an ideal gas depends only on its temperature. Because the gas ends up with the same temperature change in both methods Acme Gas uses, the change of the internal energy of the gas will be the same in both cases.

In particular, the change in the internal energy of the gas in both methods is

$$\begin{aligned}\Delta U &= KE \\ &= \frac{3}{2}nRT \\ &= \frac{3}{2}(6.0 \text{ mol})(8.31 \text{ J/K}\cdot\text{mol})(400 \text{ K} - 590 \text{ K}) \\ &\approx -14{,}200 \text{ J}\end{aligned}$$

So because the ideal gas drops in temperature, the internal energy of the gas is reduced — in this case, by 14,200 joules.

Now you can plug the value of ΔU into the work equations for the ideal gas in both methods:

$$W = Q - (-14{,}200\text{ J})$$

In method one, the gas absorbs 5,500 joules, so you have

$$W_1 = 5{,}500\text{ J} - (-14{,}200\text{ J}) = 19{,}700\text{ J}$$

And for the second method, the gas absorbs 1,500 joules, so here's how much work the gas does:

$$W_2 = 1{,}500\text{ J} - (-14{,}200\text{ J}) = 15{,}700\text{ J}$$

"In method one," you tell the president of Acme Gas, "the gas does 19,700 joules of work. In method two, the gas only does 15,700 joules of work."

"We'll use method one, then," says the president. "And stop those gases from getting lazy!"

Staying constant: Isobaric, isochoric, isothermal, and adiabatic processes

You come across a number of quantities in this chapter — volume, pressure, temperature, and so on. The ways in which these quantities vary as work is done determine the final state of the system. For example, if a gas is doing work while you keep its temperature constant, the amount of work performed and the intermediate and final states of the system will be different from when you keep the gas's pressure constant instead.

A gas performs work only if the gas expands. You can show this idea mathematically. First, note that work, W, equals force, F, times distance, s (see Chapter 9):

$$W = Fs$$

In turn, force equals pressure, P, times area, A (see Chapter 8). This means that you can write work as pressure times area times distance:

$$W = PAs$$

Finally, the area times the distance *(As)* equals the change in volume, ΔV, so here's the new work equation:

$$W = P\Delta V$$

The formula $W = P\Delta V$ makes graphs of pressure versus volume very useful in thermodynamics. The curve you draw shows how pressure and volume change in relation to each other, and the area under the curve shows how much work is done.

In this section, I cover four standard conditions under which work is performed in thermodynamics: constant pressure, constant volume, constant temperature, and constant heat. I also graph pressure and volume for each of these processes and show you what work looks like. ***Note:*** When anything changes in these processes, the change is assumed to be *quasi-static,* which means the change comes slowly enough that the pressure and temperature are the same throughout the system's volume.

At constant pressure: Isobaric

When you have a process where the pressure stays constant, it's called *isobaric* (*baric* means "pressure"). In Figure 17-1, you see a cylinder with a piston being lifted by a quantity of gas as the gas gets hotter. The volume of the gas is changing, but the weighted piston keeps the pressure constant.

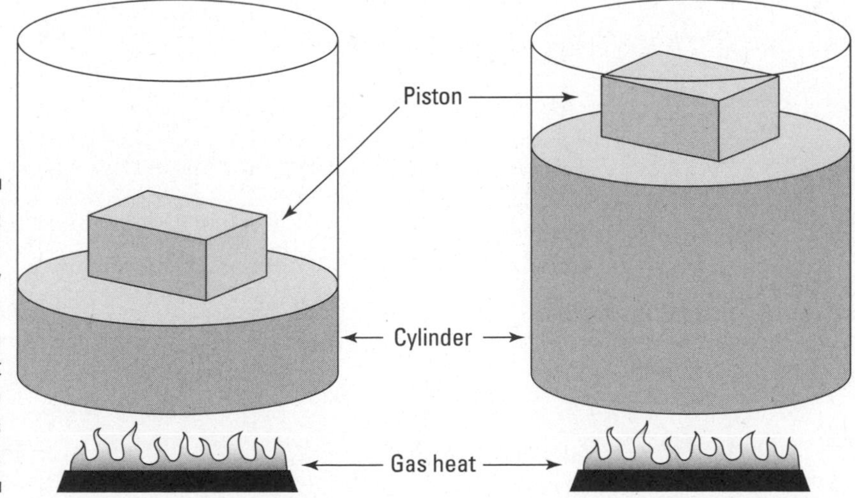

Figure 17-1: An isobaric system may feature a change in volume, but the pressure remains constant.

Graphically, you can see what the isobaric process looks like in Figure 17-2, where the volume is changing while the pressure stays constant. Because $W = P\Delta V$, the work is the shaded area beneath the graph.

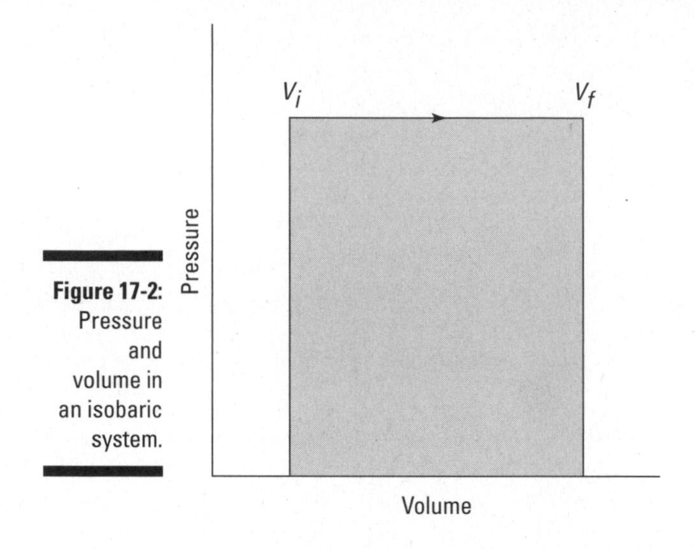

Figure 17-2: Pressure and volume in an isobaric system.

Say you have 60 cubic meters of an ideal gas at a pressure of 200 pascals. You heat the gas until it expands to a volume of 120 cubic meters (see Chapter 14 for details on gas expansion as temperature rises). How much work does the gas do? All you have to do is plug in the numbers:

$$W = P\Delta V = (200\ \text{Pa})(120\ \text{m}^3 - 60\ \text{m}^3) = 12{,}000\ \text{J}$$

The gas does 12,000 joules of work as it expands under constant pressure.

Working with constant water pressure

Suppose you're waiting for a connecting flight to the next physics conference. You look around but don't see much to amuse yourself with — just a water fountain. Proving that physicists can find fun anywhere, you take a gram of water from the fountain and put it into the pocket isobaric chamber that you always happen to carry with you. As an airport security guard looks on, you increase the pressure to 2.0×10^5 pascals and increase the temperature of the water by 62°C.

You note that the gram of water increases in volume by 1.0×10^{-8} cubic meters. "Hmm," you think. "I wonder what work was done by the water and what the change in internal energy of the water was." The process was isobaric, so the work done by the water was

$$W = P\Delta V$$

Filling in the numbers and doing the math yields:

$$W = (2.0 \times 10^5 \text{ Pa})(1.0 \times 10^{-8} \text{ m}^3) = 0.002 \text{ J}$$

So that's the work done by the water. What about the change in the internal energy of the water? The first law of thermodynamics tells you that

$$\Delta U = Q - W$$

You know W, but what is Q? Q is the heat absorbed by the water. You know the change in temperature of the water, and using the water's specific heat capacity (Chapter 15), you can find the heat actually absorbed by the water using this equation:

$$Q = cm\Delta T$$

Water's specific heat capacity is 4,186 J/kg·°C. Plugging in the numbers and doing the math gives you

$$Q = cm\Delta T = (4{,}186 \text{ J/kg·°C})(0.0010 \text{ kg})(62\text{°C}) \approx 260 \text{ J}$$

Now back to the first law of thermodynamics:

$$\Delta U = Q - W$$

Substituting in the values gives you the change in internal energy:

$$\Delta U = Q - W = 260 \text{ J} - 0.002 \text{ J} \approx 260 \text{ J}$$

Hmm, you think. The work done was a tiny 0.002 joules, while the change in internal energy was 260 joules. Interesting — very little work was done because the water didn't expand much, but you saw a fair gain in internal energy because the water's temperature went up.

Increasing steam's energy without changing the pressure

Now you decide to find the work done by something that can really expand, such as steam. Would the work done change by a lot? You decide to take a look.

Using your isobaric chamber, you raise the temperature of a gram of water until it becomes steam. Then you raise the temperature of the steam by 62°C (just as you raise the temperature of the liquid water by the same amount in the preceding section) while keeping the pressure at 2.0×10^5 pascals. This time, you note that the steam expanded by a lot more than the liquid water did — by 7.1×10^{-5} cubic meters.

How much work did the steam do? Because the expansion was in your pocket isobaric chamber, the process didn't involve a change in pressure, so you can use the following equation:

$$W = P\Delta V$$

Substituting in the numbers and doing the math gives you

$$W = P\Delta V = (2.0 \times 10^5 \text{ Pa})(7.1 \times 10^{-5} \text{ m}^3) \approx 14 \text{ J}$$

Now what about the change in internal energy of the steam? Once again, you can use the equation for the first law of thermodynamics:

$$\Delta U = Q - W$$

You know W, but what is Q? Q is the heat absorbed by the steam. You know the temperature change of the steam, so you can use the following equation:

$$Q = cm\Delta T$$

Plugging in the numbers and doing the math gives you

$$Q = cm\Delta T = (2{,}020 \text{ J/kg}\cdot{}^\circ\text{C})(0.0010 \text{ kg})(62^\circ\text{C}) \approx 126 \text{ J}$$

Going back to the first law of thermodynamics, you get the following after you plug in the numbers and do the math:

$$\Delta U = Q - W = 126 \text{ J} - 14 \text{ J} = 112 \text{ J}$$

The steam did much more work than the water did when it expanded, so less energy was available to boost the total internal energy of the steam.

"Hey buddy," says the airport security guard, indicating your pocket isobaric chamber, "What's that contraption?"

"This contraption just told me that the steam does a lot more work through expansion under isobaric conditions than liquid water does."

The security guard blinks and says, "Oh."

At constant volume: Isochoric

What if the pressure in a system isn't constant? You may see a simple closed container, which can't change its volume. In this case, the volume is constant, so you have an *isochoric* process.

In Figure 17-3, someone has neglectfully tossed a spray can onto a fire. As the gas inside the spray can heats up, its pressure increases, but its volume stays the same (unless, of course, the can explodes).

Figure 17-3: An isochoric system features a constant volume as other quantities vary.

How much work does the fire do on the spray can? Look at the graph in Figure 17-4. In this case, the volume is constant, so *Fs* (force times distance) equals zero. No work is being done — the area under the graph is zero.

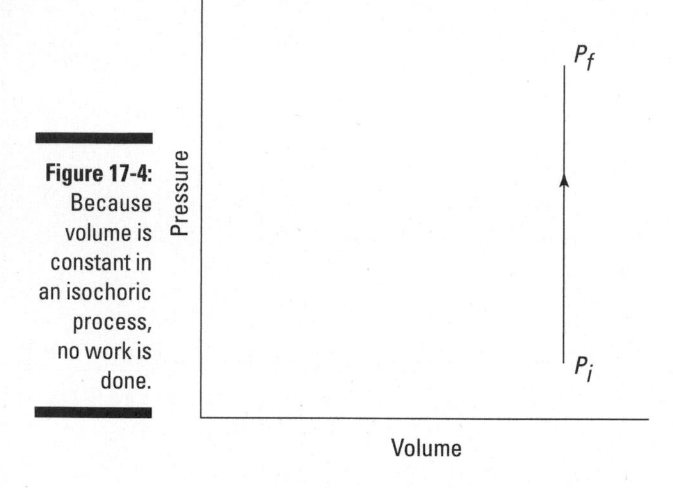

Figure 17-4: Because volume is constant in an isochoric process, no work is done.

Here's an example. The CEO of Acme Pressure Vessels approaches you and says, "We're adding 16,000 joules of energy to 5 moles of ideal gas at constant volume, and we want to know how much the internal energy changes. Can you help?"

You get out your clipboard and explain. The work done by an ideal gas depends on the change in its volume: $W = P\Delta V$ (see the earlier section "Staying constant: Isobaric, isochoric, isothermal, and adiabatic processes" for details on why). Because the volume change is zero in this case, the work done is zero.

The change in internal energy of an ideal gas is $\Delta U = Q - W$. Because W is zero, the following is true:

$$\Delta U = Q$$

You turn to the CEO and say, "You've added 16,000 joules of energy to an ideal gas at constant volume, so the change in the gas's internal energy is exactly 16,000 joules."

"What?" says the CEO. "That was too easy. We won't pay."

Handing the CEO a receipt, you say, "You already have. Thanks for your business."

At constant temperature: Isothermal

In an *isothermal system,* the temperature remains constant as other quantities change. Look at the remarkable apparatus in Figure 17-5. It's specially designed to keep the temperature of the enclosed gas constant, even as the piston rises. When you apply heat to this system, the piston rises or lowers slowly in such a way as to keep the product of pressure times volume constant. Because $PV = nRT$ (see Chapter 14), the temperature stays constant as well. (Remember that n is the number of moles of gas that remains constant, and R is the gas constant.)

What does the work look like as the volume changes? Because $PV = nRT$, the relation between P and V is

$$P = \frac{nRT}{V}$$

You can see this equation graphed in Figure 17-6, which shows the work done as the shaded area underneath the curve. But what the heck is that area?

The work done in an isothermal process is given by the following equation, where *ln* is the natural log (*ln* on your calculator), R is the gas constant (8.31 J/mol·K), V_f is the final volume, and V_i is the initial volume:

$$W = nRT \ln\left(\frac{V_f}{V_i}\right)$$

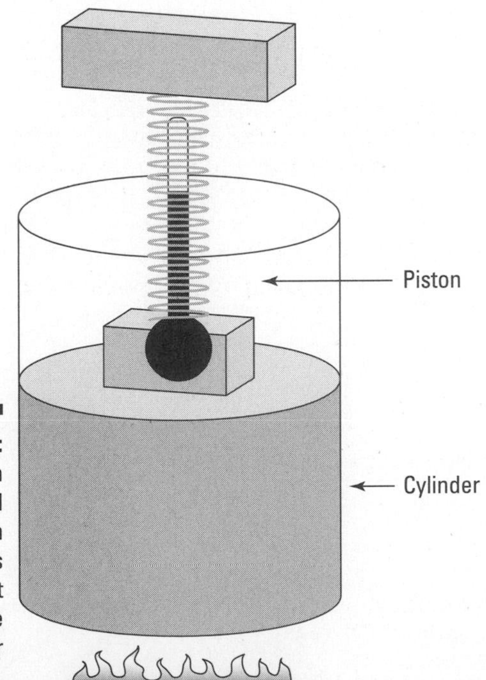

Figure 17-5: An isothermal system maintains a constant temperature amidst other changes.

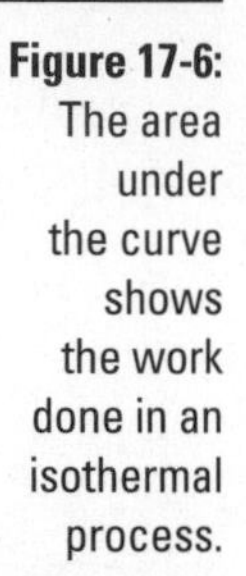

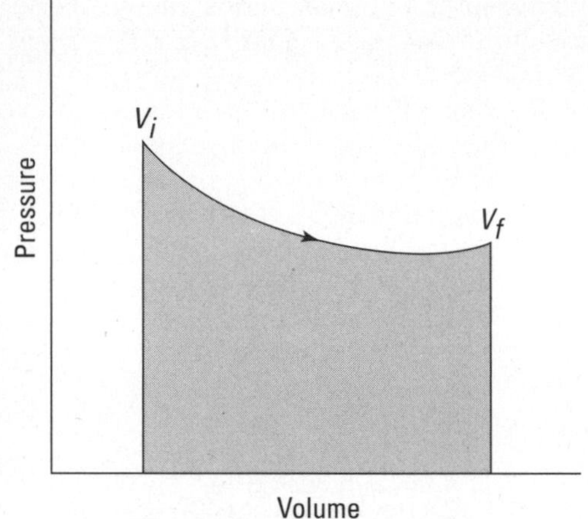

Figure 17-6: The area under the curve shows the work done in an isothermal process.

Because the temperature stays constant in an isothermal process and because the internal energy for an ideal gas equals $(3/2)nRT$ (see Chapter 16), the internal energy doesn't change. Therefore, you find that heat equals the work done by the system:

$$\begin{aligned} \Delta U &= Q - W \\ 0 &= Q - W \\ Q &= W \end{aligned}$$

If you immerse the cylinder you see in Figure 17-5 in a heat bath, what would happen? The heat, Q, would flow into the apparatus, and because the temperature of the gas stays constant, all that heat would become work done by the system.

Say that you have a mole of helium to play around with on a rainy day of temperature 20°C, and for fun you decide to expand it from $V_i = 0.010\ \text{m}^3$ to $V_f = 0.020\ \text{m}^3$. What's the work done by the gas in the expansion? All you have to do is plug in the numbers:

$$\begin{aligned} W &= nRT \ln\left(\frac{V_f}{V_i}\right) \\ &= (1.0)(8.31\ \text{J/mol}\cdot\text{K})\left[(273.15 + 20)\ \text{K}\right] \ln\left(\frac{0.020\ \text{m}^3}{0.010\ \text{m}^3}\right) \\ &\approx 1{,}690\ \text{J} \end{aligned}$$

The gas does 1,690 joules of work. The gas's change in internal energy is 0 joules, as always in an isothermal process. And because $Q = W$, the heat added to the gas is also equal to 1,690 joules.

Here's another example. Say that you're given 2.0 moles of hydrogen gas at a temperature of 600 kelvins for your birthday. Expanding the gas from a volume of 0.05 cubic meters to 0.10 meters isothermally, you wonder how much work the gas does, so you get out your clipboard. The work done by an ideal gas during isothermal expansion is

$$W = nRT \ln\left(\frac{V_f}{V_i}\right)$$

Plugging in the numbers and doing the math gives you

$$\begin{aligned} W &= nRT \ln\left(\frac{V_f}{V_i}\right) \\ &= (2.0 \text{ mol})(8.31 \text{ J/mol}\cdot\text{K})(600 \text{ K}) \ln\left(\frac{0.10 \text{ m}^3}{0.05 \text{ m}^3}\right) \\ &\approx 6{,}900 \text{ J} \end{aligned}$$

So the gas does 6,900 joules of work during its expansion.

So what about the change in the internal energy of the gas? You know that the change in internal energy is $\Delta U = (3/2)nR\Delta T$ (see Chapter 16 for details). Therefore, because ΔT equals zero in an isothermal process, ΔU is zero as well. So the change in the internal energy of the gas is zero during the isothermal expansion.

At constant heat: Adiabatic

In an *adiabatic process,* no heat flows from or to the system. Take a look at Figure 17-7, which shows a cylinder surrounded by an insulating material. The insulation prevents heat from flowing into or out of the system, so any change in the system is adiabatic.

Examining the work done during an adiabatic process, you can say $Q = 0$, so ΔU (the change in internal energy) equals $-W$. Because the internal energy of an ideal gas is $U = (3/2)nRT$ (see Chapter 14), the work done is the following:

$$W = \left(\frac{3}{2}\right) nR(T_{fi} - T)$$

where T_f represents the final temperature and T_i represents the initial temperature. So if the gas does work, that work comes from a change in temperature — if the temperature goes down, the gas does work on its surroundings.

You can see what a graph of pressure versus volume looks like for an adiabatic process in Figure 17-8. The adiabatic curve in this figure, called an *adiabat,* is different from the isothermal curves, called *isotherms.* The work done when the total heat in the system is constant is the shaded area under the curve.

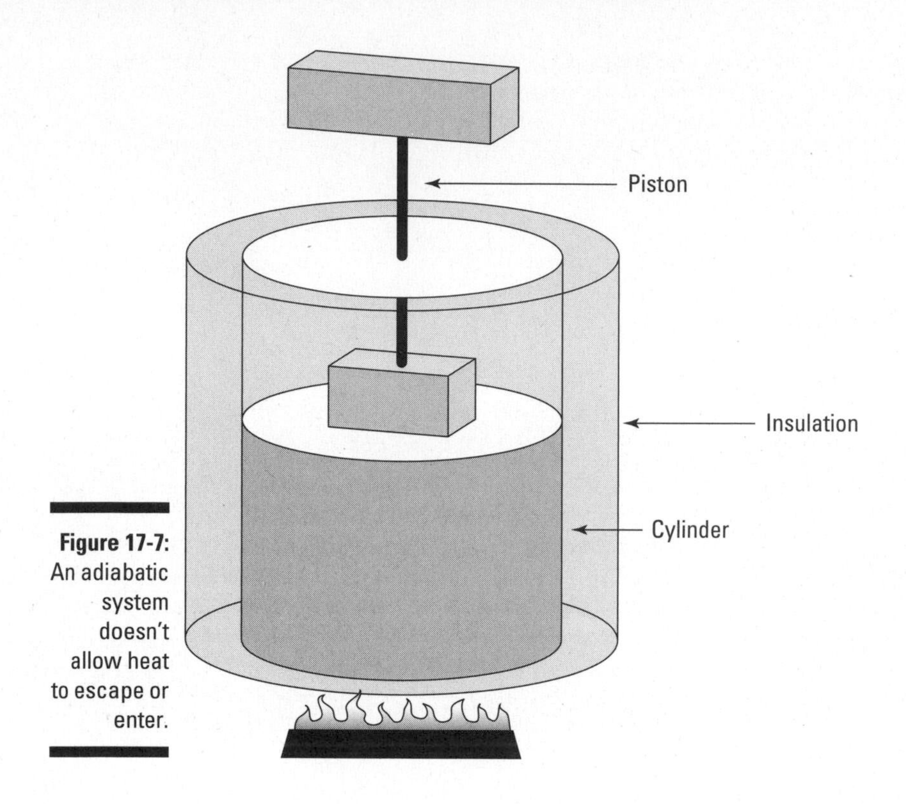

Figure 17-7: An adiabatic system doesn't allow heat to escape or enter.

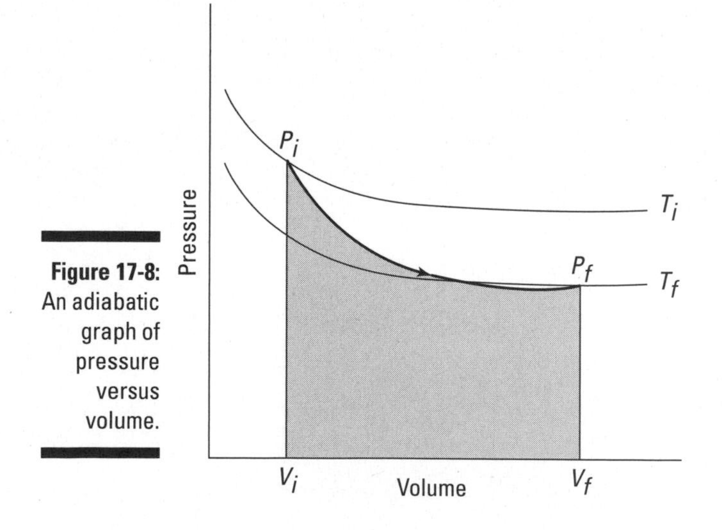

Figure 17-8: An adiabatic graph of pressure versus volume.

In adiabatic expansion or compression, you can relate the initial pressure and volume to the final pressure and volume this way:

$$P_i V_i^{\gamma} = P_f V_f^{\gamma}$$

In this equation, γ is the ratio of the specific heat capacity of an ideal gas at constant pressure divided by the specific heat capacity of an ideal gas at constant volume (*specific heat capacity* is the measure of how much heat an object can hold; see Chapter 15):

$$\gamma = \frac{C_P}{C_V}$$

How can you find those specific heat capacities? That's coming up next.

Figuring out molar specific heat capacities

To figure out specific heat capacity, you need to relate heat, Q, and temperature, T. You usually use the formula $Q = cm\Delta T$, where c represents specific heat capacity, m represents the mass, and ΔT represents the change in temperature.

For gases, however, it's easier to talk in terms of *molar specific heat capacity,* which is given by C and whose units are joules/mole-kelvin (J/mol·K). With molar specific heat capacity, you use a number of moles, n, rather than the mass, m:

$$Q = Cn\Delta T$$

To solve for C, you must account for two different quantities, C_P (constant pressure) and C_V (constant volume). Solved for Q, the first law of thermodynamics states that

$$Q = W + \Delta U$$

So if you can get ΔU and W in terms of T, you're set.

First consider heat at constant volume (Q_V). The work done (W) is $P\Delta V$, so at constant volume, no work is done; $W = 0$, so $Q_V = \Delta U$. And ΔU, the change in internal energy of an ideal gas, is $(3/2)nR\Delta T$ (see Chapter 16). Therefore, Q at constant volume is the following:

$$Q_V = \Delta U = \frac{3}{2}nR\Delta T$$

Now look at heat at constant pressure *(Q_P)*. At constant pressure, work *(W)* equals $P\Delta V$. And because $PV = nRT$, you can represent the work as nRT: $W = P\Delta V = nR\Delta T$. At constant pressure, the change in energy, ΔU, is still $(3/2)nR\Delta T$, just as it is at constant volume. Therefore, here's Q at constant pressure:

$$Q_P = W + \Delta U$$
$$= \frac{3}{2}nR\Delta T + nR\Delta T$$

So how do you get the molar specific heat capacities from this? You've decided that $Q = Cn\Delta T$, which relates the heat exchange, Q, to the temperature difference, ΔT, via the molar specific heat capacity, C. This equation holds true for the heat exchange at constant volume, Q_V, so you write

$$Q_V = C_V n\Delta T$$

where C_V is the specific heat capacity at constant volume. You already have an expression for Q_V, so you can substitute into the earlier equation:

$$\frac{3}{2}nR\Delta T = C_V n\Delta T$$

Then you can divide both sides by $n\Delta T$ to get the specific heat capacity at constant volume:

$$C_V = \frac{3}{2}R$$

If you repeat this for the specific heat capacity at constant pressure, you get

$$C_P = \frac{3}{2}R + R = \frac{5}{2}R$$

Now you have the molar specific heat capacities of an ideal gas. The ratio you want, γ, is the ratio of these two equations:

$$\gamma = \frac{C_P}{C_V} = \frac{5}{3}$$

For an ideal gas, you can connect pressure and volume at any two points along an adiabatic curve this way:

$$P_i V_i^{5/3} = P_f V_f^{5/3}$$

Finding a new pressure after an adiabatic change

Suppose you start with 1.0 liter of gas at a pressure of 1.0 atmosphere. After an adiabatic change (where no heat is gained or lost), you end up with 2.0 liters of gas. What would the new pressure, P_f, be? Putting P_f on one side of the equation gives you

$$P_f = \frac{P_i V_i^{5/3}}{V_f^{5/3}}$$

Plug in the numbers and do the math:

$$P_f = \frac{P_i V_i^{5/3}}{V_f^{5/3}} = \frac{(1.0 \text{ atm})(1.0 \text{ L})^{5/3}}{(2.0 \text{ L})^{5/3}} \approx 0.31 \text{ atm}$$

The new pressure would be 0.31 atmospheres.

Building a bigger lab: An adiabatic-change practice problem

There you are, the world-famous physicist, on vacation in Antarctica. The head of a South Pole scientific team comes running up to you and asks for your help. "We've got a big problem," the director says.

"Oh yes?" you ask.

"We put an explorer on the South Pole in a lab with vacuum chamber walls — the walls prevent any heat from being gained or lost to the environment, so he stays nice and cozy," the director says. "The problem is we pressurized it too highly. It's at 2 atmospheres and the scientist is very uncomfortable. We'd like to expand the volume of the lab so that it's at 1 atmosphere inside."

Always willing to come to the aid of a fellow scientist, you take out your clipboard. The specially constructed lab has vacuum chamber walls, so no heat is exchanged with the outside. Therefore, the expansion will be an adiabatic one, and this equation applies:

$$P_i V_i^{5/3} = P_f V_f^{5/3}$$

The scientists want to reduce the pressure from 2 atmospheres to 1 atmosphere, so

$$\frac{P_f}{P_i} = 0.5$$

Solving the pressure-volume equation for the ratio of pressures, P_f/P_i, you get

$$\frac{P_f}{P_i} = \frac{V_i^{5/3}}{V_f^{5/3}}$$

$$\frac{P_f}{P_i} = \left(\frac{V_i}{V_f}\right)^{5/3}$$

If you raise both sides of this equation to the power 3/5, you get

$$\left(\frac{P_f}{P_i}\right)^{3/5} = \left(\frac{V_i}{V_f}\right)$$

Then if you invert the terms on both sides (this is the same as raising both sides to the power of –1), you get the following

$$\left(\frac{P_i}{P_f}\right)^{3/5} = \left(\frac{V_f}{V_i}\right)$$

Finally, multiply both sides by V_i to get

$$V_f = V_i\left(\frac{P_i}{P_f}\right)^{3/5}$$

If you put in the value of the pressure ratio, you have

$$V_f = V_i 2^{3/5} \approx 1.5V_i$$

So V_f is about $1.5V_i$. You turn to the director and say, "Expand the lab's volume by 50 percent."

"Thanks," says the director. "Your usual fee?"

"No charge for a fellow scientist," you say.

Flowing from Hot to Cold: The Second Law of Thermodynamics

The *second law of thermodynamics* says that heat flows naturally from an object at a higher temperature to an object at a lower temperature, and heat doesn't flow in the opposite direction of its own accord.

The law is certainly borne out in everyday observation — when was the last time you noticed an object getting colder than its surroundings unless another object was doing some kind of work? You can force heat to flow away from an object when it would naturally flow into it if you do some work — as with refrigerators or air conditioners — but heat doesn't go in that direction by itself.

Heat engines: Putting heat to work

You have many ways to turn heat into work. You may have a steam engine, for example, that has a boiler and a set of pistons, or you may have an atomic reactor that generates superheated steam that can turn a turbine.

Engines that rely on a heat source to do work are called *heat engines;* you can see the principle behind a heat engine in Figure 17-9. A heat source provides heat to the engine, which does work. The waste heat left over goes to a *heat sink,* which effectively has an infinite heat capacity, because it can take such a large amount of heat energy without changing temperature. The heat sink could be the surrounding air, or it could be a water-filled radiator, for example. As long as the heat sink is at a lower temperature than the heat source, the heat engine can do work — at least theoretically.

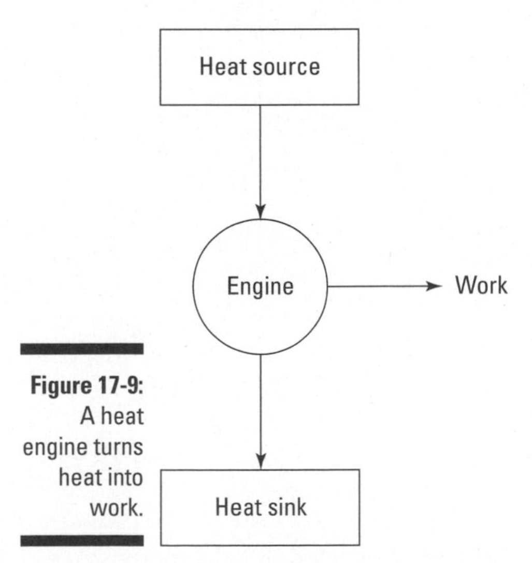

Figure 17-9: A heat engine turns heat into work.

Evaluating heat's work: Heat engine efficiency

Heat supplied by a heat source is given the symbol Q_h (for the hot source), and heat sent to a heat sink is given the symbol Q_c (for the cold heat sink).

With some calculations, you can find the efficiency of a heat engine. The efficiency is the ratio of the work the engine does, W, to the input amount of heat — the fraction of the input heat that the engine converts to work:

$$\text{Efficiency} = \frac{\text{Work}}{\text{Heat Input}} = \frac{W}{Q_h}$$

If the engine converts all the input heat to work, the efficiency is 1.0. If no input heat is converted to work, the efficiency is 0.0. Often, the efficiency is given as a percentage, so you express these values as 100 percent and 0 percent.

Because total energy is conserved, the heat into the engine must equal the work done plus the heat sent to the heat sink, which means that $Q_h = W + Q_c$. Therefore, you can rewrite the efficiency in terms of just Q_h and Q_c:

$$\text{Efficiency} = \frac{W}{Q_h} = \frac{Q_h - Q_c}{Q_h} = 1 - \left(\frac{Q_c}{Q_h}\right)$$

Finding heat from a car engine

Say that you have a heat engine that's 78.0 percent efficient and that produces 2.55×10^7 joules of energy. Perhaps this is the energy produced by the engine of a car from burning one tank of fuel. How much heat does the engine use, and how much does it reject? Well, you know that $W = 2.55 \times 10^7$ joules and that

$$\text{Efficiency} = \frac{W}{Q_h}$$

$$0.780 = \frac{2.55 \times 10^7 \text{ J}}{Q_h}$$

Solving for Q_h gives you

$$Q_h = \frac{2.55 \times 10^7 \text{ J}}{0.780} \approx 3.27 \times 10^7 \text{ J}$$

The amount of input heat is 3.27×10^7 joules. So how much heat gets left over and sent into the heat sink, Q_c? You know that $Q_h = W + Q_c$, and you can rearrange the problem to solve for Q_c:

$$Q_c = Q_h - W$$

Plugging in the numbers gives you

$$Q_c = Q_h - W = (3.27 \times 10^7 \text{ J}) - (2.55 \times 10^7 \text{ J}) = 7.2 \times 10^6 \text{ J}$$

The amount of heat sent to the heat sink is 7.2×10^6 joules.

Finding heat from your race car

You're at the race track, testing out your new physics racer. You're proud of the car, which has 25 percent efficiency. Today, you estimate, it's already produced 8,000 joules of work. Then you notice a mechanic about to put his hand on the radiator. "Don't touch that," you say. "It's got to be hot."

"How hot could it be?" the mechanic asks.

How hot indeed? The car gets rid of its excess heat through the radiator, you think. So how much heat did the car get rid of? Your car is 25 percent efficient and has done 8,000 joules of work, so the input heat is

$$Q_h = \frac{W}{\text{Efficiency}}$$

Plugging in the numbers gives you:

$$Q_h = \frac{W}{\text{Efficiency}} = \frac{8{,}000\text{ J}}{0.25} = 32{,}000\text{ J}$$

Okay, the input heat was 32,000 joules. You know that the input heat equals the work done plus the output heat $Q_h = W + Q_c$, so the output heat is

$$Q_c = Q_h - W$$

The input heat was 32,000 joules and the engine did 8,000 joules of work, so plug in the numbers and solve:

$$Q_c = Q_h - W = 32{,}000\text{ J} - 8{,}000\text{ J} = 24{,}000\text{ J}$$

"How hot could the radiator be?" you ask the mechanic. "Twenty-four thousand joules hot, that's how hot!"

"But how hot is that?" replies the mechanic? Indeed, if the radiator is 10 kilograms and it absorbs 24,000 joules of energy, how much does its temperature change? Well, if its specific heat capacity is 460 J/kg·K, then the change in temperature is the following:

$$\Delta T = \frac{24{,}000\text{ J}}{(10\text{ kg})(460\text{ J/kg}\cdot\text{K})} \approx 5.2\text{ K}$$

A difference of 5.2 kelvins is the same as a difference of 5.2°C. "Oh, only five degrees? That's fine," says the mechanic. You wonder whether you should tell him that's more than 40°F.

Limiting efficiency: Carnot says you can't have it all

Given the amount of work a heat engine does and its efficiency, you can calculate how much heat goes in and how much comes out (along with a little help from the law of conservation of energy, which ties work, heat in, and heat out together). But why not create 100-percent efficient heat engines? Converting all the heat that goes into a heat engine into work would be nice, but the real world doesn't work that way. Heat engines have some inevitable losses, such as through friction on the pistons in a steam engine.

Studying this problem, Sadi Carnot (a 19th-century engineer) came to the conclusion that the best you can do, effectively, is to use an engine that has no such losses. If the engine experiences no losses, the system will return to the state it was in before the process took place. This is called a *reversible process.* For example, if a heat engine loses energy overcoming friction, it doesn't have a reversible process, because it doesn't end up in the same state when the process is complete. You have the most efficient heat engine when the engine operates reversibly.

Carnot's principle says that no nonreversible engine can be as efficient as a reversible engine and that all reversible engines that work between the same two temperatures have the same efficiency. Here's the kicker: A perfectly reversible engine doesn't exist, so Carnot came up with an ideal one.

Finding efficiency in Carnot's engine

No real engine can operate reversibly, so Carnot imagined a kind of ideal, reversible engine. In the *Carnot engine,* the heat that comes from the heat source is supplied at a constant temperature T_h. Meanwhile, the rejected heat goes into the heat sink, which is at a constant temperature T_c. Because the heat source and the heat sink are always at the same temperature, you can say that the ratio of the heat provided and rejected is the same as the ratio of those temperatures (expressed in kelvins):

$$\frac{Q_c}{Q_h} = \frac{T_c}{T_h}$$

And because the efficiency of a heat engine is Efficiency = $1 - (Q_c/Q_h)$, the efficiency of a Carnot engine is

$$\text{Efficiency} = 1 - \frac{Q_c}{Q_h} = 1 - \frac{T_c}{T_h}$$

This equation represents the *maximum possible efficiency* of a heat engine. You can't do any better than that. And as the third law of thermodynamics states (see the final section in this chapter), you can't reach absolute zero; therefore, T_c is never 0, so the efficiency is always 1 minus some number. You can never have a 100-percent efficient heat engine.

Using the equation for a Carnot engine

Applying the equation for maximum possible efficiency (Efficiency = $1 - Q_c/Q_h = 1 - T_c/T_h$) is easy. For example, say that you come up with a terrific new invention: a Carnot engine that uses a balloon to connect the ground (27°C) as a heat source to the air at 33,000 feet (about –25°C), which you use as the heat sink. What's the maximum efficiency you can get for your heat engine? After converting temperatures to kelvins, plugging in the numbers gives you

$$\text{Efficiency} = 1 - \frac{Q_c}{Q_h} = 1 - \frac{T_c}{T_h} = 1 - \left(\frac{248\text{ K}}{300\text{ K}}\right) \approx 0.173$$

Your Carnot engine can be no more than 17.3 percent efficient — not too impressive. On the other hand, assume you can use the surface of the sun (about 5,800 kelvins) as the heat source and interstellar space (about 3.40 kelvins) as the heat sink (such is the stuff science-fiction stories are made of). You'd have quite a different story:

$$\text{Efficiency} = 1 - \frac{T_c}{T_h} = 1 - \left(\frac{3.40\text{ K}}{5{,}800\text{ K}}\right) \approx 0.999$$

You get a theoretical efficiency for your Carnot engine — 99.9 percent.

Here's another example. You're in Hawaii, taking a well-deserved vacation with other hard-working physicists. The summer has been hot, and as you lounge on the beach, you read an article about the energy crisis caused by all those whirring air conditioners. You put down the paper as the happy physicists bobbing in the surf call to you, saying you should come in for a dip.

"How warm is it?" you ask.

"Very," they say, bobbing up and down. "About 300 kelvins."

Hmm, you think. If you could create a Carnot engine and use the surface of the ocean as the input heat source (300 kelvins) and the bottom of the ocean (about 7°C, or 280 kelvins) as the heat sink, what would the efficiency of such an engine be? And how much input heat would you need to supply the entire energy needs of the United States for one year (about 1.0×10^{20} J)?

You know that Efficiency = $1 - (T_c/T_h)$, so plug in the numbers and do the math to find the efficiency:

$$\text{Efficiency} = 1 - \left(\frac{280\ \text{K}}{300\ \text{K}}\right) \approx 0.067$$

Hmm, 6.7 percent efficiency. So how much input heat would be needed to get 1.0×10^{20} joules out? You know that Efficiency = W/Q_h, so

$$Q_h = \frac{W}{\text{Efficiency}}$$

Plugging in the numbers and doing the math yields

$$Q_h = \frac{1.0 \times 10^{20}\ \text{J}}{0.067} \approx 1.5 \times 10^{21}\ \text{J}$$

How much would taking that heat out of the top meter of the Pacific Ocean change its temperature by? Assume that the top meter of the Pacific Ocean contains about 1.56×10^{14} cubic meters of water — that's 1.56×10^{17} kilograms of water.

The heat gained or lost is tied to temperature change by $Q = cm\Delta T$, so the temperature change would be

$$\Delta T = \frac{Q}{cm}$$

Plugging in the numbers and doing the math gives you a temperature change of

$$\Delta T = \frac{1.5 \times 10^{21}\ \text{J}}{(4{,}186\ \text{J/kg}\cdot{}^{\circ}\text{C})(1.56 \times 10^{17}\ \text{kg})} \approx 4.5^{\circ}\text{C}$$

So if your Carnot engine were connected from the top of the Pacific Ocean to the bottom and sucked all its heat out of the top meter of the surface water, it'd lower the temperature of that top meter of water by 4.5°C to supply all the energy needs of the United States.

Going against the flow with heat pumps

Usually, Carnot engines take heat from the hot reservoir (Q_h), do work, and then dump the leftover heat in the cold reservoir (Q_c). But what if you swapped the hot and cold reservoirs and actually did some work on the

Carnot engine (instead of having it do work on you)? Then you could "pump heat uphill," from the cold reservoir to the hot reservoir. You can do this if you connect the input of a Carnot engine to the cold reservoir and connect the exhaust to the hot reservoir.

Why on Earth would you want to move heat? Think about a cold room on an even colder day. If you connect a Carnot engine to the outside — which is colder than the inside — doing some work on the Carnot engine can drive heat into the room. This use of a Carnot engine is called a *heat pump,* because you put in work to drive heat uphill, from the cold reservoir to the hot one.

Why are heat pumps a good way to warm your house? Consider heating with electric heat instead. If you use enough electricity to add 1,000 joules to the heat inside your house, you have to pay for 1,000 joules of energy. But if you pump the heat from outside to inside, most of the heat comes from the cold reservoir, and you only have to provide the work needed to pump the heat into the hot reservoir.

A heat pump can be used to move heat in the other direction, too, to bring about cooling. In this case, mechanical work is used to pump heat from a source at a higher temperature to a lower temperature. Your refrigerator uses electrical energy to drive a compressor unit, which forms part of the refrigeration cycle.

Heating with less work

Operating heat pumps requires less work than the heat they transfer. For example, suppose you're vacationing in your woodland cabin, which is at 20°C (that is, 293 kelvins — about 68°F). You want to pump some heat in from outside, which is at 10°C (283 kelvins — about 50°F). You decide you need about 4,000 joules of heat. How much work would you need to pump 4,000 joules into your house?

You unpack a Carnot engine and connect it to the outdoors so that it will use the outside (which is colder) as its hot reservoir and the inside of your cabin (which is warmer) as its cold reservoir. To get heat to flow uphill like that, you have to do work on the Carnot engine instead of having it do work on you.

So how much work do you need to pump 4,000 joules of heat inside? You can start from this equation:

$$Q_h = W + Q_c$$

Here, Q_h is the heat dumped into the cabin and Q_c is the heat taken from the outside. W is the amount of work you need to supply to the heat pump. You're trying to solve for the work, so rearrange the equation:

$$W = Q_h - Q_c$$

For a Carnot engine, $Q_c/Q_h = T_c/T_h$. Therefore, here's the formula for the heat taken from the outside:

$$Q_c = Q_h\left(\frac{T_c}{T_h}\right)$$

Now plug this value of Q_c into the work equation ($W = Q_h - Q_c$) and simplify:

$$W = Q_h\left[1 - \left(\frac{T_c}{T_h}\right)\right]$$

You want to get 4,000 joules of heat into the room, so Q_h = 4,000 joules. Plugging in the heat and temperatures and doing the math gives you

$$W = 4{,}000 \text{ J}\left[1 - \left(\frac{283 \text{ K}}{293 \text{ K}}\right)\right] \approx 136 \text{ J}$$

So you'd need only 136 joules of work to pump 4,000 joules of heat in from the outside. See why heat pumps can be so attractive? If you were using electric heat, you'd have to pay for the full 4,000 joules.

However, as the temperature outside gets lower and lower, you have to do more work to pump heat indoors, because you have a bigger temperature difference to overcome. For example, what if the temperature outside were –20°C (that is, 253 kelvins, or –4°F)? In this case, how much work do you have to do on the Carnot engine to pump 4,000 joules of heat into the cabin?

You can use the same work equation you just derived:

$$W = Q_h\left[1 - \left(\frac{T_c}{T_h}\right)\right]$$

Plugging in the numbers and doing the math gives you

$$W = 4{,}000 \text{ J}\left[1 - \left(\frac{253 \text{ K}}{293 \text{ K}}\right)\right] \approx 546 \text{ J}$$

So when the temperature outside is 10°C, you only need 136 joules to pump 4,000 joules of heat into your cabin. But when the outside temperature is –20°C, you need 546 joules to pump the same 4,000 joules. Notice, however, that in either case, you get 4,000 joules for a lot less than the whole 4,000 joules you'd have to pay for if you were using electric heat.

Checking a heat pump's performance

The heat delivered to you from a heat pump is more than the work you put into the heat pump. You can measure how much more heat you get out of a heat pump than the work you put in using the *coefficient of performance (COP):*

$$\text{COP} = \frac{Q_h}{W}$$

The coefficient of performance tells you how much heat you get out of a heat pump per work you have to put in to it.

For something like electric heat, where you have to pay for all the heat you get, the coefficient of performance is 1. But for a heat pump, the coefficient can be a lot higher than 1, indicating that you get more heat out of the pump than the work you put in.

The coefficient of performance depends on the inside and outside temperatures. You can put the coefficient of performance into a form that makes its dependence on temperature explicit.

Because $W = Q_h - Q_c$, the coefficient of performance equation becomes

$$\text{COP} = \frac{Q_h}{W} = \frac{Q_h}{Q_h - Q_c}$$

Or if you multiply both the numerator and denominator by $1/Q_h$, you can express this like so:

$$\text{COP} = \frac{Q_h}{W} = \frac{1}{(1 - Q_c / Q_h)}$$

For a Carnot engine, $Q_c/Q_h = T_c/T_h$, so you end up with

$$\text{COP} = \frac{1}{(1 - Q_c / Q_h)} = \frac{1}{(1 - T_c / T_h)}$$

Suppose you're pumping heat from 283 kelvins to 293 kelvins. You have a coefficient of performance of

$$\text{COP} = \frac{1}{(1 - T_c / T_h)} = \frac{1}{(1 - 283\text{ K} / 293\text{ K})} \approx 29$$

So when the inside is at 293 kelvins and the outside at 283 kelvins, you pump 29 times as much energy as the work you do to make the heat transfer. Not bad.

Going Cold: The Third (And Absolute Last) Law of Thermodynamics

Absolute zero is the lower limit for the temperature of any system, and the third law of thermodynamics can be formulated in terms of this temperature. The *third law of thermodynamics* is pretty straightforward — it just says that you can't reach absolute zero (0 kelvins, or about –273.15°C) through any process that uses a finite number of steps. In other words, you can't get down to absolute zero at all. Each step in the process of lowering an object's temperature to absolute zero can get the temperature a little closer, but you can't get all the way there.

Although you can't get down to absolute zero with any known process, you can get close. And if you have some expensive equipment, you discover more and more strange facts about the near-zero world. I have a pal who discovered how liquid helium works at very, very low temperatures — below two-thousandths of a kelvin. For example, the helium will climb entirely out of containers by itself if you get it started. For these and some other observations, he and some friends got the Nobel Prize in Physics in 1996, the lucky dogs (you can read about it at `nobelprize.org`).

Part V
The Part of Tens

In this part . . .

I let physics off the leash in Part V, and it goes wild. Here, I list discoveries and ideas that had profound impacts on physics and changed the way people view their world. I also list ten great scientists and outline the contributions they made to the field of physics.

Chapter 18

Ten Physics Heroes

In This Chapter

- Looking at people who made major contributions to physics
- Lending names to famous laws and units of measurement

Through the centuries, physics has had thousands of heroes — people who furthered the field in some way or another. In this chapter, you take a look at ten physics heroes who've done their bits to make physics what it is today. And just because age has its privileges, I've arranged these in chronological order by birth date.

Galileo Galilei

Galileo Galilei (1564–1642) was an Italian physicist, mathematician, astronomer and philosopher. He was a very important person in the Scientific Revolution — at various times, people called him the father of modern observational astronomy, the father of modern physics, and even the father of science.

He's perhaps best known for his improvements to the telescopes and the consequent observations he was able to make. Among his other achievements were the confirmation of the phases of Venus, the discovery of the four largest satellites of Jupiter (now named the *Galilean moons*), and the observation and analysis of sunspots. He also studied the motion of objects undergoing constant acceleration.

Famously, he supported the *heliocentric* view of the solar system, which says the planets orbit around the sun, not the Earth. That was a tough stance to take in 1610, and he got into trouble for it with the Catholic Church, which in 1616 declared it "false and contrary to Scripture." In 1632, he was tried by the Roman Inquisition, found guilty of heresy, and forced to recant. He spent the rest of his life under house arrest. Modern physicists can be glad that kind of thing doesn't go on much anymore.

Robert Hooke

Like many early physicists, Robert Hooke (1635–1703) had his finger in many pies — he was a scientist, architect, investor, and so on. He's best known for his law of elasticity, *Hooke's law,* which says that the restoring force on an object undergoing an elastic pull is proportional to the displacement of the object and a constant, often called the *spring constant* (see Chapter 13).

Hooke experimented in many different fields, however — in fact, he was the first person to use the term *cell* to refer to the basic unit of life. Originally very poor, he grew quite wealthy through his investments. He was very active after the Great Fire of London, surveying the ruins in organized maps. He was also a well-known architect, and buildings he designed still survive in England.

Sir Isaac Newton

Sir Isaac Newton (1643–1726) was an exceptional genius. He was an English physicist, mathematician, astronomer, natural philosopher, and theologian. His accomplishments include the following:

- Laying the groundwork for most of classical mechanics
- Discovering universal gravitation
- Discovering the three laws of motion
- Building the first practical reflecting telescope
- Developing a theory of color based on prisms
- Discovering an empirical law of cooling
- Studying the speed of sound
- Sharing the credit with Gottfried Leibniz for the development of differential and integral calculus
- Demonstrating the generalized binomial theorem, an ancient mathematical problem of the expansion of the sum of two terms into a series
- Developing Newton's method for approximating the roots of a function
- Adding to the study of power series

Newton greatly influenced three centuries of physicists. In 2005, the members of Britain's Royal Society were asked who had the bigger effect on the history of science and made the greater contribution to humankind — Sir Isaac Newton or Albert Einstein. The Royal Society chose Newton.

Benjamin Franklin

Benjamin Franklin (1706–1790) is familiar to most people as one of the Founding Fathers of the United States. He was an author, printer, political theorist, politician, postmaster, scientist, inventor, statesman, and diplomat. He invented the following:

- The lightning rod
- Bifocals
- The Franklin stove
- A carriage odometer
- The glass "armonica" (a popular musical instrument of the day)
- The first public lending library in America

Franklin even created the first fire department in Pennsylvania. He was also a leading newspaperman and printer in Philadelphia (the major city of the colonies at that time). He became wealthy publishing *Poor Richard's Almanack* and *The Pennsylvania Gazette.* He played a large role in the creation of the University of Pennsylvania and was elected the first president of the American Philosophical Society. He became a national hero when he headed the effort to have Parliament repeal the unpopular Stamp Act.

As a scientist, Franklin is famous for his work with electricity. The idea that lightning is electricity may seem pretty clear today, but in Franklin's day, the largest manmade sparks were only an inch or so long. No one knows whether he really performed his most famous experiment — tying a key to a kite string and flying it during a thunderstorm to see whether it could draw sparks from the key, indicating that lightning was electricity (this experiment is so famous that I've had students who confused Franklin with Francis Scott Key). However, Franklin did write about how someone could carry out such an experiment, saying that flying the kite *before the storm actually started* would be important, or else you'd risk getting electrocuted.

Charles-Augustin de Coulomb

Charles-Augustin de Coulomb (1736–1806) is best known for developing *Coulomb's law,* which defines the electrostatic force of attraction or repulsion between charges. In fact, the MKS unit of charge, the *coulomb* (C), was named after him.

Coulomb originally came to prominence with his long-titled work *Recherches théoriques et expérimentales sur la force de torsion et sur l'élasticité des fils de metal* ("Theoretical and experimental research on the force of torsion and the elasticity of metal wire").

Throughout his life, Coulomb conducted research into many fields, but his work in electrostatics was what brought him true fame. He showed that electrostatic attraction and repulsion varied inversely as the square of the distance between the charges. There was still a lot of work to be done, though — Coulomb thought electric "fluids" were responsible for the charges.

Amedeo Avogadro

Amedeo Avogadro (1776–1856) is most well-known for *Avogadro's number,* approximately 6.022×10^{23} — the number of molecules contained within a mole (see Chapter 16 for details). He started practicing as a lawyer after getting his doctorate. In 1800, he started studying mathematics and physics and became so interested (who wouldn't be?) that he turned to it as his new career.

Avogadro was a pioneer of physics on the microscopic level with *Avogadro's hypothesis,* which says that "equal volumes of all gases under the same conditions of temperature contain the same number of molecules." Unfortunately, acceptance of the hypothesis was slow because of opposition from other scientists and a general confusion between molecules and atoms.

Fifty years later in the Karlsruhe Congress, Stanislao Cannizzaro was able to get general agreement on Avogadro's hypothesis. When Johann Josef Loschmidt calculated Avogadro's number for the first time in 1865, Loschmidt happily called it *Loschmidt's number.* But the general scientific community, in deference to the guy who first suggested that such a number existed, renamed it Avogadro's number.

Nicolas Léonard Sadi Carnot

Nicolas Léonard Sadi Carnot (1796–1832) was a French physicist and military engineer. In 1824, he published his work *Reflections on the Motive Power of Fire,* which gave the theoretical description of heat engines, now called the *Carnot cycle.* That work laid the theoretical foundations for the second law of thermodynamics (see Chapter 17).

Some people call Carnot the father of thermodynamics because he came up with concepts such as Carnot efficiency, the Carnot theorem, the Carnot heat engine, and others.

James Prescott Joule

James Joule (1818–1889) was an English physicist who set as his task to study the relationship between heat and work (steam engines were very big in his time). His studies led to laws on the conservation of energy (see Chapter 9), which led to the development of the first law of thermodynamics (Chapter 17). As a result, the MKS unit of energy was named the *joule.*

He also worked on the opposite side of the thermometer from steam, getting as close as he could to absolute zero, along with Lord Kelvin (coming up next). Joule's interests were wide-ranging — in fact, he's the one who discovered the relationship between the electrical current through a resistance and the heat generated, now called *Joule's law.*

William Thomson (Lord Kelvin)

William Thomson (1824–1907) did important work in analyzing electricity mathematically and formulating the first and second laws of thermodynamics. Like many physicists of his day, he had many interests, starting off as an electric telegraph engineer and inventor, which made him famous — and rich. With enough money to do what he wanted, Thomson turned to physics, naturally.

Physicists remember him for developing the absolute zero scale of temperature, which bears his name to this day — the Kelvin scale (see Chapter 14). Already a knight, he became a nobleman in recognition of his achievements in thermodynamics. He's also almost as well-known for his work on developing a maritime compass as on the laws of thermodynamics. Queen Victoria knighted him as Lord Kelvin for his work on the transatlantic telegraph.

Albert Einstein

Perhaps the most well-known physicist in the popular mind is Albert Einstein (1879–1955). Einstein, whose name has become synonymous with *genius,* made many contributions to physics, including the following:

- The special and general theories of relativity
- The founding of relativistic cosmology
- The explanation of the *perihelion precession* of Mercury, which is the gradual rotation of the axis of the elliptical orbit of the planet
- The prediction of the deflection of light by gravity *(gravitational lensing)*
- The first fluctuation dissipation theorem, which explained the *Brownian motion* of molecules, which is the random jittery motion of small particles suspended in a fluid, which is caused by collisions with the molecules of the fluid
- The photon theory
- Wave-particle duality
- The quantum theory of atomic motion in solids

Einstein was the scientist who, on the eve of World War II, alerted President Franklin D. Roosevelt that Germany could be creating an atomic bomb. As a result of that warning, Roosevelt created the top secret Manhattan Project, which led to the development of the atomic bomb.

In 1921, Einstein won the big one, the Nobel Prize, "for his services to Theoretical Physics, and especially for his discovery of the law of the photoelectric effect."

Einstein was affected by that absent-mindedness that scientists who habitually spend all their time thinking about their studies can suffer from. He's said to have painted his front door red so he could tell which house was his. People joke that he once asked a child, "Little girl, do you know where I live?" And the little girl answered, "Yes, Daddy. I'll take you home."

Chapter 19

Ten Wild Physics Theories

In This Chapter

- Pinpointing the smallest distance and smallest time
- Getting comfortable with uncertainty
- Exploring space for physics facts
- Uncovering the truth about microwave ovens
- Coming to grips with your bearings in the physical world

This chapter gives you ten outside-the-box physics facts that you may not hear or read about in a classroom. As with anything in physics, however, you shouldn't really consider these "facts" as actual facts — they're just the current state of many theories. And in this chapter, some of the theories get pretty wild, so don't be surprised to see them superseded in the coming years.

You Can Measure a Smallest Distance

Physicists now have a theory that a "smallest distance" exists. It's the *Planck length,* named after the physicist Max Planck. The length is the smallest division that, theoretically, you can divide space into. However, the Planck length — about 1.6×10^{-35} meters, or about 10^{-20} times the approximate size of a proton — is really just the smallest amount of length with any physical significance, given the current understanding of the universe. Smaller than this, and the whole notion of distance breaks down.

There May Be a Smallest Time

In the same sense that Planck length is the smallest distance (see the preceding section), Planck time is the smallest amount of time. The *Planck time* is the time light takes to travel 1 Planck length, or 1.6×10^{-35} meters. If the speed of light is the fastest possible speed, you can easily make a case that the shortest time you can measure is the Planck length divided by the speed of light. The Planck length is very small, and the speed of light is very fast, which gives you a very, very short time for the Planck time:

$$\text{Planck time} = \frac{1.6 \times 10^{-35}\ \text{m}}{3.0 \times 10^{8}\ \text{m/s}} \approx 5.3 \times 10^{-44}\ \text{s}$$

The Planck time is about 5.3×10^{-44} seconds, and the notion of time breaks down as times become smaller than this.

Some people say that time is broken up into quanta of time, called *chronons,* and that each chronon is a Planck time in duration.

Heisenberg Says You Can't Be Certain

You may have heard of the uncertainty principle, but you may not have known that a physicist named Heisenberg first suggested it. Of course, that explains why it's called the *Heisenberg uncertainty principle,* I suppose. The principle had its beginnings in the wave nature of matter, as Louis de Broglie suggested. Matter is made up of particles, such as electrons. But particles also act as waves, much like light waves — you just don't normally notice it because particles have such small wavelengths.

Particles have wave-like properties, and the more localized the wave is, the more certain you can be about the position of the particle. However, the wavelength of the wave is directly related to the momentum of the particle. The more definite the wavelength of a wave, the more certain you can be about the momentum of the particle. But because of the nature of waves, the more definite the wavelength, the more spread out in space it becomes. That's why the more sure you are of the momentum, the less sure you can be of the position, and vice versa. You can also say that the more precisely you measure their locations, the less precisely you know their momentums.

Black Holes Don't Let Light Out

Black holes are created when particularly massive stars use up all their fuel and collapse inwardly to form super-dense objects, much smaller than the original stars. Only very large stars end up as black holes. Stars that aren't quite massive enough to collapse that far often end up as neutron stars instead. A *neutron star* occurs when gravity has smashed together all the electrons, protons, and neutrons, effectively creating a single mass of neutrons with the density of an atomic nucleus.

Black holes go even further than that. They collapse so far that not even light can escape their intense gravitational pulls. How's that? The photons that make up light aren't supposed to have any mass. How can they possibly be trapped in a black hole?

Photons are indeed affected by gravity, a fact predicted by Einstein's theory of general relativity. Tests have experimentally confirmed that light passing next to massive objects in the universe is bent by their gravitational fields. Gravity affects photons, and the gravitational pull of a black hole is so strong that photons can't escape it.

Gravity Curves Space

Isaac Newton gave physicists a great theory of gravitation, and from him came the following famous equation:

$$F = \frac{Gm_1m_2}{r^2}$$

where F represents force, G represents the universal gravitational constant, m_1 represents one mass, m_2 represents another mass, and r represents the distance between the masses (see Chapter 7). Newton was able to show that what made an apple fall also kept the planets in orbit. But Newton had one problem he could never figure out: how gravity could operate instantaneously at a distance.

Enter Einstein, who created the modern take on this problem. Instead of thinking of gravity as a simple force, Einstein suggested in his theory of general relativity that space and time are actually different aspects of a single entity called *spacetime*. Mass and energy curve the spacetime, and this curvature *is* gravity!

Einstein's idea is that mass and energy curve space and time (and ultimately, that's where the idea of wormholes in space comes from). The curvature of space and time is *gravity.* Mathematically, you treat time as the fourth dimension when working with relativity. The vectors you use have four components: three for the x-, y-, and z-axes and one for time, t.

What's really happening when a planet orbits the sun is that it's simply following the shortest path through the curved spacetime through which it travels. The mass of the sun curves the spacetime around it, and the planets follow that curvature.

Matter and Antimatter Destroy Each Other

One of the coolest things about high-energy physics, also called *particle physics,* is the discovery of antimatter. Antimatter is sort of the reverse of matter. The counterparts of electrons are called *positrons* (which are positively charged), and the counterparts of protons are *antiprotons* (which are negatively charged). Even neutrons have an antiparticle: *antineutrons.* A neutron is made up of smaller particles called *quarks,* which have antiparticle versions, too. So the antineutron has no charge just like the neutron, but each of the quarks it's made of is the anti- version of the neutron quarks.

In physics terms, matter is sort of on the plus side, and antimatter sort of on the negative side. When the two come together, they destroy each other, leaving pure energy — light waves of great energy, called *gamma waves.* And like any other radiant energy, gamma waves can be considered heat energy, so if you have a pound of matter and a pound of antimatter coming together, you'll have quite a bang.

That bang, pound for pound, is much stronger than a standard atomic bomb, where only 0.7 percent of the fissile material is turned into energy. When matter hits antimatter, 100 percent is turned into energy.

If antimatter is the opposite of matter, shouldn't the universe have as much antimatter as it does matter? That's a puzzler, and the debate is continuing. Where's all the antimatter? The jury is still out. Some scientists say that there could be vast amounts of antimatter around that people just don't know about. Immense antimatter clouds could be scattered throughout the galaxy, for example. Others say that the way the universe treats matter and antimatter is a little different — but different enough so that the matter people know of in the universe can survive.

Supernovas Are the Most Powerful Explosions

What's the most energetic action that can happen anywhere, throughout the entire universe? What event releases the most energy? What's the all-time champ when it comes to explosions? Your not-so-friendly neighborhood supernova. A *supernova* occurs when a very massive star explodes. The star's fuel is used up, and its structure is no longer supported by an internal release of energy. At that point, the star collapses in on itself, and if the star is massive enough, the gravitational potential energy that the star had is suddenly released upon the collapse.

Among the 100 billion stars in the Milky Way, the last known supernova occurred nearly 400 years ago. (I say *known* because light takes quite a while to reach Earth; a star could've gone supernova 100 years ago, but if it's far enough from Earth, no one would know it yet.)

Most of the star that becomes a supernova explodes at speeds of about 10,000,000 meters per second, or about 22,300,000 miles per hour. By comparison, even the highest of explosives on Earth detonate at speeds of 1,000 to 10,000 meters per second.

Because the physics of how a star explodes is quite well understood, physicists can observe the apparent brightness of a supernova in a distant galaxy and work out how far away that galaxy must be. This development has led to the most-accurate measurements of the rate of expansion of the universe!

The Universe Starts with the Big Bang and Ends with the Gnab Gib

The first ideas about the large-scale nature of the universe tended to hypothesize that the universe was steady and unchanging and that it had existed for all time and would continue to do so.

Astronomer Edwin Hubble measured the velocities of galaxies and found that they were all moving apart from each other and that the more distant the galaxy, the faster it was moving away. This could only mean one thing: The universe is expanding. (The best way to imagine this is to think of the galaxies as dots drawn on a balloon that's being inflated. Each of the dots moves away from all the others as the balloon expands, and the wider the separation between the dots, the faster they move away from each other.) This means the universe yesterday was slightly smaller than it is today, and so on and so on backward in time until the universe was all concentrated into a single point! This is the point at which space and time were concentrated into what's called a *singularity*. It's from this singularity, in a single violent event called the *Big Bang,* that space, time, and the universe expanded to what it is today.

Given that the universe was "born" in the Big Bang, this raised the question of whether it might "die." Or if not, what might the ultimate fate of the universe be? Well, Einstein's theory of general relativity is useful here, because it tells how space and time curve with a given distribution of matter and energy. The theory predicts that the ultimate fate of the universe depends on the density of mass and energy in the universe. If enough mass and energy is in the universe, then that mass and energy may cause enough attraction to halt the expansion of the universe and reverse it — bringing the entire universe back to a single point in an event called the *Big Crunch.* Otherwise, the universe will continue to expand forever — getting colder and darker. Neither option seems very appealing!

Microwave Ovens Are Hot Physics

You can find plenty of physics going on in microwaves — everyday items you may have taken for granted in your pre-physics life. What really happens in a microwave oven? A device called a *magnetron* generates waves of radiation similar to those involved in the transport of thermal energy (see Chapter 15). These waves are called *electromagnetic waves,* and they have a similar form to sine waves.

Electromagnetic waves with different wavelengths have very different properties. If they have a wavelength in a particular range, then they're visible as light; in another range, at longer wavelengths, they cause water to heat up. The waves exert forces on the molecules as they pass through water, causing the molecules to oscillate in a way similar to simple harmonic motion (Chapter 13).

You may remember from chemistry that water molecules are polar because of the arrangement of the hydrogen and oxygen atoms and the distribution of the electrons. The hydrogen and oxygen atoms share electrons, but the electrons spend more time by the oxygen nucleus, which has a stronger pull. This means that one end of the molecule has a partial positive charge and the other has a partial negative charge.

A microwave is composed of an oscillating electrical field, and the water molecules, with their partial charges, rotate to align with that changing field.

The oscillating water molecules bump and jostle the surrounding molecules that constitute the food. This increased oscillating and jostling motion of the molecules is exactly what's meant by an increased temperature — and your dinner is ready! The frequency of the microwave determines the frequency of the oscillating molecules (the frequency of their simple harmonic motion), and this transfers energy to the molecules at a rate that increases with the frequency (and intensity) of the wave. The frequency of microwaves is just right for increasing the temperature at the rate required for cooking food.

Microwave ovens were invented by accident, during the early days of radar. A man named Percy Spencer put his chocolate bar in the wrong place — near a magnetron used to create radar waves — and it melted. "Aha," thought Percy. "This could be useful." And before he knew it, he had invented not only microwave ovens but also microwave popcorn (no kidding).

The universe is full of microwaves, which are a sort of leftover heat glow from the Big Bang. The discovery of this so-called *cosmic background microwave radiation* in the 1960s was a powerful confirmation of the Big Bang theory. You can read more about microwaves and other forms of electromagnetic radiation in *Physics II For Dummies* (Wiley).

Is the Universe Made to Measure?

Fundamental constants are fixed and written into the laws of physics, which describe the whole universe. These constants describe things such as the strength of gravity and the relative masses of fundamental particles. Physicists hope to develop a theory that can explain why the fundamental physical constants have the values that they do. Physicists would like their final theory of everything to be completely self-contained, leaving nothing unexplained — even the values of the fundamental constants.

Physicists have worked out what the world would be like if the constants were slightly different. What would happen if gravity were slightly weaker? What would happen if the forces holding atoms of matter together were slightly stronger? And the answer they find is that if any of the constants were only slightly different from the values that they have, then people wouldn't be able to live in this universe. For example, if gravity were slightly weaker, then stars couldn't form and we'd have no sun. And if gravity were slightly stronger, then stars would burn their fuel so quickly that life wouldn't have time to evolve! How can people explain why the constants seem to be so finely tuned?

The *anthropic principle* says that the constants have to have the values that they do because if they didn't, then we wouldn't be here to measure them. This is a very curious argument that many people are not happy with!

Another puzzle related to the constants is the question of why gravity is so weak. Gravity is exceedingly weak compared to other forces, such as electrical forces (the same kind of force that makes your hair stand on end when you rub a balloon on your shirt and bring the balloon near your head). This question has led some physicists to contemplate extra dimensions to space and time.

Glossary

Here's a glossary of common physics terms you come across in this book. ***Note:*** Words in italics appear in separate glossary entries.

absolute zero: The lower limit of physically possible temperature

acceleration: The rate of change of *velocity,* expressed as a *vector*

adiabatic: Without releasing heat into or absorbing heat from the environment

angular acceleration: The rate of change of *angular velocity*

angular displacement: The angle between the initial and final angular positions

angular momentum: The product of an object's *moment of inertia* and its *angular velocity*

angular velocity: The rate of change of *angular displacement*

Avogadro's number: The number of molecules in a *mole,* 6.022×10^{23}

blackbody: A body that absorbs all *radiation* incident upon it, reaches a thermodynamic balance with this incident energy, and radiates it all back

Boltzmann's constant: A thermodynamic constant with a value of 1.38×10^{-23} joules per kelvin; it quantifies the average amount of energy of individual particles, at a given temperature, and is given by the gas constant divided by *Avogadro's number*

buoyancy: The upward-acting force on a body immersed in a fluid that's equal in magnitude to the weight of the fluid displaced by the object

centripetal acceleration: The *acceleration* needed to keep an object in circular motion; centripetal acceleration is directed toward the center of the circle

centripetal force: The *force,* directed toward the center of the circle, that keeps an object going in circular motion

conduction: The transmission of heat through a material via direct contact

conservation of energy: The law of physics that says that the total energy of a closed system doesn't change

convection: A mechanism for transporting heat through the motion of a heated gas or liquid

conversion factor: The number that relates two sets of units

density: A quantity of mass divided by volume

displacement: The change in an object's position in terms of distance and direction

elastic collision: A collision in which *kinetic energy* is conserved (*momentum* is conserved, too, as it is in any collision)

emissivity: A property of a substance showing how well it *radiates*

energy: The ability of a system to do *work*

FPS system: The system of measurement that uses feet, pounds, and seconds

frequency: The number of cycles of a periodic occurrence per unit of time

friction: The force between two surfaces that always acts to oppose any relative movement between them

heat: The flow of thermal energy

heat capacity: The amount of *heat* needed to raise the temperature of one unit of mass of a substance by 1 degree

hertz: The *MKS* unit of measurement of *frequency* — one cycle per second

impulse: The product of the amount of force on an object and the time during which the force is applied

inelastic collision: A collision in which kinetic energy isn't conserved (though momentum is conserved, as it is in any collision)

inertia: The tendency of masses to resist changes in their motion

isobaric: At constant *pressure*

isochoric: At constant volume

isothermal: At constant *temperature*

joule: The *MKS* unit of *energy* — one newton-meter

kelvin: The *MKS* unit of temperature, equal in size to a degree Celsius; the Kelvin scale starts at *absolute zero*

kilogram: The *MKS* unit of *mass*

kinematics: The branch of *mechanics* concerned with motion without reference to *force* or *mass*

kinetic energy: The *energy* of an object due to its motion

kinetic friction: *Friction* that resists the motion of an object that's already moving

latent heat: The heat per kilogram needed to cause a change in phase in a substance

law of conservation of momentum: A law stating that the *momentum* of a system doesn't change unless influenced by an external force

linear momentum: The product of an object's *mass* times its *velocity;* momentum is a *vector*

magnitude: The size, amount, or length associated with a *vector* (vectors are made up of a direction and a magnitude)

mass: The quantitative measure of the property that makes matter resist being accelerated

mechanics: The area of physics that deals with the motions of bodies and the forces imposed upon them

MKS system: The measurement system that uses meters, kilograms, and seconds

mole: A quantity of substance that's defined to have a number of atoms (or molecules if the substance is molecular) equal to *Avogadro's number*

moment of inertia: The property of matter that makes it resist rotational acceleration

newton: The *MKS* unit of *force;* the amount of force that would accelerate a *mass* of 1 kilogram with an *acceleration* of 1 meter per second^2

normal force: The *force* a surface applies to an object, in a direction perpendicular to that surface

oscillate: Move or swing side to side regularly

pascal: The *MKS* unit of *pressure,* equal to 1 *newton* per meter^2

period: The time it takes for one complete cycle of a repeating event

phase (of matter): One of four notably distinct states of matter: solid (the molecules are relatively fixed in place), liquid (the molecules are free to flow but are bound relatively close to each other), gas (the molecules are free to flow and are far apart from each other relative to their size), and plasma (the atoms have been broken down to form a gas of subatomic particles)

potential energy: The *energy* an object has because of its internal configuration or its position when a *force* is acting on it

power: The rate at which work is done by a system

pressure: *Force* applied to a surface divided by the surface area over which the force acts

radians: The *MKS* unit of angle; 2π radians are in a circle; one radian is the angle subtended by an arc that has a length equal to the radius of the circle

radiation: A physical mechanism that transports *heat* and *energy* as electromagnetic waves

resultant: A *vector* sum

rotational inertia: See *moment of inertia*

scalar: A quantity that has magnitude but not direction (in contrast to a *vector,* which has both)

significant digits (significant figures): The number of digits that are of known value, according to the precision of the measurement and any subsequent calculations

simple harmonic motion: Repetitive motion in which the restoring *force* is proportional to the *displacement*

specific gravity: The *density* of a substance relative to a reference substance

specific heat capacity: A material's *heat capacity* per kilogram

standard pressure: One atmosphere, or 1.01×10^5 pascals

standard temperature: A temperature of 0°C

static friction: *Friction* on a stationary object

streamline: Lines in a fluid flow that are parallel to the velocity of the fluid at every point

temperature: A measure of molecular movement in a substance; when two objects are in thermal contact yet no heat flows between them, then they are defined to be at the same temperature

thermal conductivity: A property of a substance showing how well or how poorly *heat* moves through it

thermal expansion: The increase in length or volume of a material as it gets hotter

thermodynamics: The section of physics covering *heat* and *matter*

torque: The product of a *force* around a turning point and the force's perpendicular distance to that turning point

vector: A mathematical construct that has both a *magnitude* and a direction

velocity: The time rate of change of an object's position, expressed as a *vector* whose *magnitude* is speed

viscosity: The "thickness" of a fluid; the rate at which the velocity changes across a fluid flow increases with viscosity

weight: The *force* exerted on a *mass* by a gravitational field

work: *Force* multiplied by the *displacement* over which that force acts and the cosine of the angle between them; force is equal to the amount of *energy* transferred by a force

Index

• Symbols •

• A •

• B •

• C •

• D •

• E •

• F •

• G •

• H •

• I •

• J •

• K •

• N •

• O •

• P •

• Q •

• R •

• S •

• T •

• U •

• X •

• Y •

• Z •

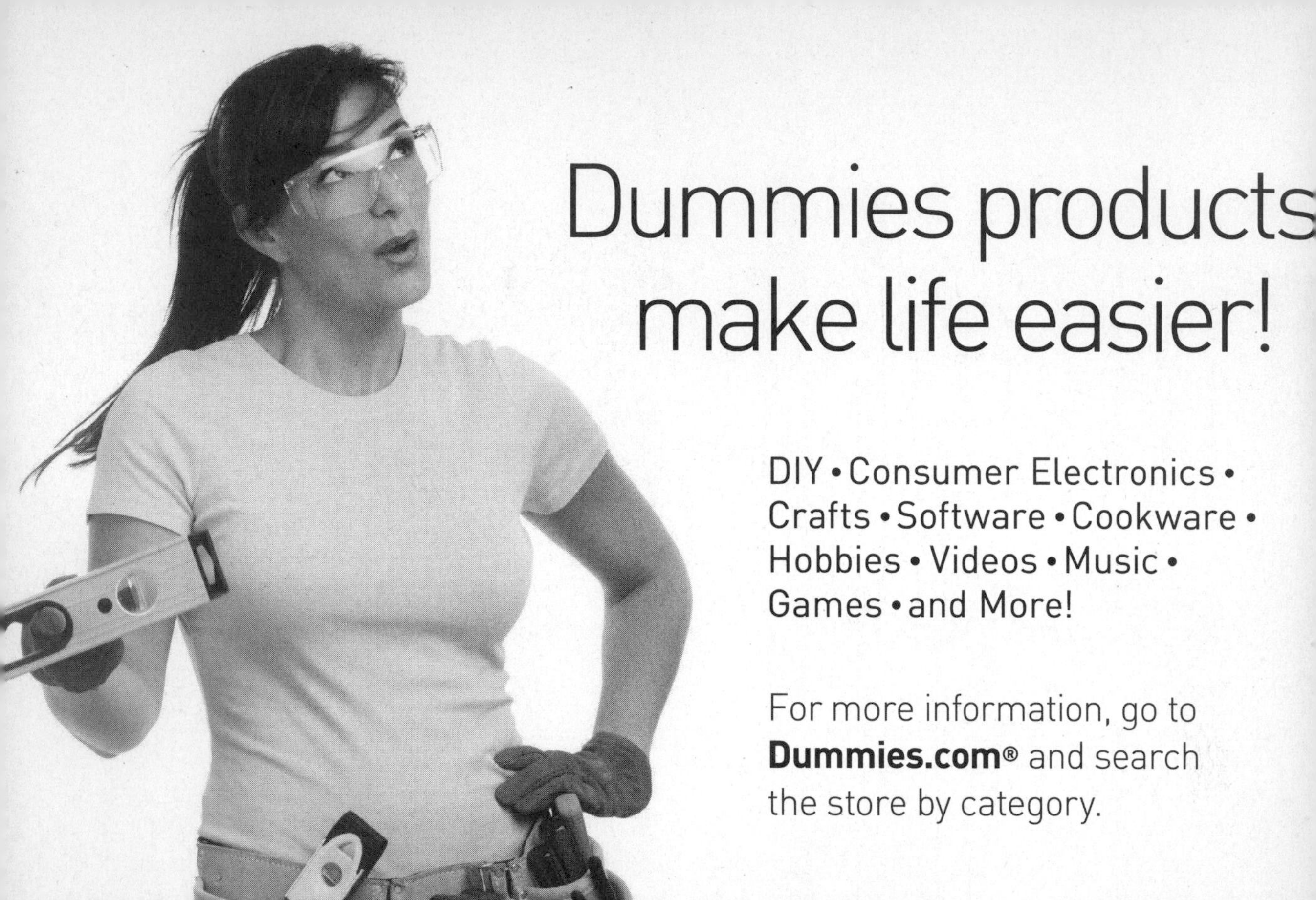
Dummies products
make life easier!
DIY • Consumer Electronics •
Crafts • Software • Cookware •
Hobbies • Videos • Music •
Games • and More!
For more information, go to
Dummies.com® and search
the store by category.

Piano
FOR
DUMMIES
The Fast and Easy Way To Start Playing!

The fun and easy way™ to
pick up the basics and start expressing yourself!
Acrylic Painting Set
FOR
DUMMIES
All you need to start painting right away!
10 Acrylic paints
6 White nylon brushes
2 Painting boards
1 Palette knife
1 Palette
1 Drawing pencil
1 Pencil sharpener
1 Eraser
1 Instruction book
An Art Set for the Rest of Us!™

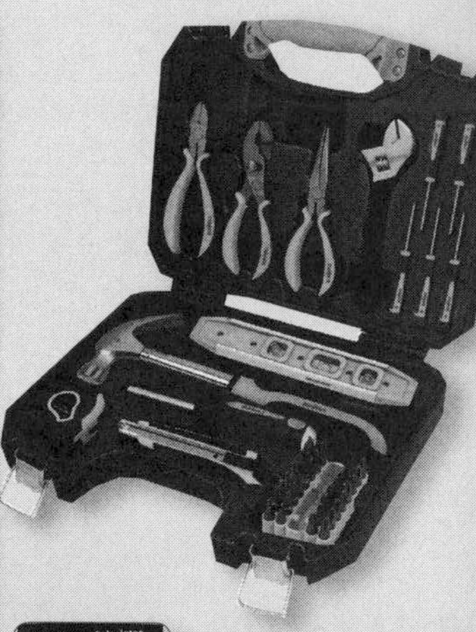

Drip Irrigation Kit
FOR
DUMMIES

Plug In and Play — Electric Guitar the fun and easy way™!
Electric Guitar
Starter Pack
FOR
DUMMIES
Electric Guitar Basics
Get ready to Rock!
KONA

DIY Project Calculator
DUMMIES

Windows 7
Transfer Kit
FOR
DUMMIES
The easiest way to move to Windows 7

GPS Navigation
DUMMIES

FOR
DUMMIES®
Making everything easier!™